Directory of Important World Honey Sources

World Honey Sources Directory

This book contains a vast quantity of precious data about plants and bees and it is marvellous to see it in print again and available to new generations. Best of all would be if people selecting trees become aware of this useful information and consult it to inform their choice: nowadays we need whenever possible to choose species and cultivars with value to bees and other insect pollinators.

The meticulous work of gathering data and compiling it into this useful book was a labour of love for Eva Crane and her team. It was the outcome of many years' scrutiny of the world's research literature - across many languages, and carefully recording any data pertaining to plants visited by honey bees. By the 1980's Eva Crane and her team had considerable data on thousands of plant species, each of them reported somewhere in the world to be major sources of honey. They then began the task of selecting from this lengthy list, those 467-plant species which could make it to the short list and be justifiably described as the 'Important world honey sources' of the book's title. For each plant species, data is provided on its economic and other uses, how much nectar / pollen and/or honeydew it provides, honey flow data (if any exist), and concerning the honey's chemical and physical properties.

I was the youngest member of staff at the International Bee Research Association when Eva Crane was compiling the data for this fantastic book, and remember well the tremendous enthusiasm of the team, and endless discussion of which plants should make it to the short list. IBRA had recently acquired its first 'WordStar' word processor, and this was perfect for organising the vast data that had been collected, and permitted it to be readily searched, e.g. to compile lists of salt-tolerant plants, or of honey that granulates quickly. Much of the research that is cited here is from meticulous 20th century field work, watching bees and recording their behaviour - long before the digital era - and of course the observations remain valid. The entries may look at first impenetrable - but persevere to learn the codes and once you begin using this book, you will become addicted to consulting it and the interesting body of work upon which it is built.

Dr Nicola Bradbear
Director *Bees for Development*
President *Apimondia Scientific Commission Beekeeping for Rural Development*

IBRA

INTERNATIONAL BEE
RESEARCH ASSOCIATION

NB
B

Directory of Important World Honey Sources

By
Eva Crane, Penelope Walker and Rosemary Day

reprint published in 2018 by
International Bee Research Association
and
Northern Bee Books

First published in 1984 by International Bee Research Association.

Reprinted and republished as a facsimile edition in 2018 by:
    International Bee Research Association
    91 Brinsea Road, Congresbury, Bristol, BS49 5JJ, UK
        and
    Northern Bee Books
    Scout Bottom Farm, Mytholmroyd, Hebden Bridge, HX7 5JS, UK

ISBN 978-0-86098-141-1

Obtainable from:
www.ibrabee.org.uk
    and
www.northernbeebooks.co.uk

CONTENTS

FOREWORD

This book is new in concept, in contents, and in method of preparation. It identifies 467 plants - from a preliminary selection list of 2569 - that are reported, somewhere in the world, to be a major source of the honey produced there. Some of these honey sources are widespread, such as lucerne (alfalfa) and many of the eucalypts; others are confined to a single area, such as plectranthus which grows only on certain slopes of the western Himalayas. Some other plants that may possibly also be important honey sources are listed in Section 4 as candidate plants.

Section 5 consists of the MAIN ENTRIES for 452 nectar-producing plants in alphabetical order of their botanical names, followed by MAIN ENTRIES for 15 honeydew-producing plants. Common names of the plants are indexed at the end of Section 8.

For each honey source, details are given (as far as is known, and quantified where possible): of the plant, its economic uses, its flowering period and nectar (or honeydew) flow, its pollen and its honey production, and of the honey's chemical composition and physical properties including flavour, aroma and granulation.

MAIN ENTRIES have been printed from master sheets typed on a BDP word processor. Certain components of each entry were coded in such a way that searches can be made for plants with particular characteristics, e.g., tolerant to drought, or with a high honey potential, or whose honeys are light in colour, or granulate rapidly - or not at all. There are 51 search fields altogether, and the lists in Section 7 were made by using some of them.

In spite of the limited financial resources available for the Directory, we believe that it provides much useful information in an accessible form, which should be effective in improving honey production in developing countries (see Section 3b). We hope that it will also stimulate scientists and beekeeping specialists to obtain the information that the Directory shows is still lacking. IBRA will run further search programmes to compile Satellite Directories listing special groups of plants with relevant information on them. Section 3c describes a practical way in which additional information can be added to the data bank and made available where it is needed.

Dr Eva Crane was in charge of the project throughout, and was the contract holder with the funding agency, International Development Research Centre, Ottawa (IDRC). For many years she has been very conscious of the need for a reference book on the world's most important honey plants, and she knew that there was a great deal of information to draw upon. A system whereby this information could be organized and made available for use, through word processor facilities, was worked out in consultation with Mike Prudence, then of Thomas Hill International.

The selection of plant and honey characteristics to be recorded for the Directory was made by all three authors together. Eva Crane and Penelope Walker planned the methods for input of the information with coded search fields, and organized the material for publication. Penelope Walker was responsible for running the search programmes devised by Mike Prudence, and throughout the work she co-ordinated the many separate operations and the sequence of detailed checks that was necessary for accuracy and consistency. Eva Crane contributed to many individual entries from her personal knowledge of honey plants, and she advised on various points of difficulty as they arose.

The Directory is Phase II of a wider project, of which Phase I was funded by the International Union of Biological Sciences (see Section 3a below). Rosemary Day collected the data for Phase I, and she made the selection of plants for the Directory and compiled the details about each plant; Penelope Walker was responsible for the information on honey; and both authors contributed data on the plants' nectar and pollen production and honey flow. Joy Fish had charge of the Bibliography references and assisted with records of nectar rating and with indexes. In the course of the work, we consulted over a thousand publications in the IBRA Library, and Inge Allen made this possible by locating and reshelving many of them for us.

Judith Dolby and Susan Henriques were responsible for work on the word processor, including production of masters from which the Directory is printed. Dr Margaret Adey read the final drafts for inconsistencies.

We are indebted to IDRC for the funding that enabled the Directory to be prepared and published. We greatly appreciate help received from many institutions, including the International Commission for Bee Botany of IUBS; Royal Botanic Gardens, Kew; Commonwealth Institute of Entomology, London; International Council for Research in Agroforestry, Nairobi. We also sincerely thank the many individuals who have assisted with specific aspects, particularly Dorothy Galton (for translating from Russian), Dr J. Louveaux, Professor R.W. Shuel, Dr A.D. Stewart, Dr G. Vorwohl - and especially Dr G.E. Wickens of the Royal Botanic Gardens, Kew, who answered our many botanical queries.

1. HOW TO USE THE DIRECTORY

In alphabetical order of botanical name, MAIN ENTRIES 001-452 are important nectar sources of honey, and 01D-15D, which follow, are important honeydew sources of honey. Abbreviations and language and country codes are listed in Section 2, and author reference codes in the Bibliography (Section 6).

(a) SUMMARY OF INFORMATION IN MAIN ENTRIES 001-452 (NECTAR SOURCES)

For each plant, information (as available) is printed in the order below. Data on similar characteristics are grouped together in paragraphs or "blocks". Underlining and bold type in this Section indicate underlining and bold type in the MAIN ENTRIES. Section 1c gives explanatory notes for the characteristics.

entry number **Botanical name of plant, authority; family**
any synonyms

common names
vegetative form of plant; floral description
Distribution; and where native. **Habitat**
Soil. Temperature. Rainfall

Economic and other uses
Food. Fodder. Fuel. Timber. Land use. Soil benefit. Other uses

Warning; alert to beekeepers
Warning. Alert to beekeepers

Nectar rating + honeybee species; blooms, nectar flow; composition
Nectar rating of a plant in a country is:
   N1 = major honey source
   N2 = medium honey source
   N3 = minor honey source
   N = honey source, importance unrated
(Section 1c explains these, and gives codes for honeybee species.)
Blooms (dates). **Nectar flow** (dates or duration). **Nectar secretion.**
Sugar concentration. Sugar value. Sugar analysis. Other characteristics.

Honey flow
Honey yield (kg/colony/season). **Honey potential** (kg/ha[/season]).

Pollen
P1, P2, P3, P (countries given, as for nectar rating).   **Yield.**
Pollen value (to bees).   **Chemical analysis.**   **Colour.**   **Pollen grain**
illustrated/described, and if under-/over-represented in honey.
Reference slide availability

Honeydew produced (countries), and any other information

Recommended for planting to increase honey production
In which country.   How to propagate.   Other notes.   Reminder to see
Warning and/or **Alert to beekeepers**

Honey: chemical composition (% by wt of honey except where stated)
Water.   **Sugars, total.**   **Glucose.**   **Fructose.**   **Sucrose.**   **Reducing**
sugars.   **Maltose.**   **Higher sugars.**   **Dextrin**
Ash.   Inorganic constituents
pH.   **Total acid** (meq/kg).   **Free acid** (meq/kg).   **Lactone** (meq/kg)
Amylase (Gothe).   **Sucrase** (units given).   **Glucose oxidase** and other
enzymes (units given).   **HMF** (ppm)
Nitrogen.   **Amino acids.**   **Protein.**   **Colloids**
Fermentation, if likely, on storage?
Vitamins
Toxicity
Other constituents

Honey: physical and other properties
Colour.   **Pfund reading** (mm)
Relative density.   **Viscosity** at 20°C (poise).   **Optical rotation**
(deg).   **Electrical conductivity** (per ohm cm).   Other physical properties
Granulation
Flavour.   **Aroma**

(b)   INFORMATION IN MAIN ENTRIES 01D-15D (HONEYDEW SOURCES)

Plants 001-452 secrete nectar from which bees produce honey.   Plants
01D-15D are hosts to certain insects (Hemiptera) which pierce the surface
and imbibe the sap thus made accessible.   The honeydew they excrete is
collected by bees and made into honey.

For 01D-15D, the following information is presented in the same style as in
001-452:
    botanical details, **Economic uses**, any **Warning/Alert**, information on
    Pollen; also data on **Honey**.
Preceding **Honey** are data on:
    Honeydew, grouped alphabetically by **insect** secreting it, giving informa-
    tion on **D1** rating (comparable with **N1**), honey yield, honey potential,
    flow (dates), honeydew analysis.   Countries from which data were
    obtained are given.

The taxonomy of honeydew-producing insects is difficult.   Names and
synonyms used (for 01D-15D and for the plants in 001-452 with **Honeydew**
information) were checked by the Commonwealth Institute of Entomology.

(c) INSTRUCTIONS AND EXPLANATORY NOTES

(i) General

Information available for a genus, or for a plant which we could not identify from the common name, has been excluded. With three exceptions (entries 306, 357, 11D), each plant entry refers to a single species.

In each entry and within each block, available information is presented in a standard order (as in Section 1a). For each characteristic, records are listed in alphabetical order of country code (see Section 2). Where country codes are not quoted, alphabetical author order is followed. Metric units (and °Celsius) are used throughout and, where necessary, appropriate conversions have been made.

Author reference codes quoted in the entries lead to full references in the Bibliography (Section 6). References are not generally cited for information on the plant itself, but references extensively used for this purpose are marked °° in the Bibliography.

Specialist terms used in botany, agriculture, etc., are kept to a minimum; they are explained in appropriate technical reference books.

(ii) Information within each entry

Explanatory notes are given here for the characteristics listed in Section 1a.   * indicates items of information coded to provide search fields, so that searches can be made for plants or honeys with certain characteristics.

Name of plant: Botanical name and authority (as used at the Royal Botanic Garden, Kew), followed by plant family, in bold type. Any recognized synonyms are printed immediately below.

Common names: Names in general use in English are given first; then those used regionally, e.g. (En/USA); then names in other languages used widely in the tropics, Es (Spanish), Fr (French), Pt (Portuguese); then names in other (indicated) languages used in countries where the plant has a nectar rating. Some common names are identified by country code rather than by language. Stringent selection was necessary for inclusion in the Directory, and transliterated names are only exceptionally used. Language and country codes are given in Section 2.

Form of plant*: Tree, shrub or herb, size, brief description of plant and of flower.

Distribution is given by climatic zones* and continents within which the plant now grows; the order in which they are listed has no significance. Plant Habitat follows.

Soil preferences include tolerance to salt*, and drainage requirements. Temperature limits include effects of frost*; "frost damage" ranges from slight injury to death of the plant. Rainfall preference includes drought tolerance* (or resistance). Terms used by the publication referred to are usually quoted verbatim.

Economic and other uses

The list for each plant is not necessarily complete, but in view of the importance of multipurpose plants in development programmes, some effort was expended in identifying other uses of plants which are important to bee-keepers as honey sources.   Uses included in the Directory are grouped under: **Food*** (for man). **Fodder*** (for animals). **Fuel***. **Timber***. **Land use** – hedges*, windbreaks*, shade*, afforestation*, amenity*. **Benefit to soil***, including erosion control*, and enrichment*. **Other uses**

Warning*; alert to beekeepers*

Warning is given for plants that are: a possible cause of injury to man or animals; invasive, or preventing growth of other plants near them; a fire hazard; especially likely to be attacked by pests. **Alert to beekeepers** is given for plants: whose nectar or pollen is toxic to bees (no plants whose honey is toxic to man are included in the Directory); whose nectar flow presents problems to beekeepers because it comes very early in the season, or may be greatly reduced by pests, or seems to make the bees "aggressive"; which produce no pollen (or pollen inadequate for brood rearing); plants whose honey has characteristics causing special problems in extraction e.g. thixotropy or very rapid granulation.   A **Warning** or **Alert** is repeated in any block where it is relevant, e.g. under **Nectar flow.**

Nectar rating + honeybee species; blooms, nectar flow; composition

The nectar rating of a plant in a country is:
  N1 = a major source of surplus honey
  N2 = a medium source of surplus honey
  N3 = a minor source of surplus honey
  N = a honey source, importance unrated

The honeybee collecting the nectar is European Apis mellifera (am) unless otherwise indicated (tm = tropical A. mellifera, ac = A. cerana, ad = A. dorsata, af = A. florea).   Square brackets indicate that we deduced the species from the context.

Most plants have two or more N1 ratings.   Each rating N1, N2, N3 is fol-lowed by a list of countries (in code, see Section 2) in which it was recorded, with the reference.   The ratings N1, N2, N3 usually refer to the amount of honey produced.   They thus involve several components, e.g. widespread occurrence of the plant; good nectar flow from a single plant (which may bloom for a few days or for several months); medium or high concentration of sugars in the nectar; accessibility of the nectar to bees.

Blooms – months (i = January, xii = December) during which the plant flowers in the country or region specified.   The reference is that cited in the nectar rating.   The months of the **Nectar flow**, or its duration, are cited similarly.

In **Nectar secretion**, numerical data in mg/fl/day (= 24h) are quoted, toge-ther with factors affecting secretion.   Many references quote a figure for the weight of nectar/flower (and/or of sugar/flower), without stating whether it is a mean daily value, or for the whole flowering period, or a mean for nectar samples extracted from several flowers on a single occa-sion.   We have not included such figures.

The **sugar concentration*** in the nectar is classed by us as [low], [medium] or [high], and the class is coded for future searches:
     low <21% by wt;   medium 21-60%;   high 61%+
     (The medium class includes all values starting with 60, e.g. 60.7%. The high class starts at 61.0%)

Sugar value* (mg/fl/day) is treated similarly:
     low <0.1;   medium 0.1-2;   high 3+

Sugar analysis – space does not permit results to be quoted, but relevant references are cited.   Other information on nectar composition may be included, for example, reports of bees collecting juice from damaged fruit.

Honey flow
Honey yield* is quoted as kg/colony/season;   the bees involved are European Apis mellifera unless otherwise indicated under **Nectar rating.**   Yields are usually from beekeepers' records, and most refer to surplus honey taken, although Russian figures are likely to be for the total amount stored.   The **honey yield** is classed by us as [moderate] or [high]:
     moderate <30;   high 30
If numerical data are lacking for a country a verbal description of yield may be quoted, such a verbatim quotation being in inverted commas.

Honey potential* is a term in common use in Eastern Europe for the estimated weight (kg) of honey that could be obtained in the course of a season from 1 hectare of land covered with the plant, assuming optimal conditions (Cra/75). The **honey potential** is classed by us as [moderate] or [high]:
     moderate <500;   high 500+

Pollen
Ratings **P1, P2, P3, P** are on a similar basis to **N1, N2, N3, N**;   where an author reference is the same for both (and there is no ambiguity), this is not repeated for pollen.   **Yield** relates to the amount produced by the flowers.   **Pollen value*** (to bees) gives further information, e.g. on nutrition/toxicity.   **Colour** is designated "of load" if the author cited states this.   Representation* of pollen grains in honey is reported here only for < 20 000 grains in 10 g honey [under-represented], or > 100 000 grains in 10 g [over-represented].   The words **Reference slide** indicate that a slide of the pollen grain is in the collection maintained by Dr G. Vorwohl at Landesanstalt für Bienenkunde, Universität Hohenheim, August-von-Hartmann-Strasse 13, 7000 Stuttgart 70, German Federal Republic.   Dr Vorwohl is actively seeking pollen samples of the plants not yet represented in his collection.   Any offers of help and any enquiries about obtaining reference slides should be sent to him.

Honeydew production*.   An entry here indicates that honeydew is produced on the plant and is collected by bees.   Following entry 452 are plants that are important sources of honeydew honey (see Section 1b).   Castanea sativa (080) and Tilia cordata (410) are important sources of both nectar and honeydew honey.

Recommended for planting to increase honey production*
The country is cited where such a recommendation has been made by the author quoted, together with other useful information.   We emphasize that this is

not a blanket recommendation to introduce an exotic plant: any proposal for introduction into a new country should be discussed with plant quarantine and other appropriate authorities before any action is taken.

Honey: composition and properties

Many quantitative data are available for the honeys from some plants, but for far too many important honeys we had to enter "no data". We hope that the Directory may draw the attention of honey chemists to this great imbalance and that, as a result, research work will be more evenly dispersed among the important world honeys.

The botanical name of a plant, and the age and previous history of honey samples studied was not always found in the reports we consulted, and we stress the importance of such details in the quantitative study of honey. We tried to exclude honeys stated to have been heat treated or to be granulated, but some of the data in the entries may not refer to monofloral, or to freshly extracted honey. Many of the values quoted were obtained by standard methods; however, some authors do not state their methods, and in some languages we have been unable to ascertain them. Where methods used several consecutive honey data were taken from one publication, the author code is cited only after the first value quoted.

Section 1a lists characteristics for which data were sought. For searches, we classed many of the following characteristics by using the limits set in the draft FAO/WHO Codex (Cod/83), or in the EEC Directive (Eur/74) - in % by wt unless otherwise stated.

water*: low <16; medium 16-20; high 21 (FAO max) and over
glucose*: low <31; medium 31-39; high 40+
fructose*: low <35; medium 35-42; high 43+
sucrose*: low <1; medium 1-4; high 5 (FAO max) and over
reducing sugars*: low <65 (FAO min)
ash*: low <0.1; medium 0.1-1.0; high 1.0 (FAO max) and over
free acids* (meq/kg): low <15.0; medium 15-39; high 40 (FAO max) and over
amylase* (Gothe): low <3 (FAO min)
HMF* (ppm): high 40 (EEC max) and over
    (HMF not included in FAO/WHO Codex)
fermentation* on storage: likely/unlikely/never
vitamins* present
toxicity* any information on adverse reactions, e.g. when fed to bees
colour: any unusual tinges*
Pfund* (mm):
    white 0-34 (includes grades water white to white)
    amber 35-114 (includes grades extra light amber, to amber)
    dark 115 + (grade dark amber)
      (Many authors do not give readings but use verbal descriptions
      of Pfund grades.)
granulation complete within:
    <2 wks = rapid; 2-52 wks = medium; >1 yr = slow
flavour* descriptions:
    bland; strong; objectionable; unusual flavour

The vocabulary used to describe honey flavours and aromas is very limited and imprecise. Also, a person often assesses a sample of honey in relation to the types he is most accustomed to. We have not quoted commonly used subjective terms such as good, pleasant or poor, but if we had used only strictly objective terms, very few honeys would have had a flavour description. Flavour is a most important characteristic of any food, and we believe that honey flavour descriptions deserve much more professional attention than they have received so far.

2. ABBREVIATIONS AND CODES

General

approx	approximately	min	minimum
C	central	mth, mths	month(s)
conc	concentration	N	north, or nitrogen
cv, cvs	cultivar(s)	no	number
deg	degrees	ppm	parts per million
	(optical rotation)	RH	relative humidity
E	east	S	south
est	estimated	syn	synonym(s)
fl, fls	flower(s)/flowering	temp	temperature
h	hour(s)	v	very
ha	hectare(s)	W	west
lvs	leaves	wk, wks	week(s)
max	maximum	wt	weight
Med	Mediterranean	yr, yrs	year(s)
meq	milliequivalents (units of acidity)		

Apis species: ac = A. cerana, ad = A. dorsata, af = A. florea, am = A. mellifera, tm = tropical A. mellifera.

Languages

En	English	Da	Danish
Es	Spanish	De	German
Fr	French	In	Indonesian
Pt	Portuguese	It	Italian
Af	Afrikaans	No	Norwegian

Other languages are named in full.

Countries

Three capital letters are used for each country (except UK), where possible the first three letters of the country's name. Territories outside the metropolitan/mainland part of the country are listed separately, since some are widely separated geographically.

AFG Afghanistan	CHN China, People's Republic	/UTT Uttar Pradesh
ALG Algeria	COL Colombia	/WBE West Bengal
ANA Angola	COS Costa Rica	INO Indonesia
ARG Argentina	CRE Crete	/JAV Java
AUS Australia (PO codes):	CUB Cuba	IRN Iran
/NSW New South Wales	CYP Cyprus	IRQ Iraq
/NT Northern Territory	CZE Czechoslovakia	IRR Irish Republic
/QD Queensland	DEN Denmark	ISR Israel
/SA South Australia	DOR Dominican Republic	ITA Italy
/TAS Tasmania	ECU Ecuador	IVO Ivory Coast
/VIC Victoria	EGY Egypt	JAM Jamaica
/WA Western Australia	ELS El Salvador	JAP Japan
AUT Austria	ETH Ethiopia	JOR Jordan
AZO Azores	EUR Europe	KEN Kenya
BAA Bahamas	FIJ Fiji	KOS Korea, South
BAL Balearic Islands	FIN Finland	KON Korea, North
BAN Bangladesh	FRA France	LEB Lebanon
BAR Barbados	FRP French Polynesia	LES Lesotho
BEG Belgium	GAM Gambia	LIY Libya
BEL Belize	GDR German Democratic Republic	LUX Luxembourg
BER Bermuda	GFR German Federal Republic	MAE Madagascar
BOL Bolivia	GHA Ghana	MAI Malawi
BOT Botswana	GRC Greece	MAK Malaysia
BRA Brazil	GUD Guadeloupe	MAQ Malta
/RG Rio Grande do Sul	GUF Guam	MAS Mariana Is.
/SC S. Catarina	GUM Guatemala	MAR Marshall Is.
/SP S. Paulo	GUS Guinea-Bissau	MAT Martinique
BUL Bulgaria	GUY Guyana	MAY Mauritius
BUM Burma	HAI Haiti	MEX Mexico
BUU Burundi	HAW Hawaii	/YUC Yucatán
CAE Cameroon	HOD Honduras	MID Midway Island
CAF Canada (PO codes):	HUN Hungary	MOR Morocco
/Alta Alberta	INI India	MOZ Mozambique
/BC British Columbia	/AND Andhra Pradesh	NAM Namibia
/Man Manitoba	/ASS Assam	NEP Nepal
/Nfld Newfoundland	/BIH Bihar	NER Netherlands
/Ont Ontario	/HAR Haryana	NEU New Caledonia
/Que Quebec	/HIM Himachal Pradesh	NEZ New Zealand
/Sask Saskatchewan	/KAR Karnataka	NIA Nicaragua
/NWT North West Territories	/KAS Jammu & Kashmir	NTR Nigeria
CAI Canary Islands	/KER Kerala	NIU Niue Island
CEN Central African Republic	/MAD Madhya Pradesh	NOW Norway
CHA Chad	/MAH Maharashtra	OMA Oman
CHG Chagos Archipelago	/ORI Orissa	PAK Pakistan
CHH Chatham Islands	/PUN Punjab	/NWFP North West
CHL Chile	/TAM Tamil Nadu	Frontier Province

PAM Panama
PAP Papua New Guinea
PAR Paraguay
PHI Philippines
POL Poland
POR Portugal
PUE Puerto Rico
REU Réunion
ROM Romania
RWA Rwanda
SEN Senegal
SEY Seychelles
SOM Somalia
SOU South Africa
/OFS Orange Free State
/TVL Transvaal
SPA Spain
SRI Sri Lanka
SUD Sudan
SUR Surinam
SWA Swaziland
SWE Sweden
SWI Switzerland
TAI Taiwan
TAN Tanzania
THA Thailand
TOG Togo
TON Tonga
TRI Trinidad and Tobago
TUN Tunisia
TUQ Turkey
UGA Uganda
URS Union of Soviet Socialist Republics
UK   United Kingdom
USA United States of America (PO codes):
/AL Alabama
/AK Alaska
/AZ Arizona
/AR Arkansas
/CA California
/CO Colorado
/CT Connecticut
/DE Delaware
/FL Florida
/GA Georgia
/ID Idaho
/IL Illinois

/IN Indiana
/IA Iowa
/KS Kansas
/KY Kentucky
/LA Louisiana
/ME Maine
/MD Maryland
/MA Massachusetts
/MI Michigan
/MN Minnesota
/MS Mississippi
/MO Missouri
/MT Montana
/NE Nebraska
/NV Nevada
/NH New Hampshire
/NJ New Jersey
/NM New Mexico
/NY New York
/NC North Carolina
/ND North Dakota
/OH Ohio
/OK Oklahoma
/OR Oregon
/PA Pennsylvania
/SC South Carolina
/SD South Dakota
/TN Tennessee
/TX Texas
/UT Utah
/VT Vermont
/VA Virginia
/WA Washington
/WV West Virginia
/WI Wisconsin
/WY Wyoming
UPP Upper Volta
URU Uruguay
VEN Venezuela
VIE Vietnam
WAK Wake Island
YEA Yemen Arab Republic
YUG Yugoslavia
ZAI Zaire
ZAM Zambia
ZIM Zimbabwe

Publications referred to

Author codes are explained at the start of the Bibliography, Section 6.

In the sections headed **Honey flow** and **Nectar**, some references are given e.g.
AA932/81; this refers to a publication (not in the Bibliography) which provides
data only for these sections; details of the publication can be found in the
journal Apicultural Abstracts, AA932/81 representing entry 932 in the 1981
volume.

3. OTHER INFORMATION

(a) BACKGROUND TO THE DIRECTORY

In the early 1970s, when planning the book "Honey: a comprehensive survey" (Cra/75), a chapter on the world's honey plants was considered essential. There was no specialist in this subject, so the editor (Eva Crane) wrote the chapter, and her first task was to try to identify the most important honey sources on a world scale. Brief characteristics of 211 plant species or genera and their honeys were finally included, together with a 17-page list of the best available publications on honey plants of individual countries, since most information is published on a national basis. In "A book of honey" (1980) the same plants, with 21 important additions, were listed in relation to their world distributions. Between 1975 and 1984 unpublished material was collected from institutions and individuals in as many countries as possible, attention being paid especially to the developing countries of the tropics and subtropics. Lists of "honey plants" which give no rating of relative importance were also filed as a basis for future investigation, but they were not used in selecting plants for the Directory.

At several International Beekeeping Congresses (especially in 1975 at Grenoble in France), discussions were initiated on the possibility of a Survey of World Honey Sources (WHOS). But in spite of much interest, and a general appreciation of the importance of the concept, there was no funding until 1979 when the International Commission for Bee Botany obtained a subvention of US$5000 from its parent body, the International Union of Biological Sciences. This was a relatively small sum, but it enabled a start to be made, and Phase I of the Survey was carried out at IBRA under the direction of Dr J. Louveaux and Dr Eva Crane (President and Vice-President, respectively, of ICBB) and of other ICBB officers. Rosemary Day, assisted by Joy Fish, recorded on paper slips a total of 2569 plant species, each reported from at least one country, and some from many countries.

Funds for Phase I were too small to allow any publication of the results, except for a list of the 160 plant families recorded, with the number of plant species in each (Cra/83a). Three families, Leguminosae (327), Myrtaceae (261) and Compositae (226), contributed nearly one-third of all species.

Phase II, the preparation and publication of the Directory, is *limited to important* honey sources and includes some plants we did not know about in Phase I. Phase III is explained in Section 3c.

(b) SPECIAL VALUE OF THE DIRECTORY TO DEVELOPING COUNTRIES

Knowledge as to which plants are important honey sources is most soundly based for technologically advanced countries of the temperate zones, and the best documented entries in the Directory are for certain plants in these countries. The wealth of such data highlights the paucity of information from many developing countries of the tropics and subtropics - the part of the world where honey production could be most dramatically increased. An important aim of raising honey production in developing countries is to add sustenance and savour to the diet of rural peoples. Apiculture utilizes food resources that would otherwise be wasted, and it does not compete with other branches of agriculture. Moreover, high populations of foraging bees can bring an added bonus by ensuring full pollination of many crops, and thus improving their yield and their quality.

Increased knowledge of honey sources and honey production can give a considerable economic reward in the tropics and subtropics: the largest exporters of honey onto the world market today are China, Mexico and Argentina.

Examples of use

(i) In tropical countries with a research capability, the Directory will show which important honey sources lack certain information, e.g. honey data.

(ii) In countries where the important honey sources are not known, the Directory will show the reader which of the plants growing in his country are major honey sources in some other country. He can then observe these plants for their local honey-producing capability.

(iii) In reforestation programmes, the Directory helps to identify suitable species for planting which are important honey sources as well as having other well known uses.

(iv) The Directory also shows which plants are recommended for planting especially because of their value in honey production, and any of these growing locally should be explored for wider planting. It also indicates non-local species that might be worth considering for importation, subject to advice from plant quarantine and other authorities.

(c) REQUEST FOR FURTHER ASSISTANCE

We are conscious of certain shortcomings of the material presented here. Some are due to our own language limitations; no-one at IBRA could make full use of literature in Arabic, Japanese, or Chinese, and we also lacked adequate on-the-spot help with Russian, in which language much has been written on honey plants. Some material is also published in languages of south-east Asia that we could not read. Language help enabling IBRA to enlarge its data bank would thus be appreciated. Sometimes we were unable to use results because we were not sure of their exact meaning, or how comparable they were with data from other countries.

We should value further relevant information that we have not located, and we hope also that readers will draw our attention to any errors of fact.

Finally, we emphasize our intention that the work should continue into Phase III, as an ongoing project which can store and utilize further information received, and print out information from searches of the enlarged data bank. This work will be of special value to environmental scientists, beekeepers and honey chemists, to the world honey trade, and especially to the developing countries of the tropics and subtropics. It will require funding, almost certainly from several sources, and we ask readers who appreciate the value of the project to put us in touch with likely sources of support.

(d) **AVAILABILITY OF THE DIRECTORY AND OF SATELLITE DIRECTORIES**

Under the terms of the contract with IDRC, a certain number of copies of the Directory is available from IBRA free of charge, on request, to institutions in developing countries that are directly concerned with apiculture.

It is intended to publish Satellite Directories of honey sources that have certain important characteristics. For instance, drought tolerant plants that give especially high honey yields are of great interest for semi-arid parts of the world. Information to be published about such plants could include botanical and growth details, economic uses, honey-producing capability, and experience in propagation specifically for honey production. We should welcome suggestions as to Satellite Directories likely to be in greatest demand. Commissioned searches for specific information can be made; enquiries about them, and correspondence about any aspect of the Survey of World Honey Sources (WHOS) should be sent to:

IBRA (WHOS), Hill House, Gerrards Cross, Bucks. SL9 0NR, UK.

4. LIST OF CANDIDATE PLANTS

The following plants do not have MAIN ENTRIES because insufficient data have been found to justify their inclusion. Some may, however, prove to be as important as certain plants in the Directory.

Acer campestre L.	Aceraceae
Achras zapota L.	Sapotaceae
Aegopodium podagraria L.	Umbelliferae
Agastache foeniculum (Pursh) Kuntze	Labiatae
Albizia gummifera (J.F. Gmelin) C.A. Smith	Leguminosae
Aleurites fordii Hemsley	Euphorbiaceae
Alhagi maurorum Medikus, syn Alhagi camelorum Fisch.	Leguminosae
Alhagi pseudalhagi (Bieb.) Desv.	Leguminosae
Allium cepa L.	Alliaceae
Allium fistulosum L.	Alliaceae
Alyssum benthami hort.	Cruciferae
Anchusa azurea Miller, syn Anchusa italica Retz.	Boraginaceae
Anchusa capensis Thunb.	Boraginaceae
Annona muricata L.	Annonaceae
Anthemis arvensis L.	Compositae
Anthemis cotula L.	Compositae
Arctium minus Bernh.	Compositae
Arctium tomentosum Miller, syn Lappa tomentosa (Miller) Lam.	Compositae
Artemisia herba-alba Asso	Compositae
Asparagus officinalis L.	Liliaceae
Baccharis latifolia Per.	Compositae
Berberis vulgaris L.	Berberidaceae
Borago pygmaea (DC.) Chater & Greuter, syn Borago laxiflora Poir.	Boraginaceae
Bryonia cretica L. subsp. dioica (Jacq.) Tutin, syn Bryonia dioica L.	Cucurbitaceae
Camellia sinensis (L.) Kuntze var. assamica (Masters) Kitamura, syn Thea assamica Masters	Theaceae
Camellia sinensis (L.) Kuntze var. sinensis, syn Thea sinensis Soem.	Theaceae
Campanula persicifolia L.	Campanulaceae
Campanula urticifolia L.	Campanulaceae
Caragana arborescens Lam.	Leguminosae
Carduus nutans L.	Compositae
Carum carvi L.	Umbelliferae
Centaurea iberica Trev. ex Sprengel	Compositae
Centaurea jacea L.	Compositae
Centaurea scabiosa L.	Compositae
Centratherum phyllolaenum Benth.	Compositae

Ceratonia siliqua L.	Leguminosae
Cichorium intybus L.	Compositae
Citrus aurata Poir.	Rutaceae
Citrus limonia Osbeck	Rutaceae
Clytostoma binatum (Thunb.) Sandw.,	
syn Bignonia purpurea Lod. ex Hook. f.	Bignoniaceae
Colchicum autumnale L.	Liliaceae
Colutea arborescens L.	Leguminosae
Consolida regalis S.F. Gray, syn Delphinium consolida L.	Ranunculaceae
Cornus mas L.	Cornaceae
Coronilla varia L.	Leguminosae
Corydalis bulbosa (L.) DC.,	
syn Corydalis cava (L.) Schweigg. & Koerte	Papaveraceae
Cotoneaster multiflora Bunge	Rosaceae
Crocus sativus L.	Iridaceae
Cydonia oblonga Mill., syn Cydonia vulgaris Pers.	Rosaceae
Daphne altaica Pall.	Thymelaeaceae
Daphne laureola L.	Thymelaeaceae
Daphne mezereum L.	Thymelaeaceae
Deutzia gracilis Sieb. & Zucc.	Philadelphaceae
Deutzia scabra Thunb., syn Deutzia crenata Sieb. & Zucc.	Philadelphaceae
Diospyros kaki L. f.	Ebenaceae
Diospyros lotus L.	Ebenaceae
Dorycnium hirsutum (L.) Ser.	Leguminosae
Echinops sphaerocephalus L.	Compositae
Echium creticum L., syn Echium australe Lam.	Boraginaceae
Ehretia tinifolia L.	Ehretiaceae
Elscholzia ciliata (Thunb.) Hyl.,	
syn Elscholzia cristata Willd.	Labiatae
Erica multiflora L.	Ericaceae
Erica scoparia L.	Ericaceae
Erythrina mysorensis Gamb.	Leguminosae
Eugenia munronii Wight, syn Jambosa munronii Walp.	Myrtaceae
Euphorbia resinifera A. Berger	Euphorbiaceae
Feijoa sellowiana Berg.	Myrtaceae
Galega officinalis L.	Leguminosae
Galeopsis ladanum L.	Labiatae
Galeopsis speciosa Mill.	Labiatae
Galium verum L.	Rubiaceae
Geranium gymnocaulon DC.	Geraniaceae
Geranium platypetalum Fisch. & Mey.	Geraniaceae
Gleditsia caspica Desf.	Leguminosae
Gleditsia macracantha Desf.	Leguminosae
Halimodendron halodendron (Pall.) Voss	Leguminosae
Harungana madagascariensis Lam. ex Poiret	Guttiferae
Helleborus niger L.	Ranunculaceae
Hippophae rhamnoides L.	Elaeagnaceae
Hylotelephium caucasicum (Grossh.) H. Ohba,	
syn Sedum caucasicum (Grossh.) A. Boriss.	Crassulaceae
Hylotelephium ewersii (Ledeb.) H. Ohba,	
syn Sedum altaicum Steph. ext.-rod.	Crassulaceae
Hylotelephium telephium (L.) H. Ohba, syn Sedum telephium L.	Crassulaceae
Justicia gendarussa Burm. f.	Acanthaceae

Lathyrus pratensis L.	Leguminosae
Lathyrus sylvestris L.	Leguminosae
Lathyrus tuberosus L.	Leguminosae
Lavatera thuringiaca L.	Malvaceae
Leontodon autumnalis L.	Compositae
Ligustrum vulgare L.	Oleaceae
Limonium gmelinii (Willd.) Kuntze,	
syn Statice gmelinii (Willd.) Kuntze	Plumbaginaceae
Lonicera altaica Pall.	Caprifoliaceae
Lonicera tatarica L.	Caprifoliaceae
Lonicera xylosteum L.	Caprifoliaceae
Malus pumila Mill.	Rosaceae
Malva moschata L.	Malvaceae
Malva pusilla Sm., syn Malva rotundifolia L.	Malvaceae
Melia azedarach L.	Meliaceae
Melilotus altissima Thuill.	Leguminosae
Melissa officinalis L.	Labiatae
Mentha rotundifolia (L.) Hudson	Labiatae
Mespilus germanica L.	Rosaceae
Mitragyna inermis (Willd.) Kuntze	Rubiaceae
Morus alba L.	Moraceae
Myosotis scorpioides L.	Boraginaceae
Myrtus communis L.	Myrtaceae
Nepeta nuda L.	Labiatae
Nigella sativa L.	Ranunculaceae
Odontites verna (Bellardi) Dumort subsp. serotina	
(Dumort.) Corb., syn Odontites serotina Dumort.	Scrophulariaceae
Oenothera biennis L.	Onagraceae
Olea europaea L.	Oleaceae
Onopordon acanthium L.	Compositae
Opuntia ficus-indica (L.) Mill.	Cactaceae
Origanum majorana L.	Labiatae
Origanum vulgare L.	Labiatae
Oyedaea verbesinoides DC.	Compositae
Palicourea crocea (Sw.) Schultes	Rubiaceae
Pappea capensis Ecklon & Zeyher	Sapindaceae
Parapiptadenia rigida (Benth.) Brenan,	
syn Piptadenia rigida Benth.	Leguminosae
Pastinaca sativa L.	Umbelliferae
Pimpinella anisum L.	Umbelliferae
Polemonium caeruleum L.	Polemoniaceae
Poncirus trifoliata (L.) Raf.	Rutaceae
Primula farinosa L.	Primulaceae
Prosopis chilensis (Molina) Stuntz	Leguminosae
Prunus armeniaca L.	Rosaceae
Prunus cerasus L.	Rosaceae
Prunus domestica L.	Rosaceae
Prunus dulcis (Mill.) D.A. Webb, syn Amygdalus communis L.	Rosaceae
Prunus persica (L.) Batsch, syn Persica vulgaris Mill.	Rosaceae
Psoralea drupacea Bunge	Leguminosae
Pulmonaria obscura Dumort.	Boraginaceae
Pulmonaria officinalis L.	Boraginaceae
Pulsatilla vulgaris Mill.,	
syn Anemone pulsatilla L. ucreinica	Ranunculaceae

Punica granatum L.	Punicaceae
Pyrus communis L.	Rosaceae
Quassia amara L.	Simaroubaceae
Quercus petraea (Mattuschka) Liebl., syn Quercus sessiflora Salisb.	Fagaceae
Ranunculus arvensis L.	Ranunculaceae
Ranunculus hyperboreus Rottb.	Ranunculaceae
Ranunculus laetus Wall. ex Hook. f. & Thom.	Ranunculaceae
Ranunculus munroanus J.R. Drum. ex Dunn	Ranunculaceae
Ranunculus muricatus L.	Ranunculaceae
Raphia vinifera Beauv.	Palmae
Reseda odorata L.	Resedaceae
Rhododendron ferrugineum L.	Ericaceae
Rhododendron ponticum L., syn Azalea ponticum L.	Ericaceae
Rhus pentaphilla (Jacq.) Desf.	Anacardiaceae
Roystonea hispaniolana Bailey	Palmae
Rubus antennifer Hook. f.	Rosaceae
Rubus biflorus Buch. Ham. ex Sm.	Rosaceae
Rubus hoffmeisterianus Kunth & Bouche	Rosaceae
Rubus irritans Focke	Rosaceae
Rubus paniculatus Sm.	Rosaceae
Ruta graveolens L.	Rutaceae
Sabal etonia Swingler ex Nash	Palmae
Sabal minor (Jacq.) Pers.	Palmae
Salix acutifolia Wlld.	Salicaceae
Salvia pratensis L.	Labiatae
Salvia verticillata L.	Labiatae
Satureia hortensis L.	Labiatae
Scrophularia vernalis L.	Scrophulariaceae
Sechium edule (Jacq.) Sw.	Cucurbitaceae
Senecio lobatus Pers.	Compositae
Solenanthus circinatus Ledeb.	Boraginaceae
Solidago virgaurea L.	Compositae
Sophora japonica L.	Leguminosae
Spiraea salicifolia L.	Rosaceae
Stachys germanica L.	Labiatae
Stachys recta L.	Labiatae
Symphoricarpos chenaultii Rehd.	Caprifoliaceae
Thevetia peruviana (Pers.) Merr.	Apocynaceae
Thymus pulegioides L.	Labiatae
Trachyspermum copticum (L.) Link	Umbelliferae
Tragopogon pratensis L.	Compositae
Trichilia arborea DC.	Meliaceae
Trichilia emetica Vahl	Meliaceae
Trifolium ambigum M.Bieb.	Leguminosae
Trifolium lupinaster L., syn Trifolium albens Fisch. ex Loud.	Leguminosae
Trifolium montanum L.	Leguminosae
Trifolium repens latum L.	Leguminosae
Trigonella caerulea (L.) Ser., syn Melilotus caerulea (L.) Desr.	Leguminosae
Trollius europaeus L.	Ranunculaceae
Turbina corymbosa (L.) Raf., syn Rivea corymbosa (L.) Hall.	*Convolvulaceae*

Tussilago farfara L.	Compositae
Ulmus laevis Pallas, syn Ulmus effusa Willd.	Ulmaceae
Ulmus minor Miller, syn Ulmus campestris sensu Sm. non L.	Ulmaceae
Vaccinium myrtillus L.	Ericaceae
Vaccinium vitis-idaea L.	Ericaceae
Viburnum opulus L.	Caprifoliaceae
Vicia cracca L.	Leguminosae
Vicia tenuifolia Roth	Leguminosae
Vincetoxicum hirundinaria Medicus, syn Cynanchum vincetoxicum (L.) Pers.	Asclepiadaceae
Wedelia trilobata (L.) Hitchc.	Compositae
Ziziphora clinopodioides Lam.	Labiatae
Ziziphora tenuior L.	Labiatae
Ziziphus lotus (L.) Desf.	Rhamnaceae

5. IMPORTANT WORLD HONEY SOURCES

MAIN ENTRIES 001–452. NECTAR PLANTS

001 Acacia berlandieri Benth.; Leguminosae

guajilla, huajilla (Es/MEX); guajillo, huajillo (USA)
Shrub, 1–4 m, spiny; fls white
Distribution subtropical N and C America; native to southern USA and north
MEX. **Habitat** desert plant growing with Prosopis and Cactus spp (USA,
Pel/76; Usa/79); forms impenetrable thickets in semi-arid steppes of
north MEX (Ord/83)
Soil some moisture needed. **Rainfall** drought tolerant (Pel/76; Usa/79)

Economic and other uses
Yields gum (Usa/79)

Nectar rating; blooms, nectar flow
N1 USA/TEX(Lov/56; Pel/76)
N2 MEX(Ord/72)
Blooms iii–v (MEX, Ord/83); ii–iv (USA/TX). **Nectar flow** – rain during
flowering stops flow (Lov/56; Pel/76)

Honey flow
Honey yield [medium] 27.0 kg/colony/season (Lov/56)

Honey: physical and other properties
Pfund white (Lov/56; Pel/76); almost water white (Ord/83)
Flavour and aroma mild (Dou/79; Lov/56; Pel/76)

002 Acacia caffra (Thunb.) Willd.; Leguminosae

common hook-thorn (En/3OU), gewone haakdoring (Af)
Shrub/tree, <12 m, one of the least prickly acacias; fls creamy-white
Distribution tropical and subtropical Africa. **Habitat** veld (BOT,
Cra/73); woodland, wooded grassland, and by rivers and streams (Pag/77);
coastal scrub
Temperature frost resistant (Pag/77); **Rainfall** v drought tolerant (Pag/77)

Economic and other uses
Fodder – lvs, pods; ?toxic (Pag/77). **Fuel.** **Timber.** **Other uses**
medicinal

Warning
Lvs and pods ?toxic to animals (Pag/77)

Nectar rating + honeybee species; blooms, nectar flow
N1 BOT[tm](Cra/73)
N RWA[tm](Bau/66)
Blooms ix-xi (southern Africa, Pag/77)

Pollen
P RWA

Honey no data

003 Acacia decurrens (Wendl.) Willd.; Leguminosae

black wattle; acácia-da-Australia, acácia negra (Pt/BRA)
Tree, <12 m, evergreen; fls yellow, slightly fragrant; similar to and
often confused with Acacia mearnsii De Wild.
Distribution tropical and subtropical Africa, Oceania, S America; native
to Australia. **Habitat** naturalized in parts of southern Africa
Soil wide range, but growth indifferent on poor soil (AUS/NSW, And/56);
salt tolerant (Kwe/78)

Economic and other uses
Fuel. Land use windbreak, shade, amenity. **Soil benefit**
stabilization. **Other uses** tannin

Nectar rating + honeybee species; blooms, nectar flow; composition
N1 RWA[tm](Bau/66)
Blooms iv-ix (BRA, Caa/72); ix-x (BRA/RG, Jul/72). **Nectar sugar**
concentration [medium] 23% (Jul/72)

Honey no data

004 Acacia greggii A. Grey; Leguminosae

catsclaw, devil's claws, paradise flower (En/USA); uña de gato (Es/MEX)
Tree/shrub, <5 m, spiny; fls pale yellow
Distribution subtropical N and C America; native to USA. **Habitat** desert
of USA/AZ (Pel/76); desert/dry steppes of north MEX (Ord/72); torrent
beds and along small streams
Soil poor dry soil preferred (Ord/83). **Rainfall** drought tolerant

Nectar rating; blooms, nectar flow; composition
N1 USA/AZ, TX(Pel/76)
N2 MEX(Ord/72; Ord/83); USA/NM(Pel/76)
N USA/AZ(Mof/81); USA/UT(Van/49)
Blooms iii-v and again in summer (USA, Lov/56; Pel/76). **Nectar flow**
heaviest in dry season after rainy autumn/winter (Ord/83); fails in
extreme heat (Pel/76). **Potassium content** and **fluorescence** (AA491/80)

Honey flow
Honey yield (kg/colony/season) [high] 72 (USA, Lov/56); [moderate] 10
(USA/TX, Pel/76)

Pollen
Pl USA/UT

Honey: chemical composition
Water [low] 14-17% (Lov/59d)

Honey: physical properties
Pfund white or extra light amber (Lov/59d); light amber (Ord/83)

005 Acacia mellifera (Vahl) Benth.; Leguminosae

blackthorn, hook-thorn (En/SOU); swarthook (Af)
Shrub/tree, 5-8 m, v spiny; fls cream/white; nectary in fl, also
?extrafloral nectaries on fl buds
Distribution tropical Africa; native to Africa.   **Habitat** dry bushveld
(Joh/73); colonizes overgrazed areas (NAM, Cla/73); on dunes in Kalahari
desert (NAM, Joh/73)
Temperature -7 to 38° (BOT/Kalahari, Cla/73).   **Rainfall** drought tolerant

Economic and other uses
Fodder - pods, twigs and fls.   **Fuel.**   **Timber.**   **Land use** shade.   **Other
uses** yields gum

Warning
Spreads rapidly by seed/vegetatively, forming spiny impenetrable thickets
(Pag/77; Usa/79)

Nectar rating + honeybee species; blooms, nectar flow
N1 BOT[tm](Cra/73)
N2 NAM[tm](Joh/73); SOU,tm(And/73; Joh/73)
Blooms ix (NAM), viii-x (SOU).   **Nectar flow** 2-3 wks (SOU); rain in ii-iv
ensures good flow the following spring (NAM); bees forage late morning to
mid afternoon when hot and dry (BOT, Cla/73)

Pollen
P2 SOU

Honey: physical properties
Colour water coloured (And/73)
Granulation slow

006 Acacia modesta Wall.; Leguminosae

phulai (PAK)
Tree
Distribution subtropical Asia.   **Habitat** plantations in PAK

Nectar **rating + honeybee species; blooms, nectar flow**
Nl PAK,ac(Pak/77, Shh/77)
Blooms iv-v (PAK).   **Nectar flow** can be adversely affected by sudden rise in temperature (Pak/77)

Pollen
P PAK

Honey no data

007 **Acacia polyphylla DC.; Leguminosae**

guarucaia (Pt/BRA)
Distribution tropical S America.   **Habitat** upland forests of coffee-growing areas (latitude 19-23 degrees S, BRA, Smt/60)

Nectar rating + honeybee species; blooms, nectar flow
Nl BRA[tm](Mun/54; Smt/60)
Blooms iii (BRA/SP)

Honey no data

008 **Acacia senegal (L.) Willd.; Leguminosae**
syn Acacia verek Guill. & Perr.

gum acacia, gum arabic tree; gommier (Fr/SEN)
Tree/shrub, 5-15 m, spiny; fls whitish spikes
Distribution tropical and subtropical Africa and Asia; native to Africa and Asia.   **Habitat** arid areas; survives hot dry winds and sandstorms; altitudes 100-1700m in E Africa
Soil poor; rocky sand/clay; no waterlogging; pH 5-8 (Usa/80).
Temperature -4 to 48° (INI, Usa/80); frost tolerant.   **Rainfall** 200-800 mm, 300-450 mm optimum; max dry period 8-11 mths (Nap/83); drought tolerant (Usa/80)

Economic and other uses
Food - pods.   **Fodder** - pods and lvs.   **Fuel.   Timber.   Land use** hedges, shade.   **Soil benefit** N-fixation, erosion control, reclamation of refractory sites.   **Other uses** yields gum arabic; rope from root fibres; medicinal; tannin

Warning
Forms spiny thickets, can become a pest.   Noxious weed in AUS and SOU (Usa/80).   Susceptible to browse damage

Nectar rating + honeybee species
Nl SEN,tm(Dou/70)

Pollen
Pollen grain illustrated and described (Smt/56a)

Honey: physical and other properties
Pfund amber (Dou/70)
Granulation rapid
Aroma v mild

009 Acacia seyal Del.; Leguminosae

mimosa épineux, (Fr/SEN)
Tree/shrub, <12 m, spiny, deciduous; fls yellow, fragrant
Distribution tropical Asia, Africa; subtropical Africa; native to Africa,
W Asia. **Habitat** drier woodland and grassland savanna; may occur on river
banks (Usa/80)
Soil wide range, even heavy clay (Usa/79); free lime ?not tolerated
(Hor/81); inundation tolerated better than by other acacias (Usa/80).
Rainfall >350 mm; drought tolerant (Usa/80)

Economic and other uses
Fodder - lvs, pods, fls (Usa/80).   **Fuel.**   **Timber.**   **Land use** shade.
Other uses yields gum arabic

Nectar rating + honeybee species
N1 SEN[tm](Dou/70; Ndi/74)

Pollen
Pollen grain illustrated and described (Smt/56a)

Honey: physical and other properties
Pfund white (Dou/70)
Aroma v mild

010 Acacia tortilis (Forssk.) Hayne; Leguminosae

umbrella thorn (En/SOU); haak-en-steek (Af); mos'arwa, musa (BOT); semra
(OMA)
Tree, 3-20 m, thorny, fls white/cream/pale yellow, fragrant
Distribution tropical Africa, Asia.   **Habitat** low altitude dry areas in
variety of woodland (Pag/77); veld (BOT, Cra/73)
Soil sandy loam, dunes and rocky soil if well drained; alkaline soil
preferred (Usa/80).   **Temperature** <50°; hardy (Pag/77); protect young
plants from frost (Usa/80).   **Rainfall** 100-1000 mm with 10-12 mths dry
period (Nap/83); drought resistant (Pag/77; Usa/80)

Economic and other uses
Fodder - lvs and pods (Pag/77); but ?toxic to animals (Usa/79).   **Fuel.**
Timber.   **Land use** windbreak, shade, afforesting dry rocky areas,
amenity.   **Soil benefit** sand stabilization, N-fixation

Warning
?Toxic to animals (Usa/79). Thorny, can become a nuisance in humid/sub-humid areas; lateral roots cause difficulties in shallow soil (Usa/79)

Nectar rating + honeybee species; blooms, nectar flow
N1 BOT[tm](Cra/73); OMA,[af+am?](Dut/79); OMA[af+am?](Fil/80)
N2 YEA,am(Fie/80)
N OMA, af(Dut/77)
Blooms xi-i (SOU, Pag/77); iv-vi (OMA)

Honey flow
Honey yield [moderate] 2-3 kg/colony/season (af, Dut/79)

Pollen
Pollen grain illustrated and described (Smt/56a)

Honey no data

011 Acer circinatum Pursh; Aceraceae

vine maple (En/USA)
Tree, <12 m, sprawling, vine-like with crooked limbs
Distribution temperate N America

Nectar rating; blooms, nectar flow; composition
N1 CAF/BC(Con/81)
N2 USA/OR,WA(Pel/76)
Blooms iv, early source of nectar (USA/OR, WA).   **Sugar concentration** [medium] 42% (Pel/76)

Honey flow
Honey yield [moderate] exceptional season 22.6 kg/colony/season (USA, Pel/76)

Pollen
P CAF/BC

Honey: physical and other properties
Pfund amber (Pel/76)
Flavour distinctive (Con/81)

012 Acer macrophyllum Pursh; Aceraceae

Pacific maple; Oregon maple (En/CAF)
Tree, <30 m, deciduous; fls yellow, fragrant
Distribution temperate N America.   **Habitat** Sierra Nevada; coastal ranges of USA/CA to south AL (Pel/76)

Economic and other uses
Timber

Nectar rating; blooms, nectar flow; composition
N1 CAF/BC(Con/81)
Blooms iv-v (CAF/BC). **Nectar flow** cut short by rain (Pel/76). **Sugar
concentration** [medium] 52% (Pel/76)

Pollen
P CAF/BC

Honey no data

013 Acer platanoides L.; Aceraceae

Norway maple; faux sycomore (Fr); Spitzahorn (De); spisslønn (No)
Tree, <30 m, deciduous; fls yellow, fragrant, monoecious, open before lvs
Distribution temperate Europe, W Asia, N America, Oceania; native to
Europe. **Habitat** mountains/uplands of N hemisphere; altitudes <1000 m
Soil chalky, alkaline, cool and deep, also acid sandy soil (Hor/81)

Economic and other uses
Timber. **Land use** shade, amenity

Nectar rating; blooms, nectar flow; composition
N1 URS(Ave/78; Fed/55)
N2 BEG(Cra/84)
N EUR(Maz/82); NOW(Lun/71)
Blooms iv-v (mid EUR). **Nectar secretion** (mg/fl/day) 0.95 (Han/80); 0.42-
0.95 (Maz/82). Also total production: 4.13 mg/male fl, 6.32 mg/female fl
(AA602/79). **Sugar concentration** [medium] 50% (Cir/80; Han/80); 30-50%
(Maz/82); 30.02, 34.48% (2 yrs, AA602/79). **Sugar value** (mg/fl/day)
[medium] 0.47 (Han/80); 0.12-0.47 (Maz/82). Also total production: 1.42
mg/male fl, 1.60 mg/female fl (AA602/79). **Potassium, sodium, calcium
contents** (AA361/68)

Honey flow
Honey potential (kg/ha) [moderate] 200 (GDR, Bec/67); 100-200 (ROM,
Apc/68; Cir/80)

Pollen
P NOW. **Yield** average, 8000 grains/fl (Maz/82). **Colour** of load dark green-
brown (Han/80); brown (Hod/74). **Pollen grain** illustrated and described
(Ayt/71; Ert/63; Hod/74). **Reference slide**

Honeydew
Honeydew produced and collected by bees: from **Periphyllus aceris** (L.),
previously Chaetophorella aceris, Chaitophoridae - rated **D1** (mid EUR,
Hag/66; Klo/65; and from **Periphyllus caracinus** (Koch), previously
Chaitophorinus coracinus (Koch), Chaitophoridae (mid EUR, Klo/65).
Honeydew also produced by **Drepanosiphum platanoidis** (Schrank) Callaphididae
(mid EUR, Klo/65); also in ALG (insect not specified, Ske/72); ROM
(Cir/80)

Honey no data

014 Acer pseudoplatanus L.; Aceraceae

great maple, plane, sycamore, sycamore maple;   sycomore (Fr);   Bergahorn (De)
Tree, <30 m, deciduous;   fls yellowish green, monoecious
Distribution temperate Europe (except N), W Asia, S America;   native to S
and central Europe.   **Habitat** woods, hedges, streamsides and on mts;
widely naturalized in UK;   grows on exposed sites where other spp fail
Soil wide range, but dry well drained soil preferred.   **Temperature** hardy

Economic and other uses
Timber.   **Land use** windbreak, shade, amenity

Nectar rating;   blooms, nectar flow;   composition
N1 URS(Glu/55)
N2 FRA(Lou/81);   GFR(Gle/77);   UK(How/79);   URS(Fed/55)
N EUR(Maz/82)
Blooms v-vi (central EUR).   **Nectar flow** 2-3 wks (How/79).   **Nectar
secretion** (mg/fl/day) 1.04 (Han/80);   0.90-1.16 (Maz/82);   0.410
(Sim/80).   Also total production: 4.87 mg/male fl, 5.92 mg/female fl
(AA602/79).   **Sugar concentration** [medium] 47% (Han/80);   37-46% (Maz/82);
35.5% (Sim/80);   male 39.0%, female 43.2% (AA602/79).   **Sugar value**
(mg/fl/day) [medium] 0.49 (Han/80);   0.31-0.54 (Maz/82);   0.58 (Sim/80).
Also total production: 2.56 mg/male fl, 1.90 mg/female fl (AA602/79).
Sugar analysis (Maz/82)

Honey flow
Honey potential (kg/ha) [moderate] 200 (GDR, Bec/67;   ROM, Apc/68;   Cir/80)

Pollen
P2 FRA.   **P3** GFR.   **P** EUR;   URS.   **Yield** sparse, 23 500 grains/fl
(Maz/82).   **Pollen value** average (Maz/82).   **Chemical analysis** (Maz/82).
Colour of load greenish yellow (Hod/74);   load green, large (av 7.9 mg)
(Maz/82).   **Pollen grain** illustrated and described (Hod/74;   Saw/81).
Reference slide

Honeydew
Honeydew produced and collected by bees: from **Periphyllus aceris** (L.)
Chaitophoridae - rated **D1** (mid EUR, Hag/66);   from **Periphyllus testudi-
naceus** (Fernie), previously P. villosus (Hartig), Chaitophoridae - honeydew
collected   from early or mid v in yrs when tree flowers (mid EUR,
Klo/65).   Honeydew also produced (insect not specified) in: ALG
(Ske/72);   ROM (Apc/68;   Cir/80);   UK (How/79)

Honey: chemical composition
Water [medium] 17.3% (Kir/60)
Dextrin 1.85% of dry matter
Ash [medium] 0.79%
pH 4.37
Colloids 1.5% of dry matter

Honey: physical and other properties
Colour greenish tinge, possibly due to presence of honeydew (How/79).
Pfund light amber (Cra/75;   Kir/60);   amber (How/79)
Optical rotation -5.58 deg (Kir/60)
Granulation slow, fine (Cra/75);   slow, coarse (Kir/60)

Flavour rank when fresh, but improves with age (How/79); like heather (Kir/60). **Aroma** unremarkable (Cra/75)

015 Acer tataricum L.; Aceraceae

Tatarian maple
Shrub/tree, <10 m, deciduous; fls greenish-white, bisexual and male fls
Distribution temperate Europe, Asia; native to SE Europe

Nectar rating + honeybee species; blooms, nectar flow; composition
N1 URS(Ave/78; Fed/55)
N3 POL(Dem/64a)
Blooms v-vi (ROM, Apc/68; Cir/80). **Nectar secretion** (mg/fl/day) male fl 0.23, bisexual fl 0.41 (Jar/60); 0.241 (Sim/80). **Sugar concentration** [medium] male fl 50.67%, bisexual fl 48.40% (Jar/60); 33.4% (Sim/80); 38.8% (AA161/79). **Sugar value** (mg/fl/day) [medium] male fl 0.13, bisexual fl 0.96 (Jar/60); 0.197 (Sim/80)

Honey flow
Honey potential (kg/ha) [high] 300-600 (ROM, Apc/68; Cir/80); [moderate] 100 (BUL, AA161/79); 84.80 (POL, Jar/60); 50-100 (POL, Dem/64a); 100 (URS, Fed/55)

Pollen
P URS. **Yield** good (Cir/80). **Colour** yellow (Cir/80). **Reference slide**

Honeydew produced ROM (Apc/68, Cir/80)

Honey no data

016 Actinodaphne angustifolia Nees; Lauraceae

pisa (INI)
Tree, 4.5 m; fls yellow
Distribution tropical Asia. **Habitat** semi-evergreen forest, W peninsular INI only

Economic and other uses
Non-edible oils and waxes

Nectar rating + honeybee species; blooms, nectar flow
N1 INI/MAH,ac(Pha/67)
Blooms xii-iii (INI/KAR). **Nectar flow** annual (Pha/67)

Honey: chemical composition
Water [medium] 18.72% 8 samples, Pha/67; same data also in Naa/70)
Glucose [medium] 35.55%. **Fructose** [medium] 38.15%. **Melezitose, raffinose** present (Pha/70). **Dextrin** 0.669%
Ash [medium] 0.494% (Pha/67); Na, K, Ca, Mg, Fe, P, Si contents (Pha/70)

Free acid [medium] 36.8 meq/kg (Pha/67);   citric, malic, succinic acids
present (Pha/70)
Amino acids 9 identified including glycine (Pha/70).   **Protein** 1.111%
(Pha/67)
Vitamins in "pisa" honey (plant source may be this sp or A. hookerii, Kai/65)

Honey: physical and other properties
Colour black (Pha/67)
Relative density 1.405.   **Viscosity** of "pisa" honey (Raj/70).   **Optical
rotation** -42 minutes (Pha/67)
Granulation medium, 4-6 mths
Flavour like molasses

017 Actinodaphne hookerii Meissn.;   Lauraceae

pisa (INI)
Distribution tropical Asia.   **Habitat** evergreen hill forest (INI/MAH, Sub/79)

Nectar rating + honeybee species;   blooms, nectar flow
N1 INI/MAH[ac](Chu/65;   Tha/62)
Blooms xii-i (INI/MAH)

Honey flow
Honey yield 10% of total (INI/MAH, Chu/65)

Pollen
P1 INI/MAH.   **Pollen value** [high] stimulates brood rearing and swarming
(Chu/65).   **Reference slide**

Recommended for planting to increase honey production
INI/MAH (Sub/79)

Honey: chemical composition
Water [medium] 19.37% (Pha/62)
Sugars, total 73.50%.   **Glucose** [medium] 33.73%.   **Fructose** [medium]
35.71%.   **Reducing sugars** 69.44%.   **Dextrin** 1.98%
Ash [medium] 0.478%
Free acid [medium] 24.5 meq/kg
Protein 1.135%
Vitamins present in "pisa" honey (plant source may be this sp or A.
angustifolia, Kai/65)

Honey: physical and other properties
Colour burnt amber (Chu/65);   dark black (Pha/62)
Relative density 1.405 (Pha/62).   **Viscosity** of "pisa" honey (Raj/70).
Optical rotation -54 minutes
Flavour like molasses (Chu/65)

018 **Adhatoda vasica Nees; Acanthaceae**

bhaikar (PAK)
Shrub, evergreen, lush, much branched (Usa/80); fls white/purple
Distribution tropical and subtropical Asia; native to Asia.   **Habitat**
waste places of plains and submontane tracts of INI; dry/moist deciduous
forests; river banks, dry slopes, forest margins (Usa/80)
Soil weathered diabase, oolitic limestone, etc (Usa/80).   **Temperature**
frost tender (Usa/80).   **Rainfall** 500-1650 mm; in dry areas must be
watered well until established (Usa/80)

Economic and other uses
Fuel.   **Land use** hedges, amenity.   **Soil benefit** green manure.   **Other uses**
medicinal, pesticides, dyes, herbicides

Warning
Unpalatable to livestock, and if unchecked readily invades new areas (Usa/80)

Nectar rating + honeybee species; blooms, nectar flow
N1 PAK,ac(Pak/77)
N3 INI/MAH[ac](Chu/80)
Blooms viii-x (INI/MAH); xii-iii (PAK)

Pollen
P3 INI/MAH.   **P** PAK

Honey no data

019 **Aegiceras corniculatum (L.) Blanco; Myrsinaceae**

river mangrove, small black mangrove (En/AUS)
Shrub/tree; fls pure white, fragrant
Distribution tropical Asia, Oceania.   **Habitat** landward side of mangrove
swamps and up rivers to tidal limits (AUS/QD, Bla/72); tidal swampy forest
(INI/WBE, Chk/72)
Soil mangrove swamps; salt tolerant

Economic and other uses
Sticks used in oyster culture

Nectar rating + honeybee species; blooms, nectar flow
N1 AUS/QD(Bla/72)
N INI/WBE,ad(Chk/72)
Blooms x-xi (AUS/QD); iii-iv (INI/WBE)

Honey flow
Honey yield [high] 54 kg/colony/season (AUS/QD, Bla/72)

Pollen
P1 AUS/QD.   **Pollen value** [high].   **Colour** greyish (Bla/72).   **Pollen grain**
described (Err/69)

Honey: physical and other properties
Pfund extra white (Bla/72)
Granulation rapid in cool weather
Flavour distinctive

020 Aesculus hippocastanum L.; Hippocastanaceae

horse-chestnut; faux chataignier, marronier (Fr); gemeine Rosskastanie
(De); castagno d'India, ippocastano (It)
Tree, <40 m, deciduous, resinous buds; fls white spotted pink and yellow
Distribution temperate Europe, Asia, Oceania; native to Europe, central
Balkan peninsula and E Bulgaria. **Habitat** mountain woodland, extensively
planted elsewhere
Soil some moisture preferred. **Temperature** -37° tolerated

Economic and other uses
Fodder - mashed seed for cattle (Hor/81). **Fuel. Timber. Land use**
shade, amenity. **Other uses** medicinal

Warning; alert to beekeepers
Warning leaf litter can be a nuisance in built-up areas. **Alert to
beekeepers** nectar and pollen ?toxic to bees (Jon/52; Scu/66; Shw/50a)

Nectar rating; blooms, nectar flow; composition
N1 IRN(Cra/73); URS(Fed/55)
N2 FRA(Lou/81)
N3 GFR(Gle/77); ITA(Ric/78)
Blooms iv-v (UK, How/79). **Nectar secretion** (mg/fl/day) 1.0-5.0 (Han/80);
1.371 (Sim/80); 1.4-1.6 (AA536/80). **Sugar concentration** [high] 69%
(Beu/49); 60-76% (Han/80); [medium] 50% (Cir/80); 50-60% (Jon/52); 49%
(Pel/76); 53.2% (Sim/80); 29.6-45.7% (3 dates, AA160/58); 36.9%
(AA161/79); 40.19-42.44% (AA536/80). **Sugar value** (mg/fl/day) [high]
3.572 (Sim/80); [medium] 1.1 (Beu/49); 0.7-2.7 (Han/80); 1.671
(Sim/75); 0.58-0.65 (AA536/80). **Sugar analysis** (Maz/59; Pec/61;
Pry/44; Wyk/52). **Alert to beekeepers** nectar ?toxic to bees (Jon/52;
Scu/66; Scu/67)

Honey flow
Honey potential (kg/ha) [moderate] 30 (GDR, Bec/67); 30-100 (ROM, Apc/68;
Cir/80); also 0.097 kg/tree (age 8 yrs, BUL, AA161/79)

Pollen
P1 URS. **P2** FRA. **P3** GFR; ITA. **Alert to beekeepers** pollen ?toxic to
bees (Jon/52; Shw/50a). **Colour** of load dull yellow (Cir/80); load dark
red to violet (Han/80); load reddish-brown (Hod/74; How/79). **Pollen
grain** illustrated and described (Ayt/71; Saw/81); chemical composition
(Sta/74). **Reference slide**

Recommended for planting to increase honey production
USA (Hay/81). Propagate by seed; grows vigorously, 60 cm per year. See
Warning; alert to beekeepers

Honey: chemical composition
Water [medium] 18.7% (Dus/67)
Sucrase 26.7.   Catalase 0.6.   **Peroxide number**   (µg/g/h) 57.2 (Dus/67);
307.0 (139.5 after 10 minutes in sunlight, 177.2 [sic] after 6 h, Dus/72)

Honey: physical properties
Colour transparent, almost colourless (Fed/55)
Viscosity "thin".   **Electrical conductivity** 0.00013/ohm cm (Dus/67)
Granulation rapid (Fed/55)

021 Aesculus turbinata Bl.;   Hippocastanaceae

Japanese horse-chestnut;   tochi (JAP)
Tree, 30 m, deciduous;   fls white/cream, with red spots
Distribution temperate Asia;   native to Japan.   **Habitat** mountains and
ravines throughout JAP (Inu/57)

Economic and other uses
Land use amenity

Nectar rating + honeybee species;   blooms, nectar flow;   composition
N1 JAP,ac,also am(Inu/57;   Sak/82)
Blooms vi (JAP).   **Nectar secretion** 0.89 mg/fl/day (AA536/80).   **Sugar
concentration** [medium] 59.60% (AA536/80).   **Sugar value** [medium] 0.53
mg/fl/day (AA536/80)

Honey flow
Honey yield (kg/colony/season) [high] 30, every 2 yrs (JAP, Inu/57);
[medium] 24, in some yrs (JAP, Wah/74)

Pollen
P2 JAP.   **Colour** red

Honey: chemical composition
Water [medium] 19.55% (Aso/60);   19.2% (Ech/75)
Sugars, total 80.8% (Ech/75).   **Glucose** [medium] 33.30% (Aso/60);   32.11%
(Ech/77).   **Fructose** [low] 33.60% (Aso/60);   36.34% (Ech/77).   **Sucrose**
[medium] 3.53% (Aso/60);   1.20% (Ech/77).   **Reducing sugars** 69.63%.
Maltose 4.05% (Ech/77).   **Higher sugars** 1.21% of total sugars (Ech/75)
pH 3.8 (Ech/75)
Amylase 86 900 units/100 ml honey (Ech/75);   435 units/mg protein
(Ech/77).   **Sucrase** 2.37 units/mg protein.   **Glucose oxidase** 13 200
units/100 ml honey (Ech/75);   66.0 units/mg protein (Ech/77).   **Catalase**
0.01 unit/mg protein.   **Acid phosphatase** 2.06 units/mg protein
Protein 0.18% (Ech/75)
Vitamins pantothenic acid 0.7-11.5 ppm (Waa/56)

Honey: physical properties
Colour yellow (Aso/60)

022 Agave americana L.; Agavaceae

American agave; American aloe (En/SOU); century plant, mescal (En/USA);
maguey (Es/HOD); agave (It)
Herb, 1-2 m, spiny, rosette of tough rigid lvs; fls yellowish-green, many
on 8-9 m stalk of plants of approx 10 yrs; plant dies after flowering
Distribution subtropical N America, tropical and subtropical C America,
(Med) Europe; native to MEX. **Habitat** arid and semi-arid areas of W
hemisphere; when cultivated for ornament full sun is required
Soil limestone; ?salt tolerant. **Rainfall** drought tolerant but irrigation
improves growth

Economic and other uses
Food - sap fermented in MEX to produce alcoholic drinks. **Land use** hedges,
amenity. **Other uses** fibres from lvs

Alert to beekeepers
Bees "cross" on this flow (USA, Pel/76)

Nectar rating + honeybee species; blooms, nectar flow
N1 USA/AZ(Pel/76)
N3 ITA(Ric/78); SOU,tm(And/73); USA/AZ(Lov/56)
N HOD(Ord/63)
Blooms xi-iii (SOU); vi-viii in Med area. **Alert to beekeepers** bees
"cross" on this flow (USA, Pel/76)

Honey flow
Honey yield [high] 41 kg/colony/season (MEX, Pel/76)

Pollen
P3 SOU. **Chemical analysis** (Sta/74). **Pollen grain** illustrated and
described (Heu/71); extremely under-represented because of large grain
size (Ric/78). **Reference slide**

Honey: physical and other properties
Colour dark (And/73; Lov/56); v dark (Pel/76)
Flavour strong, "poor" (Lov/56). **Aroma** of sour grain mash (Lov/57a)

023 Aloe davyana Schonl.; Liliaceae

davyana aloe (En/SOU)
Herb, small inconspicuous succulent, grows in clumps; fls pink
Distribution subtropical Africa. **Habitat** NE Kalahari thornveld, in the
shelter of trees and scrub (Cri/57)
Soil sandy (Cri/57). **Temperature** fls damaged by frost (SOU, Joh/75)

Alert to beekeepers
Bees reported "v aggressive" on this flow (Cra/73a)

Nectar rating + honeybee species; blooms, nectar flow
N1 SOU,tm(And/73; Mou/72)

Blooms vii–viii (SOU, Cri/57).   **Nectar flow** fls occasionally damaged by frost and fire (SOU, Joh/75).   **Alert to beekeepers** bees reported "v aggressive" on this flow (Cra/73a)

Pollen
Pl SOU.   **Pollen value** [high], and reported to lead to swarming in SOU (Cri/57; Joh/75).   **Colour** reddish-brown (Cri/57)

Honey: chemical composition
Glucose [medium] 39.1% (Joh/76a).   **Fructose** [medium] 36.2%.   **Sucrose** [medium] 3.5%
pH 3.5.   **Total acid** 42.2 meq/kg
Amylase 13.3.   **HMF** 1.3 ppm

Honey: physical and other properties
Colour v light, almost colourless (And/73); clear (Cri/57; Joh/75).
Pfund water white
Optical rotation –23.8 deg (Joh/76a)
Granulation rapid, fine (And/73); accelerated by low night and high day temps (Joh/75); soft creamy consistency like lard (Cri/57)
Flavour rather sickly sweet, not pronounced (Cri/57); mild (Joh/76a).
Aroma none (And/73); little (Joh/75)

024 Aloe dichotoma Masson; Liliaceae

quiver tree (En/SOU); keetmanshoop, kokerboom (Af)
Tree/shrub, 3–5 m, exceptionally 7 m
Distribution subtropical and tropical Africa; native to Africa.   **Habitat** dry desert and semi-desert areas, on and among rocky hills (Pag/77)
Soil not too damp (Pag/77).   **Temperature** not cold areas (Pag/77).
Rainfall drought tolerant

Nectar rating + honeybee species; blooms, nectar flow
N1 SOU,tm(Joh/73)
Nectar flow abundant after good rainy season (Joh/73)

Pollen
P NAM

Honey no data

025 Aloe mutans Reynolds; Liliaceae

mutans aloe (En/SOU)
Herb
Distribution tropical and subtropical Africa; native to Africa.   **Habitat** banks of Olifants River (SOU, Cri/57)

Alert to beekeepers
Colonies on this flow inclined to swarm (Mou/72)

Nectar rating + honeybee species; blooms, nectar flow; composition
N1 SOU,tm(And/73; Mou/72)
Blooms vii (SOU).   **Nectar flow** short and sharp (SOU).   **Alert to beekeepers** colonies on this flow inclined to swarm (Mou/72).   **Sugar concentration** [low] 18-20% (Joh/76a)

Pollen
Pl SOU

Honey: physical and other properties
Colour v pale straw (And/73);   clear (Cri/57).   **Pfund** light amber (Cri/57)
Granulation rapid, fine creamy (And/73);   colour then slightly yellowish (Cri/57)
Flavour and aroma mild (And/73)

026 Aloysia gratissima (Gill. & Hook.) Troncoso;   Verbenaceae
syn Aloysia ligustrina Small

beebrush, whitebrush (En/USA);   huele de noche, jozmincillo, vera dulce (Es/MEX)
Shrub; fls white
Distribution subtropical N America.   **Habitat** semi-desert
Rainfall drought tolerant but plant dies back during dry spell

Economic and other uses
Land use hedges, amenity

Nectar rating; blooms, nectar flow
N1 USA/TX(Lov/56)
N2 MEX(Ord/83)
Blooms vi-xii.   **Nectar flow** heaviest following rain (USA/TX, Lov/56d)

Honey flow
Honey yield [moderate] 2-3 kg/colony/season (MEX, Ord/83)

Honey: chemical composition
Water [low] can be 12-13% in dry deserts (Lov/56d)

Honey: physical and other properties
Pfund white (Lov/56)
Viscosity "heavy body"
Granulation rapid (Lov/56d)
Flavour mild (Lov/56).   **Aroma** delicate

027 Aloysia virgata (Ruiz & Pav.) A.L.Juss.; Verbenaceae
syn Lippia urticoides (Cham.) Steud.

acerillo (Es/BOL); lixa (Pt/BRA)
Shrub; fls white, fragrant
Distribution subtropical N America, Caribbean, subtropical and tropical S
America; native to Brazil. **Habitat** stony sloping farmland; does not
thrive in shade (Ord/83)
Soil rocky soil; optimum pH 6-7; not flooded ground (Ord/83).
Temperature cold tolerated (USA/FL, Ord/83)

Economic and other uses
Medicinal

Nectar rating + honeybee species; blooms, nectar flow; composition
N1 BOL[tm](Kem/71); BRA/SP[tm](Ker/60)
N BRA/SP[tm](Fle/63)
Blooms 2 mths(BRA); all yr (Caribbean). **Nectar flow** heaviest on moist
soil; higher in morning than afternoon (Ord/83). **Sugar concentration**
[medium] 24-46% (Caa/72); 33-36% (Kem/80)

Recommended for planting to increase honey production
Tropical America (Ord/83). Propagate by seed/cuttings; when mature does
not need watering/weeding; v resistant to pests

Honey: physical properties
Colour light (Ord/83)

028 Ampelopsis arborea (L.) Koehne; Vitaceae

crossvine, peppervine, pepperidge, snowvine (En/USA)
Shrub, vigorous climber; fls greenish, v small
Distribution subtropical N America. **Habitat** swamps, streams in S USA;
thickets and on fences in low land (USA/FL)
Soil rich, moist

Nectar rating; blooms, nectar flow; composition
N1 USA(Pel/76)
Blooms vi-viii (USA, Lov/66a); 2 mths USA/LA. **Nectar flow** continues in
rain (Pel/76). **Sugar concentration** [medium] 34% (AA906/76)

Honey flow
Honey yield (kg/colony/season) [high] 23-36 (USA, Pel/76); [moderate] 10-
23 (USA, Lov/66a)

Pollen
Pollen grain illustrated (Lie/72). **Reference slide**

Honey: chemical composition
Water [high] 20-30% (Lov/60)
Fermentation on storage likely (Mot/64; Lov/60)

Honey: physical and other properties
Pfund amber (Lov/60); dark amber (Mot/64)
Granulation rapid (Mot/64)

029 Anacardium excelsum (Bert. & Balbis ex Kunth) Skeels; Anacardiaceae

espavé (Es/PAM)
Tree, <35 m; fls whitish
Distribution tropical C and S America. **Habitat** forest in regions <800 m,
often forming pure stands (Ord/83)

Economic and other uses
Food fruit. **Timber.** **Land use** shade/windbreak, coffee plantations COL

Nectar rating + honeybee species
N1 PAM[tm](Ord/83)

Honey no data

030 Anacardium occidentale L.; Anacardiaceae

cashew nut; marañon (Es/COL); jambu mété, monyet (In)
Tree, <10 m, evergreen, straggly; fls pink, may be striped yellow, small,
fragrant
Distribution tropical and subtropical regions; native to BRA. **Habitat**
low country, altitudes <1200 m
Soil wide range including sandy soil, eroded or other poor sites; not low
waterlogged sites or rock. **Rainfall** 500-700 mm; drought tolerant

Economic and other uses
Food - roasted nuts. **Fuel.** **Timber.** **Soil benefit** cover and conservation. **Other uses** ink from bark; insect-repellent oil from nut
shell; liquid in husk for insulating medium; tannin

Warning
Husk toxic to man

Nectar rating + honeybee species; blooms, nectar flow
N1 COL(Cor/76); GUY[tm](Cra/73); INI/KER[ac](Dev/71; Hol/65)
N2 INO[ac](Bee/77)
Blooms ii-iii, also v-vi (tropical America, Ord/83)

Pollen
P3 INO. **Pollen grain** illustrated and described (Smt/56a). **Reference slide**

Honey no data

031 Anchusa officinalis L.; Boraginaceae

common alkanet, true alkanet; buglosse, langue de boeuf (Fr)
Herb, 20-80 cm, perennial/biennial, erect; fls violet/reddish rarely white/yellow
Distribution temperate Europe but not extreme N nor much of W; wasteland, vineyards

Economic and other uses
Medicinal

Nectar rating; blooms, nectar flow; composition
N1 POL(Dem/64a)
Blooms vi-vii (ROM, Apc/68; Cir/80). **Nectar secretion** (mg/fl/day) 1.73, 2.22 (2 yrs, Dem/63a); 1.3-4.4 (Maz/82). **Sugar concentration** [medium] 47.6, 55.6% (Dem/63a); 40-58% (Han/80; Maz/82). **Sugar value** (mg/fl/day) [medium] 0.823, 1.237 (Dem/63a); 0.6-2.5 (Maz/82). **Sugar analysis** (Wyk/52)

Honey flow
Honey potential (kg/ha) [high] 515 (Dem/63a); [medium] 400-500 (Maz/82); 57-100 (GDR, Bec/67); 407 (POL, Dem/63a); 50-100 (ROM, Apc/68; Cir/80)

Pollen
Pollen grain illustrated and described (Ert/63). **Reference slide**

Honey: chemical composition
Sugars (as % of total, Maz/64): **glucose** 37.5%; **fructose** 47.6%; **sucrose** 7.5%; **maltose** 3.7%; **fructomaltose** 3.7%

Honey: physical and other properties
Electrical conductivity 0.000145/ohm cm (Vor/64)
Granulation slow (Dem/64)

032 Andira inermis (Wright) DC.; Leguminosae

angelin; yaba (Es/CUB); almendro (Es/HOD, NIA); cola de pescado (Es/HOD)
Tree, <30 m; fls pink-violet/purple
Distribution tropical C and S America, Caribbean, W Africa. **Habitat** dry/rain forests, marshes and river banks, altitudes <900 m; humid/sub-humid zones

Economic and other uses
Timber, Land use shade in coffee plantations (PUE, Ord/83)

Warning
Honey "possibly harmful" [?]to man (Ord/83)

Nectar rating; blooms, nectar flow; composition
N1 HOD(Ord/63); NIA(Ord/63a); PUE(Ord/72)
N CUB(Ord/83)
Blooms vi-vii (tropical America, Ord/83). **Sugar concentration** [medium] 25% in fls 2.5 m from ground, 40% at 11 m (AA336/79)

Honey: chemical composition
Toxicity "possibly harmful" [?]to man (Ord/83)

Honey: physical properties
Pfund amber

033 Angelica archangelica L.; Umbelliferae
syn Archangelica decurrens Ledeb.; Archangelica officinalis (Moench.) Hoffm.

angelica
Herb, 2 m, biennial/perennial; fls greenish, in large umbels
Distribution temperate Europe, Asia. **Habitat** damp places; taiga region
of Siberia; often cultivated, frequently naturalized

Economic and other uses
Food - aromatic young stalks and petioles for confectionery and liqueurs;
roots/seeds for essential oil

Nectar rating
N1 URS/Siberia(Ave/78; Fed/55)
N2 URS/Krasnoyarsk(Kov/65)
N URS(Clu/65)

Honey flow
Honey yield [high] 5.5-6.0 kg/colony/day in best conditions (Ave/78).
Honey potential [moderate] 90 kg/ha (ROM, Apc/68; Cir/80)

Pollen
P1 URS. Reference slide

Honey: chemical composition
Sugars (as % of total, Maz/64): **glucose** 30.6%; **fructose** 57.3%; **sucrose**
2.3%; **maltose** 9.8%

Honey: physical properties
Colour dark (Dem/64); reddish (Glu/55)
Electrical conductivity 0.000287/ohm cm (Vor/64)

034 Antigonon leptopus Hook. & Arn.; Polygonaceae

Mexican creeper (En/BAR); confederate vine, pink-vine (En/USA); coral
creeper (En/ZIM); liane antigone (Fr/MAY); bellísima (Es/COL); coralillo
rosada (Es/CUB); bellacima (Es/DOR); fulmina (Es/MEX); rosa de montana
(USA)
Shrub, climber 8-15 m, deciduous perennial, tendrils at ends of racemes;
fls rose-pink
Distribution tropical and subtropical S and C America, subtropical N
America, Caribbean, tropical Africa and Asia; native to W MEX and N
central America. **Habitat** cultivated throughout the tropics as an ornamental

Economic and other uses
Food - tubers. **Land use** amenity

Nectar rating + honeybee species; blooms, nectar flow; composition
N1 BAR(Rin/79); INI/BIH[ac?](Nai/76); MAY(Bro/82)
N2 BRA/SC[tm](Wie/80); COL(Cor/76); CUB(Ord/44; Ord/56; Ord/83);
DOR(Ord/64); MEX(Ord/83); USA/FL(Pel/76)
N ZIM[tm](Pap/73)
Blooms usually v, vii-viii, and at other times (BRA/SP, Sao/54); all yr,
especially in spring and summer (tropical America, Ord/83). **Nectar
secretion** higher on cloudy humid days (Ord/54). **Sugar concentration**
[medium] 22-28% (Kem/80); 44-50% (Sao/54); 28-40% (Wie/80). **Sugar
analysis** (Vah/72)

Pollen
P1 INI/BIH. **P** TRI (Lau/76a); ZIM. **Reference slide**

Recommended for planting to increase honey production
CUB, USA/FL (Ord/54). Propagate by seed; needs careful weeding and
watering when young; "the only vine in the tropics capable of keeping
colonies active all through the yr" (Ord/54)

Honey: chemical composition
Water [medium] 16.3, 17.2% (2 samples, age 14, 2 mths (Whi/62); also
[high] over 20% in Cuba (Ord/83)
Glucose [low] 28.68, 28.24% (Whi/62). **Fructose** [low] 34.87, 34.84%.
Sucrose [low] 0.60, 0.62%. **Maltose** 6.16, 6.05%. **Higher sugars** 3.05,
3.01%. **Melezitose** 3.15, 3.06%
Ash [medium] 0.616, 0.567%
pH 4.35, 4.30. **Total acid** 54.81, 55.41 meq/kg. **Free acid** [high] 45.51,
46.91 meq/kg. **Lactone** 9.30, 8.50 meq/kg
Nitrogen 0.039, 0.074%

Honey: physical and other properties
Colour light (Cra/75); often black (Lov/56); dark brown (Mot/64); almost
white (Pel/76). **Pfund** light to dark amber (Ord/83); 104-114 mm, amber
(Whi/62)
Granulation [slow] none in 4 yrs (Mot/64)
Flavour like aster honey, or like molasses with a tang. **Aroma** character-
istic (Cra/75); strong (Pel/76)

035 Asclepias syriaca L.; Asclepiadaceae

milkweed; silkseed (En/USA)
Herb, 1-1.5 m, perennial; fls pink
Distribution temperate Europe, N America; native to America. **Habitat** dry
grassland; has been cultivated as a food plant for bees

Economic and other uses
Fibre

Alert to beekeepers
Sticky pollinia trap bees (USA, Hol/75)

Nectar rating; blooms, nectar flow; composition
N1 POL(Dem/63); ROM(Cir/80); URS(Fed/55); USA/NH(Hol/75)
N2 USA(Pel/76)
N USA(Lov/56)
Blooms vii-viii (ROM, Apc/68; Cir/80); early summer (USA/NH). **Nectar flow** each fl 4-6 days (AA581/63). **Nectar secretion** (mg/fl/day) 7.01-10.39 (3 yrs, Dem/63a); 3.087 (Sim/80); total production 7.0, 6.6 mg/fl (normal yr, dry yr, AA581/63). **Sugar concentration** [medium] 27.4-32.3% (Dem/63a); 25.2% (Shw/53); 39.2% (Sim/80); 14.5-25.6% (Mog/58); 29.5, 58.2% (normal yr, dry yr, AA581/63); 32.3% (AA526/82). **Sugar value** (mg/fl/day) [high] 3.633 (Sim/80); [medium] 2.22-2.85 (Dem/63a); also 8-12 mg/fl over 3-5 days, max 0.23-0.75 mg/fl/h in afternoon and evening (AA165/83). **Sugar analysis** (AA526/82)

Honey flow
Honey yield (kg/colony/season) [high] 45 (USA, Pel/76); 50 (USA/MI, Pel/76). **Honey potential** (kg/ha) [high] 187-576 (3 yrs, POL, Dem/63a); 400-600 (ROM, Apc/68; Cir/80); 300 (URS, Fed/55)

Pollen
Alert to beekeepers sticky pollinia trap bees (USA, Hol/75). **Pollen yield** good (Cir/80). **Colour** of load yellow (Cir/80)

Honey: chemical composition
Sugars (as % of total, Maz/64): **glucose** 33.4%; **fructose** 48.2%; **sucrose** 4.5%; **maltose** 7.1%; **fructomaltose** 6.8%

Honey: physical and other properties
Colour v light (Cra/75); yellow (Lov/56)
Viscosity "thick and heavy" (Roo/74)
Granulation slow, may take yrs (Cra/75)
Flavour characteristic (Lov/56); fruity, rather like quince, with light tang (Roo/74). **Aroma** of the flowers (Cra/75)

036 Astragalus sinicus L.; Leguminosae

Chinese milk vetch, milk vetch
Herb/shrub, short-lived; fls red-purple
Distribution temperate and sub-tropical Asia; native to China
Soil - sp is cultivated on rice soils as a soil improver (Zev/75)

Economic and other uses
Soil benefit erosion control, green manure

Nectar rating; blooms, nectar flow; composition
N1 CHN(Mad/81); JAP(Inu/57; Sak/82); TAI(Kin/79)
Blooms spring, iv (JAP). **Sugar concentration** in nectar [medium] 32.3% (Ech/77). **Sugar analysis** (Ech/77; Ech/77)

Honey flow
Honey yield [high] 30 kg/colony/season (JAP, Inu/57)

Pollen
Pl JAP.   **Chemical analysis** (AA188/76).   **Reference slide**

Recommended for planting to increase honey production
JAP(Sak/82).   Propagate by seed; grows rapidly, short-lived (Why/53)

Honey: chemical composition
Water [high] 22.0% (3 samples, Aoy/68); 26.20% (Aso/60); 20.4% (Ech/75).
Sugars, total 79.6% (Ech/75).   **Glucose** [medium] 30.77% (Aso/60);   33.06%
(Ech/77).   **Fructose** [medium] 35.56% (Aso/60); 39.63% (Ech/77).   **Sucrose**
[medium] 3.85% (Aso/60); 0.80% (Ech/77).   **Reducing sugars** 66.33%
(Aso/60).   **Maltose** 3.23% (Ech/77)
Ash [low] 0.03-0.06% (Aoy/68)
pH 3.20-3.75 (5 samples, Aoy/68); 3.8 (Ech/75).   **Total acid** 29.1-61.3
?meq/kg (Aoy/68).   **Free acid** [medium] 25.7-58.0 ?meq/kg.   **Lactone** 3.3-4.0
?meq/kg.   Results for 9 organic acids (Ech/77)
Amylase 9.1-30.0 (6.6-27.2 after 2 mths, 8.1-25.0 after 4 mths, Aoy/68);
results also in Ech/75, Ech/77, Mau/71.   **Sucrase** 0.47 unit/mg protein
(Ech/77);   trace, 2.55 g/100g/h (Mau/71).   **Glucose oxidase** 800 units/100
ml honey (Ech/75);   4.0 units/mg protein (Ech/77).   **Catalase** 0.015 unit/mg
protein.   **Acid phosphatase** 1.21 unit/mg protein (Ech/77);   22.6, 62.8 µ
moles/100g/h (Mau/71)
Nitrogen 0.022-0.044% (Aoy/68).   **Amino acids** 0.07% (Ech/77);   contents of
individual acids (Aoy/68; Ech/77).   **Protein** 0.21% (Ech/75);   0.10, 0.16%
(Mau/71)
Other constituents - compounds probably contributing to flavour (Wab/80)

Honey: physical properties
Colour pale brown (Aso/60).   **Pfund** extra light amber (Mad/81);   water
white (Pel/76)

037 Avicennia germinans (L.) L.; Avicenniaceae
syn Avicennia nitida Jacq.

black mangrove; blacktree, blackwood (En/USA); mangrove (Fr/MAT);   mangle
(Es/HOD, NIA); mangle prieto (Es/CUB, DOR, MEX);   courida (Es/GUY)
Tree/shrub, <18 m, evergreen; fls whitish, small
Distribution tropical Africa, S and C America, Caribbean;   subtropical N
America.   **Habitat** low marshy sea coasts of USA/FL, Gulf coast to TX and
tropical America, often forms thick groves around bays and river mouths
(Ord/83)
Soil salt tolerant (Pel/76).   **Temperature** plant damaged or killed by frost
(USA, Lov/62d; Pel/76)

Economic and other uses
Fuel.   Timber

Nectar rating + honeybee species; blooms, nectar flow; composition
N1 CUB(Ord/44); DOR(Ord/64); GUY(Bee/76; Cra/73; Cra/79); HOD
(Ord/63); MEX(Ord/72); NIA(Ord/63a); SUR(Bee/76); SEN[tm](Dou/70)
N2 CUB(Ord/83); USA/FL,TX(Mor/56)
N3 USA/FL,TX(Pel/76)
N GUS[tm](Sve/80); MAT(Bal/76); SUR(Cra/79); TRI(Lau/76)
Blooms vi-vii. **Nectar flow** 6-8 weeks USA/FL; trees can become salt-
coated in dry weather, which discourages bees; v sensitive to weather
(Pel/76). **Sugar analysis** (Vah/72)

Honey flow
Honey yield [high] mean 23-27, max 90 kg/colony/season (USA/FL, Lov/62d)

Pollen
P DOR. Reference slide

Honey: chemical composition
Water [high] 24% (Lov/58a)
Glucose [high] (Lov/61)
Fermentation on storage likely (Cra/75)

Honey: physical and other properties
Colour light (Cra/75; Pel/76); dark but light if not unifloral (Mot/64);
usually dark but light in Cuba (Ord/83). **Pfund** white to light amber
(Lov/56); water white or extra light amber (Lov/62d)
Viscosity "thin bodied" (Lov/56)
Granulation rapid (Mot/64; Ord/83); fine grain (Lov/61)
Flavour mild (Cra/75; Pel/76); slightly salty or brackish (Lov/56;
Mot/64); can be sweet (Lov/61). **Aroma** sometimes "swampy" (Lov/61)

038 Avicennia marina (Forssk.) Vierh. var. resinifera (Forst.) Bakh.; Avicenniaceae

grey mangrove, white mangrove
Tree/shrub, <12 m, peg-like root branches project above mud; fls deep
yellow/orange
Distribution tropical, subtropical Oceania. **Habitat** coastal; muddy
estuaries/backwaters/banks of tidal streams (AUS/QD, Bla/72)
Soil mud/silt; salt tolerant

Economic and other uses
Food - fruit can be eaten. **Fodder** lvs palatable to stock. **Timber**

Nectar rating; blooms, nectar flow
N1 AUS/SA(Boo/72)
N3 AUS/QD(Bla/72)
Blooms ii (Aus/QD, Bla/72)

Honey flow
Honey yield [moderate] max 18 kg/colony/season (AUS/QD, Bla/72); good
every 2 yrs (AUS/SOU, Boo/72)

Pollen
P2 AUS/QD.   **P3** AUS/SA.   **Yield** low to moderate (Boo/79).   **Pollen value**
good (Boo/79)

Honey: physical and other properties
Pfund extra light amber to light amber (Bla/72; Boo/72); light amber
(Wal/78)
Viscosity "light bodied" (Wal/78)
Granulation slow (Bla/72); soft grain (Wal/78)
Flavour strong

039 Azadirachta indica A. Juss.; Meliaceae

nim
Tree, <11 m, evergreen except in extreme drought; fls white, fragrant
Distribution subtropical and tropical Africa, Asia.   **Habitat** dry areas
Soil dry stony, clay, shallow or nutrient-deficient soils;   optimum pH
6.2;   not laterite outcrops;   not on waterlogged/saline soil (Usa/80).
Temperature range 0° (occasional) to 44° (shade);   seedlings killed by
frost (Usa/80).   **Rainfall** 130-1150 mm, >450 mm preferred;   drought tolerant

Economic and other uses
Food ?fruit.   **Fodder** - lvs (Asia but not in W Africa).   **Fuel.**   **Timber**
termite resistant.   **Land use** windbreak, shade, afforestation, amenity.
Soil benefit lvs for mulch;   reclamation of arid waste land.   **Other uses**
medicinal;   insecticide;   oil for lamps/lubrication/soap;   gas generation
(Usa/80)

Warning
Tree may be "aggressively" invasive (Usa/80)

Nectar rating + honeybee species; blooms, nectar flow
N1 INI/UTT,ac(Kap/57; Koh/58)
N2 INI/MAH[ac](Chu/80)
N INI/TAM[ac](Ram/37)
Blooms iii-iv (INI/UTT); iii-v (W Africa, Pam/77)

Pollen
P3 INI/MAH.   **Pollen grain** illustrated and described (Nak/65)

Recommended for planting to increase honey production
INI/MAH (Sub/62).   Propagate by fresh seed (Usa/80);   seed production
starts at 5 yrs;   grows rapidly.   See Warning

Honey: chemical composition
Water [high] 22.88% (Sig/62)
Sucrose [high] ?7.46%
Ash [low] 0.06%
Free acid [medium] 20.8 meq/kg

Honey: physical and other properties
Colour light golden (Koh/58)

Viscosity "thin"
Flavour slightly bitter

040 Baccharis dracunculifolia DC.;   Compositae

alecrim, vassoura (Pt/BRA)
Shrub (under-shrub), perennial; fls cream
Distribution subtropical and tropical S America;   native to BRA/RS.
Habitat agricultural areas;   fields, wasteland, edges of woodland scrub
(Jul/72)

Nectar rating + honeybee species;   blooms, nectar flow;   composition
N1 BRA/SP[tm](Wie/80)
N BRA[tm](Fle/63)
Blooms iii-iv (BRA/RG);   iv-vi (BRA/SC);   xi-ii (BRA/SP).   **Sugar
concentration** [high] 40-42% (Jul/72);   [medium] 25-30% (Wie/74);   [low] 12-
16% (Ama/79)

Pollen
P BRA/SP.   **Reference slide**

Honey no data

041 Baikiaea plurijuga Harms;   Leguminosae

Rhodesian teak
Tree, 8-16 m, deciduous;   fls mauve, large
Distribution tropical Africa.   **Habitat** open deciduous woodland (Pag/77);
ZAM, ZIM forms extensive and often pure stands in association with
Julbernardia and Brachystegia spp
Soil characteristic of areas of deep Kalahari sand

Economic and other uses
Timber.   **Other uses** tannins from bark and heartwood

Nectar rating + honeybee species;   blooms, nectar flow
N1 ZAM[tm](Smt/59)
Blooms xii-iii (southern Africa, Pag/77)

Honey no data

042 Banksia serrata L.f.;   Proteaceae

red honeysuckle, saw banksia (En/AUS)
Tree, <15 m, usually small and gnarled;   fls faintly bluish-grey then
yellow, cone shaped

Distribution temperate and subtropical Oceania; native to AUS/NSW, TAS, VIC
Soil poor, sandy (AUS, And/56)

Economic and other uses
Fuel.   Timber

Nectar rating; blooms, nectar flow
N1 AUS/VIC(Gom/73)
N2 AUS/NSW(Goo/47)
Blooms xiii-iv (AUS/VIC)

Honey flow
Honey yield heavier every 2 yrs (Gom/73)

Pollen
P1 AUS/VIC.   **P** AUS/NSW.   **Yield** abundant (Gom/73).   **Pollen value** [high]

Honey: chemical composition
Volatiles identified: acetoin (major), 5-hydroxy-methyl-2-furaldehyde and
methyl syringate (minor), and 7 others (traces) (Grd/79)

Honey: physical and other properties
Colour medium to dark (Goo/73)
Granulation rapid

---

043 Barbarea vulgaris R. Br.;   Cruciferae

rocket gentle, winter cress; yellow rocket (En/USA)
Herb, <1 m, biennial/perennial; fls yellow
Distribution temperate N America, Europe; native to Europe.   **Habitat**
usually wet/damp places (Tut/64); widely distributed weed (USA, Pel/76)

Economic and other uses
Food - salad/garnish

Nectar rating; blooms, nectar flow
N1 GDR(Koc/65);   URS(Glu/55)
N2 USA/OH(Lov/56)
Blooms iv-vii (GDR, Bec/67); v-vi (ROM, Cir/80)

Honey flow
Honey potential (kg/ha) [moderate] 30-35 (GDR, Bec/67);   30-40 (ROM,
Cir/80);   30 (URS, Fed/55);   51-100 (Cra/73)

Pollen
P1 URS.   **Chemical analysis** (Shp/80)

Honey: physical and other properties
Colour greenish yellow (Glu/55);   rich yellow (Lov/56)
Viscosity "thin" (Glu/55)
Flavour strong

044 Berchemia scandens (Hill) K. Koch;   Rhamnaceae

rattan vine, supple jack (En/USA)
Shrub, long climbing/trailing vine;   fls greenish, small
Distribution subtropical N America;   native to USA.   **Habitat** low thickets
throughout southern USA (Pel/76);   abundant along streams in eastern TX
(Pel/76)

Nectar rating;   blooms, nectar flow
N1 USA/LA(Lie/72)
N2 USA/AL,TX(Pel/76)
Blooms late spring (southern USA, Pel/76)

Honey flow
Honey yield [moderate] 7-14 kg/colony/season (USA/TX, Lov/56)

Pollen
Pollen grain illustrated (Lie/72)

Honey: physical and other properties
Pfund amber (Lov/56; Lov/60b);   dark amber (Pel/76; Lov/61a)
Viscosity "good body" (Lov/56)
Granulation slow, almost none after 10 yrs (Lov/60b)

045 Bidens pilosa L.;   Compositae

Spanish needle (En/USA);   romerillo blanco (Es/CUB);   clavelito del monte,
romerillo (Es/DOR);   aceitilla (Es/MEX);   picão (Pt/BRA)
Herb, <1 m, annual, erect;   fls creamy-white
Distribution subtropical N America;   tropical Caribbean, C America.
Habitat weed of cultivated land, disturbed ground, gardens;   altitudes
<2000 m
Soil wide range except very wet (Ord/83)

Economic and other uses
Food - lvs

Nectar rating + honeybee species;   blooms, nectar flow;   composition
N1 tropical America (Ord/52)
N3 DOR[tm](Ord/64);   USA/FL(Mor/56)
N Antilles(Ord/83);   MEX(Ord/83);   USA/FL(Ord/83)
Blooms most of yr (tropical America, Ord/52).   **Sugar concentration**
[medium] 24% (Ama/79)

Honey flow
Honey yield "high", but only where plant is abundant;   some honey exported
as B. pilosa may be partly derived from other plants (Ord/52)

Pollen
P1 CUB.   **P** Antilles;   DOR;   MEX;   USA/FA.   **Yield** "an inexhaustible supply
of pollen during the entire yr" (CUB, Ord/44).   **Pollen value** high
(Ord/83).   **Colour** orange-brown (Lau/73).   **Pollen grain** illustrated and
described (Smt/54a).   **Reference slide**

Honey: chemical composition
Water [low] 14% (Lov/57b)

Honey: physical and other properties
Colour reddish (Lov/57b; Mot/64); yellow (Lov/58c); reddish-golden
(Ord/83)
Granulation [slow] none under normal conditions (Mot/64; Ord/83)
Flavour rich (Lov/58; Lov/58c); mild (Mot/64); delicate (Ord/83).
Aroma fragrant (Mot/64)

046 Bombax ceiba L.; Bombacaceae
syn Bombax malabaricum DC.; Salmalia malabarica (DC.) Schott. & Endl.

red cotton-tree, silk-cotton tree; semal (INI)
Tree, <23 m, deciduous, v thick trunk; fls white/bright crimson with purple-
tipped stamens, appear before lvs
Distribution tropical/subtropical Asia, Oceania; native to tropical
Asia. **Habitat** moist deciduous forests; plantations; all over INI (Sig/62)

Economic and other uses
Food - fls. **Fodder** - native animals, eg deer, may eat fls; seed for
stock feed. **Land use** afforestation. **Timber.** **Other uses** cotton for
stuffing; good for oil/soap!

Nectar rating + honeybee species; blooms, nectar flow; composition
N1 INI/BIH,ac(Nai/76); INI/UTT,ad(Rae/80)
N2 INI/MAH[ac](Chu/80)
Blooms ii-iii (INI); i-ii (INI/KAR, KER, Kha/59). **Sugar concentration** in
nectar [low] approx 6% (Zma/80)

Pollen
P2 INI/MAH. **P** INI/BIH

Recommended for planting to increase honey production
INI/MAH, coastal areas; forestry department has commercial plantations
(Sub/79)

Honey no data

047 Borago officinalis L.; Boraginaceae

borage; bourrache (Fr/ALG); Borretsche (De); borragine (It)
Herb, 15-70 cm, annual; fls bright blue, rarely white
Distribution temperate Europe, Oceania; native to (Med) Europe. **Habitat**
dry wasteland; widely cultivated, often naturalized in warmer parts of
Europe

Economic and other uses
Food - lvs and fls. **Land use** amenity. **Other uses** medicinal

Nectar rating; blooms, nectar flow; composition
N1 GFR(Gle/77);   IRN(Cra/73);   URS(Fed/55)
N2 POL(Dem/63)
N3 ITA(Ric/78)
N ALG(Ske/72);   NEZ(Wal/78);   URS(Glu/55)
Blooms vi-viii (GDR, Bec/67);   vi-vii (ROM, Cir/80);   summer to frost (UK, How/79).   **Nectar secretion** 5.0-8.1 mg/fl/day (Han/80; Maz/82);   only at temps 6-32°, max at noon (AA105/67).   **Sugar concentration** [medium] 31.7% (Haa/60);   19-52% (Han/80; Maz/82).   **Sugar value** (mg/fl/day) [medium] 1.3 (Cra/75);   0.4-3.5 (Han/80; Maz/82).   **Sugar analysis** (Bat/73a;   Pec/61; Wyk/52;   AA753/75)

Honey flow
Honey potential (kg/ha)[moderate] 200-300 (GDR, Bec/67);   250-300 (ROM, Apc/68;   Cir/80);   200 (URS, Fed/55);   59-211 (Maz/82)

Pollen
P2 GFR.   **P3** ITA.   **P** ALG;   NEZ.   **Colour** of load beige (Hod/74);   bluish-grey/almost white (How/79).   **Pollen grain** illustrated (Hod/74).
Reference slide

Recommended for planting to increase honey production
USA (Pel/76).   Propagate by seed

Honey: chemical composition
Sugars (as % of total, Maz/64):   **glucose** 31.6%;   **fructose** 54.8%;   **sucrose** 3.8%;   **maltose** 7.5%   **fructomaltose** 2.3%

Honey: physical and other properties
Colour whitish with yellow-grey tint (Cra/75);   light (Fed/55);   v dark (Lov/56);   black (Pel/76)
Electrical conductivity 0.000193/ohm cm (Vor/64)

048 Borreria verticilata (L.) G. Meyer;   Rubiaceae

vassoura branca (Pt/BRA)
Shrub;   fls white
Distribution tropical Caribbean;   subtropical S America;   native to BRA

Nectar rating + honeybee species;   blooms, nectar flow
N1 BRA/SC[tm](Wie/80)
Blooms viii-x (BRA/SC)

Honey flow
Honey yield "contributes to most honey produced" in BRA/SC (Wie/80)

Pollen
P BRA/SC.   **Colour** white (Lau/76a).   **Pollen grain** illustrated and described (Bah/70)

Honey: physical and other properties
Colour chestnut, may be v dark (Bah/70);   ?dark brown (Bah/73)
Flavour rather sharp, unpleasant

049 Brachystegia floribunda Benth.; Leguminosae

Tree
Distribution tropical Africa. **Habitat** wooded areas
Rainfall >1000 mm

Nectar rating + honeybee species; blooms, nectar flow
N1 ZAM[tm](Zam/79)
Blooms vii-x (ZAM)

Pollen
P ZAM

Honey no data

050 Brachystegia laurentii (De Wild.) Louis ex Hoyle; Leguminosae

eko (ZAI)
Tree
Distribution tropical Africa. **Habitat** rain forest

Nectar rating + honeybee species; blooms, nectar flow
N1 ZAI,tm + stingless bees(Ich/81)
Blooms vi-vii (ZAI, Ich/81)

Honey flow
Honey yield (kg/colony) 4.5-10, tm (ZAI, Ich/81); 6, stingless bees (ZAI,
Ich/81)

Honey: physical and other properties
Flavour "subtle flavour of eko flowers" (Ich/81)

051 Brachystegia longifolia Benth.; Leguminosae

Tree, <25 m, semi-deciduous; fls greenish white
Distribution tropical Africa. **Habitat** frequent to locally dominant in all
types of miombo woodland and locally commmon on some types of Kalahari
woodland

Nectar rating + honeybee species; blooms, nectar flow
N1 ZAM,tm(Sil/76a; Smt/59; Sto/82; Zam/79)
N2 TAN,tm(Smt/57)
Blooms ix-xi (ZAM, Zam/79)

Pollen
P ?ZAM. **Pollen grain** illustrated and described (Smt/56a)

Honey: physical and other properties
Relative density >1.4129 (Sil/76a)
Granulation coarse, colour then brown

052 Brachystegia spiciformis Benth.;   Leguminosae

panda (Pt/ANA);   mumanga, ussamba (ANA)
Tree, <28 m usually 12-15 m, semi-deciduous, growth dwarfed at high
altitudes;   fls greenish-white, not showy, fragrant
Distribution tropical Africa.   **Habitat** dominant tree in ZAM, locally
common in miombo and chipya woodland on the plateau, sometimes on rocky
hills/escarpments/woodland on Kalahari sands;   also in MOZ
Soil wide range tolerated, deep sand preferred.   **Rainfall** higher rainfall
areas

Economic and other uses
Food - seed.   **Fuel.   Timber.   Land use** shade, amenity.   **Other uses** rope
from inner bark;   tannin from bark;   bark for bark hives;   medicinal

Nectar rating + honeybee species;   blooms, nectar flow
N1 ANA,tm(Ros/60);   ?TAN[tm](Smt/57);   ZAM,tm(Mou/72;   Zam/79)
Blooms vii-xi (southern Africa, Pag/77).   **Nectar flow** good but irregular
(ZAM, Sto/82)

Pollen
P ZAM.   **Pollen grain** illustrated and described (Smt/56a)

Honey no data

053 Brachystegia tamarindoides Welw. ex Benth.;   Leguminosae

mussamba (ANA)
Tree
Distribution tropical Africa

Nectar rating + honeybee species
N1 ANA,tm(Ros/60)

Honey no data

054 Brassica campestris L.;   Cruciferae

field cabbage, field mustard, rape;   black Argentine rape (En/CAF);   common
yellow mustard (En/USA);   nabo, naviza (Es/ARG);   sarisha (BAN)
Herb, 50-100 cm, annual/biennial/perennial;   fls yellow.
Use of common names without botanical names in the literature has led to
confusion between B. campestris and B. napus (and cvs)
Distribution temperate N America, S America, Europe, Asia, Oceania;
subtropical N America, S America, Asia;   tropical Asia;   native to Europe
and Asia.   **Habitat** cultivated crop plant;   waste places, and frequent weed
of cultivated land

Economic and other uses
Food - oil from seed.   **Fodder** - meal from seed

Nectar rating + honeybee species; blooms, nectar flow; composition
N1 ARG(Lut/63); BAN,ac(Dew/79); CAF/ALTA(Hen/77); CAF/MAN(Smi/72);
CAF/SASK(Pal/49); CHN,ac,am(Mad/81); JAP(Inu/57; Sak/82); PAK,ac(Shi/77);
TAI(Fan/52)
N2 PAK[ac](Shr/48); ?USA(Pel/76)
N USA/MA(Shw/50a)
Blooms xi-i (BAN); spring (JAP); xi-ii (TAI); ii-iii (USA/CA).   **Nectar**
secretion 0.2-0.4 mg/fl/day (Han/80).   **Sugar concentration** [high] 45-71%
(Han/80); [medium] 38-46% (Caa/72).   **Sugar value** [medium] 0.12-0.27
mg/fl/day (Han/80).   **Amino acid analysis** (Bak/77)

Honey flow
Honey yield (kg/colony/season) [moderate] 25 (JAP, Inu/57); 14 (TAI,
Fan/52); 7-14 (Mcg/76)

Pollen
P1 INI/PUN; JAP.   **P** BAN; PAK.   **Colour** pale yellow (Wal/78).   **Pollen**
grain illustrated and described (Nak/65)

Honey: chemical composition
Water [medium] 18.1% (1 sample, age 17 mths, Whi/62)
Glucose [low] 26.43%.   **Fructose** [medium] 37.26%.   **Sucrose** [low] 0.45%.
Maltose 11.11%.   **Higher sugars** 1.68%
Ash [medium] 0.324%
pH 4.38.   **Total acid** 34.55 meq/kg.   **Free acid** [medium] 30.00 meq/kg.
Lactone 4.55 meq/kg
Amylase 18.8
Nitrogen 0.070%

Honey: physical and other properties
Colour white or yellowish white (Fan/52); clear, yellowish (Pel/76).
Pfund 70-85 mm, light amber (Whi/62)
Granulation ?rapid (Fan/52)
Flavour like wine (Fan/52); rather like the plant (Pel/76)

055 Brassica campestris L. var. dichotoma Prain; Cruciferae

brown sarson, Indian rape, toria
Herb
Distribution subtropical Asia.   **Habitat** cultivated crop plant
Soil sandy.   **Rainfall** - irrigation required in INI/HAR, HIM, PUN

Economic and other uses
Food - oil from seed

Nectar rating + honeybee species; blooms, nectar flow; composition
N1 INI/PUN[ac](Chd/77)
N3 PAK,ac(Shr/48)
Blooms x-xii (INI/PUN).   **Sugar concentration** [medium] 31.5-64.0% (Shr/58)

Pollen
P1 INI/PUN

Honey no data

056 Brassica campestris L. var. sarson Prain; Cruciferae

sarson
Herb; fls yellow
Distribution tropical and subtropical Asia.   **Habitat** cultivated crop plant

Economic and other uses
Food - oil from seed

Alert to beekeepers
Crop spraying has killed bees (PAK, Pak/77)

Nectar rating + honeybee species; blooms, nectar flow; composition
N1 INI/UTT,ac(Kap/57; Raw/80); PAK,ac(Pak/77)
N2 INI/MAH[ac](Chu/80); INI/PUN[ac](Chd/77); PAK,ac(Shr/48)
Blooms ix-iii (INI/MAH); xi-xii (INI/PUN); xii-i (INI/UTT); i-ii
(PAK).   **Alert to beekeepers** crop spraying has killed bees (PAK, Pak/77).
Sugar concentration [high] 32-69% (Shr/58)

Honey flow
Honey yield [moderate] 5-6 kg/colony/season (PAK, Shr/48)

Pollen
P1 INI/PUN; INI/UTT.   **P2** INI/MAH.   **P** PAK.   **Colour** yellow (Kap/57)

Honey no data

057 Brassica juncea (L.) Cosson; Cruciferae

Chinese mustard, edible mustard, Indian mustard, leaf mustard, trowse
mustard; moutarde (Fr); raya (PAK)
Herb, <1 m, annual; fls bright yellow
Distribution temperate Europe; subtropical Asia, S America.   **Habitat**
cultivated crop plant; elsewhere, casual or established weed

Economic and other uses
Food - lvs; oil from seed.   **Fodder**

Nectar rating + honeybee species; blooms, nectar flow; composition
N1 PAK,ac(Pak/77); URS(Ave/78)
N2 INI/PUN[ac](Chd/77); PAK,ac(Shr/48); URS(Glu/55)
Blooms iii-iv (INI/KAS, Sha/76); i-ii (PAK).   **Sugar concentration**
[medium] 38-46% (Caa/72); 28-36% (Sao/54); 22-65% (Shr/58).   **Amino acid**
analysis (Bak/77).   **Potassium content** and **fluorescence** (AA491/80)

Honey flow
Honey yield [moderate] 5-7 kg/colony/season, mixed with honey from B. campestris var. sarson (PAK, Shr/48).   **Honey potential** [moderate] 50-60 kg/ha (south-east URS, Ave/78)

Pollen
Pl INI/PUN.   **P** PAK.   **Colour** yellow (Sha/76).   **Reference slide**

Honey no data

058 Brassica napus L.;   Cruciferae

rape
Herb, <1.5 m, annual/biennial; fls yellow.   B. napus L. taxonomy is difficult, and records stating that B. napus is grown as an oilseed crop (and records of B. napus from Europe) are included with B. napus var. oleifera.   Also, the use of common names without botanical names in the literature has led to confusion between B. napus and B. campestris (and cvs)
Distribution temperate Europe, Asia.   **Habitat** cultivated crop plant; elsewhere open ground/banks/streams/ditches;   often naturalized

Warning
Green tops of plants ?toxic to animals (Pol/69)

Nectar rating + honeybee species
N1 CHN,ac(Tse/54);   IRN(Cra/73)

059 Brassica napus L. var. napobrassica (L.) Reichenb.;   Cruciferae

shaljam (PAK)
Herb, biennial; fls yellow
Distribution subtropical Asia, temperate Europe;   native to Europe
Temperature plant overwinters in PAK

Economic and other uses
Food - root.   **Fodder**

Nectar rating + honeybee species;   blooms, nectar flow
N1 PAK,ac(Pak/77)
Blooms iii-v (PAK)

Pollen
P PAK

Honey no data

060 Brassica napus L. var. oleifera DC.; Cruciferae

oilseed rape, rape, swede rape, Swedish rape; colza (Fr); Raps (De);
colza (It); raps (No)
Herb, annual/biennial; fls yellow. Records stating that B. napus is
grown as an oilseed crop and (all records of B. napus from Europe) are
included here. The use of common names without botanical names in the
literature has led to confusion between B. napus and B. campestris (and cvs)
Distribution temperate Europe, Asia, N. America, Oceania. **Habitat**
cultivated crop plant
Soil wide range but well drained soil preferred; optimum pH 6.5 (Cab/81)

Economic and other uses
Food – oil from seed. **Fodder** – meal from seed

Warning; alert to beekeepers
Warning green tops ?toxic to animals (Pol/69). **Alert to beekeepers** Bees
continue to forage on rape at very end of flowering when pest control
measures may be necessary. Blooms early in season so beekeepers must make
colonies strong to harvest honey (UK, Cab/81). Honey granulates v rapidly
– sometimes within a few days in the comb (Cab/81)

Nectar rating + honeybee species; blooms, nectar flow; composition
N1 CZE(Svo/58); DEN(Jon/54); FRA(Lou/81; Mar/81); GFR(Gle/77);
ITA(Ric/78); UK(How/79); URS(Fed/55)
N2 GDR(Koc/65); HUN(Kos/77); POL(Dem/64a); ROM(Cir/77); USA/WI(Pel/76)
N NEZ(Wal/78); NOW(Lun/71; Thy/65); ROM(Int/65)
Blooms iv-v (FRA, AA1246/80); iv-v (ROM, Cir/80; Apc/68); v-vii
(Europe). **Alert to beekeepers** bees continue to forage on rape at very end
of flowering, when pest control measures may be necessary; blooms early in
season so beekeepers must make colonies strong to harvest honey (UK,
Cab/81). **Nectar flow** x-xi (NEZ). **Nectar secretion** 0.6 mg/fl/day
(Maz/82); effects of fertilizers (AA761/75); effects of various factors
(Frj/70). **Sugar concentration** [high] at RH 55-65%, inner nectaries 68.4%,
outer 84%, but at RH 80-90%, [medium] 21.8 and 23.4%, respectively
(AA344/83); also [medium] 45-60% (Frj/70); 28.7% (Haa/60); 36.28%
(Pet/77); [low] 5-7% (5 cvs, AA1246/80). **Sugar value** [medium] 0.29-0.90
mg/fl/day (Maz/82). **Sugar analysis** (Cab/81; Frj/70; Maz/59; Wan/64;
Wyk/52; AA678/66). **Amino acid analysis** (Bak/77; Moo/65)

Honey flow
Honey yield [moderate] 0-30 kg/colony/season (EUR, Uni/83). **Honey
potential** (kg/ha) [moderate] 35-100 (ROM, Cir/77); 40-50 (ROM, Cir/80);
100-500 under optimum conditions but usually 50 (UK, Cab/81); 50 (URS,
Fed/55)

Pollen
P1 FRA; ITA; UK. **P3** GFR. **P** NEZ; NOW; ROM. **Pollen value** high
(Cab/81). **Chemical analysis** (Sta/74). **Colour** of load lemon yellow
(Cir/80; Han/80); yellow (Wal/78). **Pollen grain** over-represented in
honey (ITA, Ric/78); >100 000 grains in 10 g (Sta/74); see also Dem/77

Recommended for planting to increase honey production
UK (Cab/81). Propagate by seed. See **Warning; alert to beekeepers**

Honey: chemical composition

Water [medium] 17.6–18.8% (4 samples, Dus/67); can be >18% (Lou/80)
Glucose [medium] 35.22% ("rape" honey, Ech/77); 39.6–42.6% (6 samples, Gon/79); 35.13% (Mur/76); also "high" (Cab/61). **Fructose** [medium] 37.25% (Ech/77); 36.9–40.2% (Gon/79); 39.69% (Mur/76). **Sucrose** [low] 0.40% (Ech/77). **Reducing sugars** 70.15% (Mur/76). **Maltose** 5.80% (Ech/77). Contents of individual sugars as % of total sugars (Maz/59; Maz/64)
Ash [low] 0.53% (Mur/76). Fe, Cu, Mn, Mg contents (36 samples, Pos/72)
pH 4.6 (Dus/72); approx 4 (Lou/80). **Free acid** [medium] approx 15 meq/kg
Sucrase 7.2–15.4 (Dus/67). **Catalase** 0–2.1 (Dus/67). **Peroxide number** (μg/g/h) 41.8–125 (Dus/67); 73.3 (25.1 after 10 minutes in sunlight, Dus/72)
Amino acids 0.00037% dry wt, also contents of individual acids (Moo/65).
Protein 0.220% (Ber/75)
Other constituents – formaldehyde, acetaldehyde, acetone, isobutyraldehyde, diacetyl present (Hoo/63)

Honey: physical and other properties

Colour bright yellow (Dus/72); pale yellow (Fed/55; Mur/76); golden (Pel/76); greyish white (Ric/78). **Pfund** water white (Cra/75; Wal/78); <35 mm, white (Lou/80); light amber when liquid (Pal/59); Pfund-Lovibond grade (Aub/83)
Relative density 1.412 (Mur/76). **Electrical conductivity** (per ohm cm) 0.00017–0.00021 (Dus/67); 0.00012–0.00027 (Pou/70); 0.000137 (Vor/64)
Granulation v rapid (Cab/81; Dem/64); fine, homogeneous (Cab/81; Ric/78); colour then white (Pal/59). **Alert to beekeepers** honey granulates v rapidly – within a few days, sometimes in the comb (Cab/81)
Flavour sweet (Cra/75); mild (Lou/80); little when granulated (Pal/59); delicate (Wal/78). **Aroma** of flowers (Lou/80); varies from none to rather unpleasant, characteristic (Wal/78)

061 Brassica nigra (L.) Koch; Cruciferae

black mustard, brown mustard; lahi (INI)
Herb, <1–3 m, annual; fls bright yellow
Distribution temperate Europe, Asia; subtropical Asia, N America; native to Europe. **Habitat** cultivated crop; waste places/ditches/cliffs

Economic and other uses

Food – mustard from seed; seedlings/green tops for salad. **Fodder. Soil benefit** green manure. **Other uses** oil for medicine/soap-making/lubricants

Nectar rating + honeybee species; blooms, nectar flow; composition

N1 INI/UTT,ad(Rae/80); INI/UTT,ac(Raw/80); PAK,ac(Pak/77); UK(How/79)
N2 URS(Glu/55); USA/CA(Pel/76)
Blooms vi–ix (Europe); xi (INI/UTT); iii–v (PAK). **Nectar flow** fails some seasons (USA/CA, Van/41). **Sugar concentration** [medium] 48.6% (nectar from honey sacs of bees, Ord/83). **Amino acid analysis** (Bak/77)

Honey flow

Honey yield (kg/colony/season) [moderate] >27 (INI/UTT, Koh/58); 4.25, although crop failed (INI/UTT, Raw/80)

Pollen

P PAK; USA/CA. **Pollen value** high (Ord/83). **Colour** yellowish
(Ord/83). **Reference slide**

Honey: chemical composition

Water content may be high
Fermentation reported likely (Lov/56)

Honey: physical and other properties

Colour light, bright (How/79); yellow cast (Lov/56). **Pfund** light amber
(Lov/56; Ord/83)
Flavour usually mild (Lov/56); strong like mustard when freshly extracted
(Ord/83). **Aroma** strong when fresh (How/79)

062 Brassica rapa L. subsp. oleifera DC.; Cruciferae

winter rape; navette (Fr); ryps (No)
Herb, annual
Distribution temperate Europe. **Habitat** cultivated crop plant

Nectar rating; blooms, nectar flow; composition

N1 FRA(Lou/81)
N2 ROM(Cir/77)
N NOW(Lun/71)
Blooms iv-v (ROM). **Sugar analysis** (Pec/61)

Honey flow

Honey potential [moderate] 30-100 kg/ha (ROM, Cir/77)

Pollen

P1 FRA. P NOW; ROM

Honey no data

063 Bucida buceras L.; Combretaceae

júcaro negro (Es/CUB); guaraguao (Es/DOR); cacho de toro (Es/HOD); pucté
(Es/MEX)
Tree; fls white, small
Distribution subtropical N America, tropical C America and Caribbean.
Habitat shores/marshes/river mouths

Economic and other uses

Timber hard, resistant. **Land use** amenity

Nectar rating; blooms, nectar flow

N1 BEL(Mul/79)
N3 DOR(Ord/04)
N CUB(Ord/83); DOR(Ord/83); HOD(Ord/63); MEX(Ord/83)

Blooms i-iv (tropical America, Ord/83).   **Nectar flow** cannot be relied on every yr (Ord/83)

Pollen
P DOR

Honey no data

064 Butea monosperma (Lam.) Taub.;   Leguminosae
syn Butea frondosa Roxb.

bastard teak;   dhak, prayash (INI)
Tree;   fls crimson
Distribution subtropical Asia

Economic and other uses
Other uses medicinal;   food for lac insects;   red dye from fls

Nectar rating + honeybee species ;   blooms, nectar flow
N1 INI/MAD[ac](Khn/48)
N3 INI/PUN[ac](Chd/77)
Blooms i-ii (INI/MAD);   iv-v (INI/PUN)

Pollen
P INI/MAD

Honey: physical properties
Pfund ?light amber (Khn/48)

065 Byrsonima crassifolia (L.) DC.;   Malpighiaceae

chaparro manteco (Es/VEN)
Shrub/tree, 3-6 m;   fls yellow/red
Distribution tropical Caribbean and C America.   **Habitat** savannas of Antilles and Venezuela;   forms large areas of scrub, "nancitales", along coasts of C America (Ord/83)

Economic and other uses
Food - fruit

Nectar rating;   blooms, nectar flow;   composition
N1 VEN(Cra/73)
Blooms summer (tropical America, Ord/83);   vi (COS, Ord/83).   **Amino acid analysis** of nectar (AA336/79)

Pollen
Reference slide

Honey no data

066 Caesalpinia coriaria (Jacq.) Willd.; Leguminosae

dividivi (Es); guatapana (Es/DOR)
Tree, small to medium size, deciduous; fls yellow/whitish
Distribution tropical S America, C America, Caribbean; subtropical C
America.   **Habitat** deciduous forest (Ord/83)
Soil dry preferred (Ord/83).   **Rainfall** drought tolerant (Ord/83)

Economic and other uses
Timber.   **Other uses** - tannin from pods

Nectar rating; blooms, nectar flow
N1 DOR(Ord/66; Ord/72)
N2 DOR(Ord/64; Ord/83)
Blooms ix-ii (C America, Ord/83)

Honey flow
Honey yield "much dividivi honey in hives in October" (Caribbean area,
Ord/83)

Pollen
P DOR

Honey: physical properties
Pfund light amber (Ord/83)

067 Cajanus cajan (L.) Millsp.; Leguminosae

pigeon pea, red gram; feijão boere (Pt/MOZ); cajan, arthar (INI)
Shrub, <3 m, annual/biennial/perennial; fls yellow spotted red
Distribution tropical Oceania, Africa, Asia, Caribbean, S America, C
America; subtropical Africa, Asia; native to NE Africa and Asia.
Habitat cultivated crop plant; wide range from arid to humid areas; grows
well on "difficult" sites; some cvs do not crop in shade or salt spray;
some cvs for altitudes >3000 m
Soil infertile/arid; light sand or deep loam preferred; waterlogging not
tolerated; some cvs salt tolerant.   **Temperature** mean <35°, optimum 18-
29°; killed by frost.   **Rainfall** 400-2500 mm with 5-6 mths max dry
period; drought tolerant; not suited to wetter areas of tropics

Economic and other uses
Food - pods; seeds for dhal (INI).   **Fodder** - pods/husks/lvs for cattle;
lvs for silkworms/lac insects.   **Fuel.**   **Land use** hedges, windbreaks,
temporary shade.   **Soil benefit** erosion control; N-fixation.   **Other uses**
gunpowder from charcoal; thatching; basketry; gum

Nectar rating + honeybee species; blooms, nectar flow
N1 INI/BIH,ac(Nai/76);   MOZ,tm(Cra/73)
N TRI(Lau/76)
Blooms xii (INI/BIH); winter (tropical America, Ord/83)

Pollen
Pollen grain illustrated and described (Sao/61)

Honey: physical and other properties
Colour distinctive greenish hue in the comb (Lau/76)

068 Caldcluvia paniculata D. Don; Cunoniaceae

tiaca (Es/CHL)
Tree, <10 m; fls yellowish-white, fragrant
Distribution temperate S America; native to Chile.   **Habitat** by streams;
part shade
Soil poor

Nectar rating; blooms, nectar flow
N1 CHL(Car/38; Fis/38; Kar/60)
Blooms xii-i (CHL/Osorno).   **Nectar secretion** prolific

Pollen
Pollen grain illustrated and described (Heu/71)

Honey: physical and other properties
Colour light (Car/38)
Granulation fine
Aroma v aromatic

069 Calea urticifolia (Miller) DC.; Compositae

jalacate (Es/HOD)
Shrub, 1-3 m; fls bright yellow
Distribution subtropical C America, tropical C America.   **Habitat** common in
Guatemala at edges of forests; also on Pacific coast of C America
especially in the Fronseco Gulf area of Honduras (Ord/83)

Nectar rating; blooms, nectar flow
N1 HOD(Ord/83)
N2 NIA(Ord/63a)
N MEX(Ord/83)
Blooms x-xii (HOD)

Honey flow
Honey yield [high] 34 kg/colony/season (22 hives, Ord/83)

Pollen
PI NIA.   Reference slide

Honey no data

070 Calliandra calothyrsus Meissn.; Leguminosae

red calliandra
Shrub, 5-10 m; fls red
Distribution tropical C America, Asia; native to C America. **Habitat** humid/sub-humid zones; altitudes 150-1500 m
Soil wide range; good tolerance to flooding. **Rainfall** 1000-2000 m with 3-4 mths max dry period

Economic and other uses
Fodder lvs. **Fuel** "excellent fast-growing source" (Usa/80). **Land use** hedges, afforestation, amenity, firebreak, weed suppression. **Soil benefit** erosion control; N-fixation; organic manure; mulch

Warning
?Invasive, spreads rapidly by seed (Usa/80)

Nectar rating + honeybee species; blooms, nectar flow
N1 INO/JAV,ac(Peu/80)
Blooms all yr (INO/JAV)

Honey flow
Honey yield [moderate] 1.3 kg/colony/mth (INO/JAV, Peu/80)

Pollen
P1 INO/JAV

Recommended for planting to increase honey production
INO/JAV, Peu/80. Propagate by seed/large cuttings; grows rapidly, 2.5-3.5 m in 6-9 mths (Usa/80). See **Warning**

Honey: physical and other properties
Flavour bitter-sweet (Usa/80)

071 Callistemon citrinus (Curt) Skeels; Myrtaceae

crimson bottle brush (En/AUS)
Shrub/tree, 4-9 m; fl stamens bright red, resembling a bottle-brush
Distribution temperate and subtropical Oceania; native to Australia
Soil damp soil; also poor dry soil; some salt tolerance

Economic and other uses
Land use amenity

Nectar rating + honeybee species; blooms, nectar flow
N1 PAK,ac(Pak/77)
N2 AUS/VIC(Gom/73)
Blooms xi-xii (AUS/VIC)

Pollen
P2 AUS/VIC. **P** PAK

Honey no data

072 Calluna vulgaris (L.) Hull; Ericaceae

heather, ling; brande, bruyère commune, callune vulgaire (Fr); hedelyng
(Da); Besenheide, Heidekraut (De); brentoli, brugo (It); røsslyng (No)
Shrub, 15-80 cm, evergreen; fls pinkish-lilac
Distribution temperate Europe (rare in Med region and SE), also in small
isolated places where introduced in Africa, Asia, N America, Oceania (Bei/-
40). **Habitat** moors, heaths, open woods, sand-dunes, etc; often dominant
over large areas; altitudes in UK <1000 m, optimum 100-300 m (Wht/54)
Soil damp conditions preferred but also on sandy and rocky soils; lime not
tolerated. **Rainfall** – chief Calluna areas in UK have high rainfall (Wht/54)

Economic and other uses
Fodder grazing for sheep/cattle/deer/grouse. **Fuel.** **Land use** amenity.
Other uses tannins; rope; dyes

Alert to beekeepers
Honey is considered unsuitable as sole winter food for bees (UK/Scotland,
How/79). It is thixotropic (Cra/75), due to protein content (Pry/50), so
difficult to extract from combs

Nectar rating; blooms, nectar flow; composition
N1 DEN(Lou/77); ?FIN(Enb/54); FRA(Lou/81); GDR(Lou/77); GFR(Lou/77);
NER(Mod/52); NOW(Lun/71; Thy/65); POL(Lou/77); SWE(Lou/77);
UK(How/79); URS(Ave/78; Fed/55)
N2 BEG(Cra/84); DEN(Fre/57; Jon/54); FIN(Kay/79); GFR(Gle/77);
IRR(Hil/68)
N3 ITA(Ric/78); USA(Lov/56)
N AZO(Hab/72); CZE(Svo/58); NEZ(Wal/78); SPA(Sau/82b)
Blooms vii-late ix. **Nectar flow** viii-early ix; influenced by soil, rain-
fall, ?altitude; temp must be >13.9° (UK, Wht/54). **Nectar secretion**
(mg/fl/day) 0.14-0.58 (Han/80; Maz/82); 0.29-0.47 (Szk/73); greatest
when fls are newly opened and on 1-2 yr old shoots following burning (UK,
Wht/54). **Sugar concentration** [medium] 23-24% (Han/80); 23-39% (Maz/82);
43.4-52.2% (Szk/73); 35-45% (Wht/54). **Sugar value** (mg/fl/day) [medium]
0.12-0.14 (Han/80); ?0.12 (Lou/77); 0.12-1.2 (Maz/82); 0.148-0.213
(Szk/73). **Sugar analysis** (Wan/64); nectar composition (Wht/54). **Amino
acid analysis** (Moo/65)

Honey flow
Honey yield 2-39 kg/ha (POL, Maz/82). **Honey potential** [moderate] 100-200
kg/ha (FRA, Lou/77); 200 (ROM, Apc/68); 100-150 (ROM, Cir/80); 200 (URS,
Ave/78)

Pollen
P1 FRA. **P3** GFR; ITA. **P** AZO; NEZ; NOW. **Yield** low, 17 000 grains/fl
(Maz/82). **Colour** dull white to light slate grey (Wht/54); load light to
dark grey, small, 5.8 mg (Maz/82); illustrated (Hod/74). **Pollen grain**
illustrated (Hod/74; Saw/81); described (Hod/74; Saw/81; Wht/54); ?over-
represented in honey (Wht/54). **Reference slide**

Honey: chemical composition

Water [high] 19.2-21.6% (3 samples, Dus/67);   19.69-20.45% (Feo/71);   17.6-
26.7% (27 samples, Mic/54);   16.0-24.8% (25 samples, Mic/55)
Sugars, total 72.0, 76.7% (Ver/65).   **Glucose** [medium] 33.3-34.0%
(Feo/71);   31.2-32.9% (5 samples, Gon/79).   **Fructose** [medium] 37.5-39.3%
(Feo/71);   41.4-42.6% (Gon/79).   **Sucrose** [medium] 0.5-0.9% (Feo/71);
3.39, 1.78% (Ver/65).   **Reducing sugars** 71.5-72.9% (Feo/71).   **Maltose,
lactose, galactose** present (Wan/64).   **Dextrin** 4.8-5.9% (Feo/71)
Ash [medium] 0.18-0.69% of dry wt (Mic/54);   see also Feo/71
pH 4.5 (Dus/72);   4.1-4.3 (Feo/71);   4.0-4.6 (Lou/77);   3.69-12.9
(Mic/54);   4.20-5.36 (25 samples, Mic/55).   **Total acid** 2.3-3.5 (degrees,
by Polish standard method, Feo/71).   **Alkalinity number** 11.3-13.4 (Feo/71)
Amylase 21.0-36.9 (Polish standard method, Feo/71);   40 (Lou/66).   **Sucrase**
31.0-35.5 (Gontarski 1957 method, Dus/67);   24.7 (Dus/72a);   45.8-49.5
(Feo/71).   **Catalase** 0.1-3.6 (Dus/67).   **Peroxide number** (μg/g/h) 29.0-33.6
(Dus/67);   111.8 (Dus/71);   132.2 (103.2 after 10 minutes in sunlight,
Dus/72)
Nitrogen 0.97-1.09% (Feo/71);   0.13-0.38% (17 samples, Mic/54).   **Amino
acids** 0.000526% dry wt, also contents of 9 individual acids given
(Moo/65);   12, 13 individual acids identified in 2 samples (Kum/74).
Protein 1.3-1.8% (How/79).   **Colloids** 4.1-8.8% (Feo/71);   0.48-3.34% of dry
wt (Mic/54);   1.35-3.9% (Mic/55)
Vitamins C 40-52 ppm (Feo/71)
Alert to beekeepers honey considered unsuitable as sole winter food for
bees (UK/Scotland, How/79)

Honey: physical and other properties

Colour golden (Bab/61);   light, dark or reddish brown (Cra/75);   reddish
(Lou/77; Lov/56).   **Pfund** amber (Lou/77; Lov/56);   variable, sometimes
dark amber (Mic/54);   8.5 mm, extra white (Uni/83)
Viscosity "very heavy body" (Lov/56).   **Alert to beekeepers** thixotropic
(Cra/75), due to protein content (Pry/50), so difficult to extract from
combs;   thixotropy ratios in Mic/54 and Mic/55.   **Optical rotation** 14.35 to
15.03 deg (Feo/71). **Electrical conductivity** (per ohm cm) 0.00075-0.00080
(Dus/67);   0.00075-0.00095 (Lou/77);   0.0008 regarded as lowest level for
Calluna honey (Hao/75)
Granulation slow, forming large crystals (Lou/77);   slow, forming hard
spherical grains (Uni/83)
Flavour characteristic, slightly bitter (Lou/77);   aromatic (Lov/56);   mild
but pronounced (Wal/78).   **Aroma** pronounced, characteristic (Cra/75);
strong, floral (Lou/77)

073 Calycophyllum candidissimum (Vahl) DC.;   Rubiaceae

lemon wood;   surra (Es/COS);   dagame (Es/CUB);   salamo (Es/HOD);   canelo,
palo camarón (Es/MEX);   madrono (Es/NIA);   alazano (Es/PAM)
Tree, 6-8 m;   fls white (bracts)
Distribution tropical C America, Caribbean, S America.   **Habitat** deciduous
forests of plains/hills especially along Pacific coast of C America;   low
altitudes, also mountains in east (CUB)
Soil chalky/rocky soil;   calcareous/gravelly soil

Economic and other uses
Timber valuable.   **Land use** shade for coffee trees

Nectar rating; blooms, nectar flow
N1 CUB(Ord/44);   HOD(Ord/63);   MEX(Ord/72); NIA(Ord/63a)
N2 CUB(Ord/52)
N COS, CUB, MEX, NIA, PAM, VEN (Ord/83)
Blooms xi-i (CUB); mid-x to i (tropical America, Ord/83)

Honey flow
Honey yield "a species of major importance" in tropical America (Ord/83)

Recommended for planting to increase honey production
CUB (Ord/44).   Recommended for planting by roadsides

Honey: physical and other properties
Pfund water white, beyond Pfund scale (Ord/83)
Granulation medium

074 Canthium coromandelicum (Burm. f.) Alston; Rubiaceae
syn Canthium parviflorum (Lam.)

karegida (INI)
?Herb; fls yellow
Distribution tropical Asia

Nectar rating + honeybee species; blooms, nectar flow
N1 INI/KAR,KER[ac](Kha/59)
Blooms iv-v (INI/KAR, KER)

Pollen
P INI/KAR, KER

Honey no data

075 Carica papaya L.; Caricacae

papaya, pawpaw, tree melon
Tree, 6-8 m, evergreen, dioecious; fls yellow, small
Distribution tropical Africa, Asia, S America, Oceania; native to tropical
America.   **Habitat** cultivated crop plant; sheltered site
Soil wide range but must be well drained

Economic and other uses
Food – fruit; papain from fruit for brewing; lvs for tenderizing meat.
Other uses medicinal

Nectar rating + honeybee species; blooms, nectar flow; composition
N1 ETH[tm](Cra/73);   MOZ[tm](Cra/73);   PHI[ac](Row/76)

N3 INO[ac](Bee/77)
Nectar flow nectar in male fls inaccessible to bees (Mcg/76).   **Sugar concentration** [medium] 24-34% (Mcg/76);   26% (Row/76);   24-34% (AA516/70).   **Sugar analysis** (Row/76).   Juice from damaged fruit collected by bees (GHA, Gor/64);   perhaps some N1 entries may refer to juice not nectar?

Pollen
P3 INO.   **Pollen value** important in BOL, OMA.   **Pollen grain** illustrated and described (Mag/78).   **Reference slide**

Honey no data

076 Carnegiea gigantea (Engelm.) Britton & Rose;   Cactaceae

pitahaya, saguaro (Es/MEX)
Shrub (columnar cactus), post-like, 6-18 m high, <0.6 m diameter;   fls white, open at night and lasting into the day
Distribution subtropical N America, C America;   native to N America and C America.   **Habitat** desert zones in Sonora (MEX), adjacent areas of USA
Rainfall very drought tolerant

Economic and other uses
Food - fruit;   seeds;   alcoholic drink from fruits.   **Land use** amenity

Nectar rating;   blooms, nectar flow;   composition
N1 MEX(Ord/83)
N2 MEX(Ord/72)
Blooms iv-vi (MEX).   **Nectar secretion** 5 ml or more/fl (Mcg/59).   **Sugar concentration** [medium] 25% (Mcg/59).   Juice from ripe fruit also collected and stored by bees, resulting in red patches in combs (Ord/83).

Pollen
P1 USA/AZ.   **Yield** 12 or more bee loads/fl (Mcg/59).   **Pollen value** good (Mcg/59).   **Colour** of load cream (Mcg/59)

Honey: physical properties
Viscosity "very thick" (Ord/83)

077 Carvia callosa (Nees) Brem.;   Acanthaceae

garikalu, karvi (INI)
Shrub, unbranched shoots 2.6 m, plietesial, gregarious;   fls bluish-purple
Distribution tropical Asia;   native to India (W peninsular, rarely C India/MAH, KAR).   **Habitat** evergreen forests of Western Ghats (Wak/81); altitudes 600-1500 m (INI/MAH, Pha/64)
Soil thin laterite on steep rocky slopes (INI/MAH, KAR, Sur/78).
Temperature 10-34° (Pha/64).   **Rainfall** 6500-7500 mm with monsoon vi-ix (Pha/64)

Economic and other uses
Fuel. **Timber** sticks for wall construction.   **Other uses** medicinal, insect
repellent, aromatic (pandadi) oil, pulp for paper-making

Warning; alert to beekeepers
Warning lvs toxic to man and animals (Sur/78).   **Alert to beekeepers** honey
is thixotropic (Chu/65)

Nectar rating + honeybee species; blooms, nectar flow; composition
N1 INI/KAR,ac(Diw/72; Pha/64); INI/MAH,ac(Chu/65; Ded/53; Pha/64;
Sur/78)
N3 INI/KAR,ac,ad,Trigona(Diw/64)
Blooms vii-ix (INI/KAR); viii-x (INI/MAH); flow stimulates brood rearing
and swarming (INI/MAH, Chu/65; Sur/78).   **Nectar flow** once in 8 yrs; all
plants flower in same yr then die (Pha/64).   **Nectar secretion** peak in
early afternoon (Sur/78); highest if no excessive rains during blooming
(Diw/72).   **Sugar concentration** [medium] 20-37% (Pha/62; Sur/78); [low]
<5% when RH is high (Pha/64); bees stop foraging when concentration <15%
(Sur/78).   **Sugar analysis** (Wak/81)

Pollen
P1 INI/KAR,MAH.   **P3** INI/KAR.   **P** INI/KAR.   **Yield** mostly collected between
08.00 and 10.00 h (Pha/64).   **Colour** of load light pink (Sur/78).   **Pollen
grain** illustrated and described (Chu/65; Ini/77; Sur/78)

Recommended for planting to increase honey production
INI/MAH (Sub/79).   Propagate by cuttings or by seed sown during monsoon;
blooms after 8 yrs from seed, same yr as parent plant from cuttings
(Sur/78).   Cold storage of seed gives annual succession of blooming (Sur/-
78).   See **Warning; alert to beekeepers**

Honey: chemical composition
Water [medium] 18.33% (Pha/62); 19.08% (8 samples, Pha/67; also Pha/70)
Sugars, total 77.82% (Pha/62).   **Glucose** [medium] 28.17% (Pha/62); 31.21%
(Pha/67).   **Fructose** [medium] 38.29% (Pha/62); 35.24% (Pha/67).   **Sucrose**
[high] >10% (Pha/67).   **Reducing sugars** 66.46% (Pha/62); 66% (Pha/67).
Maltose, melezitose, raffinose present (Pha/67).   **Dextrin** 2.14% (Pha/62);
1.92% (Pha/67)
Ash [medium] 0.298% (Pha/62); 0.248% (Pha/67).   Na, K, Ca, Mg, Fe, P, Si
contents (Pha/67)
Free acid [medium] 36 meq/kg (Pha/62; Pha/67).   **Citric, malic, succinic
acids** present (Pha/70)
Amylase 6.22 (Sur/78).   **Sucrase** 24.38
Amino acids glutamic acid, tyrosine, valine, leucine present (Naa/70).
Protein 0.817% (Pha/62); 0.734% (Pha/67); 1.02% (Sur/78)
Vitamins C 113.5 ppm (Sur/78)
Indian Standard for this honey, see Ini/77

Honey: physical and other properties
Colour dark red, becoming burnt amber (Chu/65); greenish amber (Pha/62);
greenish fluorescence observed when freshly extracted (Sur/78).   **Pfund**
dark amber (Pha/64; Pha/67; Sur/75).   **Opacity** values (Pha/67)
Relative density 1.41 (Pha/62); 1.404 (Pha/67).   **Viscosity (alert to
beekeepers)** thixotropic (Chu/65; Sur/75); "thick" (Sur/78).   **Optical
rotation** +2 deg 18 minutes (Pha/62); +03 minutes (Pha/67)

Granulation slow, 5-6 yrs (Pha/67)
Aroma characteristic (Pha/64); strong (Sur/75)

078 Cassia siamea Lam.; Leguminosae

yellow cassia; casia de Siam (Es/VEN)
Tree, 15-20 m, evergreen; fls yellow
Distribution tropical Asia, Caribbean, C America, Africa; subtropical N
America; native to SE Asia from INO to SRI. **Habitat** wide range from arid
to humid areas; plantations, river banks, irrigated land, etc; lowlands
<1200 m
Soil deep, relatively rich soil preferred; laterite and limestone tolerated; poor tolerance to waterlogging. **Temperature** tropical heat tolerated; frost not tolerated. **Rainfall** monsoon areas preferred; in dry
areas, roots need access to deep soil moisture; 500-1000 mm/yr with max
dry period 4-5 mths

Economic and other uses
Fodder for cattle/sheep; seeds/pods/lvs highly toxic to pigs (Usa/80).
Fuel. Timber. Land use windbreak, afforestation, amenity. **Soil
benefit** soil conservation; organic manure. **Other uses** host plant for
sandalwood (Santalum album)

Warning
Wood may contain irritant yellow powder (Usa/80). Seeds/pods/lvs highly
toxic to pigs (Usa/80)

Nectar rating; blooms, nectar flow
N1 VEN(Cra/73; Ste/71)
Blooms vi-x (VEN)

Pollen
Pollen grain illustrated and described (Smt/54a)

Honey no data

079 Castanea pubinervis (Hassk.) C.K. Schn.; Fagaceae
syn Castanea crenata Sieb. & Zucc.

sweet chestnut; kuri (JAP)
Tree/shrub, 10-30 m, deciduous; fls yellowish-white?
Distribution temperate Asia; native to Japan. **Habitat** foothills

Economic and other uses
Food - nuts. **Timber. Land use** amenity

Nectar rating; blooms, nectar flow
N1 JAP(Inu/57; Sak/82)
Blooms early summer (JAP); mid-vi to early vii (JAP, Wah/74)

Pollen
P1 JAP

Honey: chemical composition
Mineral content high (Wah/74)

Honey: physical and other properties
Colour black or brown (Inu/57); v dark (Wah/74)
Flavour characteristic, bitter (Wah/74). **Aroma** "unpleasant" (Inu/57);
characteristic (Wah/74)

080 **Castanea sativa Mill.; Fagaceae**

chestnut, Spanish chestnut, sweet chestnut; châtaignier (Fr); castagno (It)
Tree, <30 m, deciduous; fls greenish-yellow catkins, monoecious
Distribution temperate Europe, Asia, N Africa; also Oceania but does not
thrive there; native to southern Europe. **Habitat** woods and mountain
slopes; widely cultivated
Soil well drained; usually calcifuge; deep, rich moist, preferably
alluvial. **Rainfall** moist climate preferred but dry periods tolerated

Economic and other uses
Food – nuts. **Timber** – especially for wine storage vessels. **Land use**
amenity

Nectar rating; blooms, nectar flow; composition
N1 BEG(Cra/84); FRA(Lou/81; Mar/81); GRC(Adm/54); ITA(Ric/78);
URS(Fed/55); YUG(Adm/54; Shl/81)
N2 HUG(Haz/55)
N EUR(Maz/82)
Blooms vi-vii (HUG). **Nectar flow** 3-4 wks, fills gap between Robinia
pseudacacia and Stachys annua flows (HUG, Haz/55). Honeydew flow may
occur with nectar flow (Klo/65). **Sugar concentration** [medium] 37.1%
(Pek/80a). **Sugar analysis** (Bat/72; Bat/73a; Maz/82)

Honey flow
Honey yield (kg/colony/season) [moderate] 25-28 (ITA/Umbria, Ric/78); 10-
17, exceptionally 25 (YUG, Shl/81). **Honey potential** (kg/ha) [moderate] 30
(GDR, Bec/67); 50-120 (ROM, Cir/80); 400-500 (URS/Caucasia, AA1500/81);
26-50 (Cra/75)

Pollen
P1 FRA; ITA; URS. **P** YUG. **Pollen value** excellent (Sta/74). **Chemical
composition** (Maz/82). **Colour** of load dull yellow to yellowish green
(Han/80); load bright yellow (Ric/78); illustrated (Hod/74). **Pollen
grain** illustrated and described (Haz/55; Saw/81); pollen heavily over-
represented in honey (Mal/77; Ric/78); >100 000 grains in 10 g, and up to
1 million (Maz/82); 490 000 grains in 10 g (Pes/80). **Reference slide**

Honeydew
Honeydew produced and collected by bees: from **Lachnus roboris** (L.),
previously L. longipes (Dufour), Lachnidae – flow 1–2 wks during flowering
(S EUR, Klo/65); also from **Myzocallis castanicola** Baker, Callophididae –
(S EUR, Klo/65); and **Parthenolecanium rufulum** (Cockerell), previously
Eulecanium rufulum (Cockerell), Coccidae – bees can only use this flow
early in morning and/or? at high RH (S EUR, Klo/65). Honeydew produced by
L. roboris and M. castanicola (ROM (Cir/80). Also produced (insect not
specified) in ITA (Ric/78); YUG (Kul/59)

Honey: chemical composition
Water [medium] 17.6–19.1% (5 samples, Dus/67); 17.1–19.3% (10 samples,
Iva/78); 15.7% (Kir/60); 16.0% (Mal/77); 18–19% (Shl/81).
Sugars, total 71.46–75.30% (Iva/78); 70.2–71.2% (Shl/81); 76.8, 79.6%
(Ver/65). **Glucose** [medium] 32.4–33.5% (5 samples, Gon/79); 32.15% (Mur/-
76); 25.22–30.80% (Shl/81); 22.29% (2 samples, Tou/80). **Fructose** [med-
ium] 42.5–44.4% (Gon/79); 36.74% (Mur/76); 35.00–44.68% (Shl/81); 39.19,
39.37% (Tou/80). **Sucrose** [medium] 0.00–6.17% (Iva/78); 0.35, 0.33%
(Tou/80); 2.21, 2.62% (Ver/65). **Reducing sugars** 67.80–72.50% (Waa/78);
69.64% (Mur/76). Contents of individual sugars expressed as % of total
sugars (Bat/73; Maz/59); also **maltose, isomaltose, trehalose, gentiobiose**
(Bat/73). **Dextrin** 2.71% (Kir/60)
Ash [medium] 0.70–1.20% (Iva/78); 1.28% (Kir/60); 0.77% (Mal/77); 0.28%
(Mur/76); 0.45–0.98% (20 samples, Pes/80); 0.25–0.42% (Shl/81)
pH 4.5 (Dus/72); 5.45 (Had/63); 5.15 (Kir/60); 5.87 (Mal/77). **Total**
acid (meq/kg) 10.6 (Had/63); 8.5–13.0 (Iva/78); 12.94 (Mal/77). **Free**
acid (meq/kg) [low] 6.8 (Had/63); 10.69 (Mal/77); 25–35 (Shl/81).
Lactone (meq/kg) 3.8 (Had/63); 2.25 (Mal/77). **Citric acid** 0.017% (Tou/-
80). **Malic acid** 0.064%
Amylase 13.0–18.4 (Iva/78); 17.7 (Mal/77). **Sucrase** 19.0–32.8 (Dus/67);
28.7 (Dus/72a). **Catalase** 0.0–0.2 (Dus/67). **Peroxide number** (μg/g/h) 100–
285 (Dus/67a); 391.5 (294.0 after 10 minutes in sunlight, Dus/72);
results for honeydew honey (Dus/67a; Dus/71). **HMF** 0.38 ppm (Mal/77)
Nitrogen 0.036% dry wt (Bos/78); 0.12% (Kir/60). **Amino acids**, free
0.201%, protein 0.118% (Bos/78); 11, 15 individual acids identified in 2
samples (Kum/74). **Protein** 0.375% (Ber/75). **Colloids** 0.48% (Kir/60)
Fermentation on storage unlikely, low yeast count (Maa/73)
Honey from this sp may be derived from both nectar and honeydew, if the
flows coincide (Klo/65)

Honey: physical and other properties
Colour often reddish (Cra/75); light brown (Fos/56); may be slightly
fluorescent if honeydew present (Haz/55); yellow (How/79); darker if
honeydew present (Klo/65); pale yellow (Mur/76); dark (Pia/81); dark
brown (Ric/78); honeydew honey almost black (Ric/78). **Pfund** light/dark
amber (Cra/75); amber (Kir/60); Pfund–Lovibond grade (Aub/83)
Relative density 1.4241 (Mur/76). **Viscosity** "thin" (Fed/55). **Optical**
rotation (deg) –33.91 (Bat/73); –2.25 to –3.50 (Iva/78); –3.99
(Kir/60). **Electrical conductivity** (per ohm cm) 0.0006–0.0008 (Dus/67a);
0.000578–0.000742 (Iva/78); 0.001553 (Mal/77); 0.00916–0.001844
(Pes/80); for honeydew honey, >0.009 (Dus/67a)
Granulation slow (Fos/56; Pia/81); fine (Cra/75); slow, irregular
(Pes/80); absent or v irregular (Ric/78)
Flavour pronounced, may be bitter (Cra/75; Maz/82); sharp (Fos/56); v
objectionable, bitter (Kir/60); rather bitter (Ric/78). **Aroma** sharp, and
like the flowers (Cra/75; Maz/82); v aromatic (Fos/56; Pes/80); pungent
(Pia/81)

081 Catunaregam spinosa (Thunb.) Tirvengadum; Rubiaceae
syn Randia dumetorum Retz. Poir.; Xeromphis spinosa (Thunb.) Blume

gela, karegida (INI)
Tree, <5 m, thorny; fls yellow-green
Distribution tropical Asia; native to Asia.   **Habitat** evergreen forests,
edges of forests (INI/KAR, Diw/64); dry/moist deciduous forests

Economic and other uses
Medicinal

Nectar rating + honeybee species; blooms, nectar flow
Nl INI/KAR[ac](Diw/64; Kha/59); INI/KER[ac](Kha/59); INI/MAH[ac](Chu/-
65; Ded/53)
Blooms iv-v (INI/MAH).   **Nectar flow** adversely affected by cloudy weather
(INI/MAH, Chu/65)

Honey flow
Honey yield 25% of annual honey harvest (INI/MAH, Chu/65)

Pollen
P INI/KAR, KER.   **Pollen grain** illustrated and described (Chu/65).
Reference slide

Honey: chemical composition
Water [medium] 17.23% (Naa/70); 16.09% (Pha/62); 17.19% (Raj/70)
Sugars, total 79.89% (Pha/62).   **Glucose** [medium] 5.70% (Naa/70); 35.26%
(Pha/62).   **Fructose** [medium] 41.17% (Naa/70); 42.54% (Pha/62).   **Reducing**
sugars 77.80% (Pha/62).   **Maltose, melezitose, raffinose** present
(Pha/70).   **Dextrin** 1.80% (Naa/70); 1.89% (Pha/62)
Ash [medium] 0.169% (Naa/70); 0.167% (Pha/62).   Contents of Na, K, Ca,
Mg, Fe, P, Si (Pha/62)
Free acid [medium] 12.9 meq/kg (Pha/62); 16.4 meq/kg (Naa/70).   **Citric,**
malic, succinic acids present (Pha/70)
Amino acids glutamic acid, tyrosine, leucine present (Pha/70); glycine,
tyrosine, serine, proline not present (Kai/64).   **Protein** 0.486% (Naa/70);
0.419% (Pha/62)
Vitamins 8 μg thiamine/100 g; riboflavin (in bound form), ascorbic acid,
niacin also present (Kai/65)

Honey: physical and other properties
Colour golden yellow (Chu/65); pale yellow (Pha/62)
Relative density 1.416 (Chu/65); 1.425 (Pha/62).   **Viscosity** at 26° over
100 poise (est), at 45°, 20 poise approx (Raj/70).   **Optical rotation** -2
deg 4 minutes (Naa/70); -2 deg 12 minutes (Pha/62)
Granulation medium (Chu/65); 8-12 months (Naa/70)

082 Ceiba pentandra (L.) Gaertn.; Bombacaceae

kapok, silk-cotton tree; ceiba (Es/ELS, VEN); faux kapokier (Fr/CEN);
fromager (Fr/CEN, IVO); kapokier de Java (Fr/IVO); randu (In)
Tree, 30-40 m, deciduous, spiny trunk with buttresses at base; fls
white/pink, large, showy

Distribution tropical C America, S America, Africa, Asia; subtropical N
America; native to tropical America, Amazon region. **Habitat** important
plantation crop; forest conditions; low altitudes; naturalized on coast
(western INI, Sig/62)
Soil - sites preferred where water is on/near surface. **Rainfall** areas of
heavy rainfall preferred

Economic and other uses
Food - seeds. **Fodder** - cattle food from pressed cake. **Timber.** **Land
use** hedges, shade, amenity. **Other uses** fibre (kapok) from fruit capsules

Nectar rating + honeybee species; blooms, nectar flow; composition
N1 ?CEN,tm(Dou/79); ELS(Woy/81); INO,ac(Bee/77); VEN(Cra/73)
N2 THA,ac(Pre/82)
N IVO[tm](Dou/80); SEN,tm(Ndi/74)
Blooms xii-i (ELS); ii-iii (INI, Sig/62); xii-iv when tree has no lvs,
not every yr (tropical America, Ord/83); i-iv (VEN). **Nectar secretion**
copious, tends to run out of corolla at anthesis (Frj/70). **Sugar concentra-
tion** [medium] 50% (Zma/80)

Pollen
P1 ELS; INO. **Yield** heavy (Ord/83). **Colour** orange (Ord/83). **Pollen
grain** sticky (Frj/70)

Honey: physical and other properties
Colour light (Dou/79). **Pfund** amber (Mot/64; Woy/81); light amber (Ord/83)
Flavour characteristic (Mot/64; Ord/83)

083 Centaurea cyanus L.; Compositae

blue cap, blue-bottle, cornflower; bachelor's button (En/USA); bleuet,
centaurée bleuet (Fr); Kornblume (De)
Herb, 30-60 cm, annual; fls blue (pink/purple/white); nectary in fl, also
extrafloral nectaries on seed heads and flower buds (Dus/69)
Distribution temperate Europe, N America, N Africa; native to SE Europe

Nectar rating; blooms, nectar flow
N1 GDR(Koc/65); URS(Glu/55)
N2 FRA(Lou/81)
N3 GFR(Gle/77)
N ALG(Ske/72); EUR (Maz/82); URS(Fed/55); USA/MA(Shw/50a)
Blooms v-x (EUR); v-vii (GDR). **Nectar secretion** 0.43 mg/fl/day; total 5
mg/fl head (Maz/82). **Sugar concentration** [medium] 30-40% (Cir/80); 31-
35% (Han/80); may be >60% (Maz/82); 33.9% (Mog/58); 16% at 07.00 h to
35% (max) at 17.00 h (AA969/79); 31.0-34.5% (extrafloral nectaries,
Dus/69). **Sugar value** [medium] 1.15 mg/fl/day (Han/80; Maz/82). **Sugar
analysis** (Dus/69; Maz/82)

Honey flow
Honey potential (kg/ha) [moderate] 60 (GDR, Bec/67); 50-60 (ROM, Cir/77);
51-100 (Cra/75)

Pollen
P1 URS.   P2 FRA.   P3 GFR.   P URS.   **Pollen yield** 7.8 mg/fl head
(Maz/82).   **Pollen value** high (Sta/74).   **Colour** of load light grey (Han-
/80; Maz/82); load beige (Hod/74).   **Pollen grain** illustrated (Hod/74);
illustrated and described (Saw/81); chemical composition (Sta/74); pollen
under-represented in honey (Cra/75; Dus/69); but >100 000 grains in 10 g
honey sample [over-represented] (Sta/74).   **Reference slide**

Honey: chemical composition
Glucose 37.4% of total sugars (Maz/64).   **Fructose** 44.0%.   **Sucrose** 6.3%.
Maltose 7.9%.   **Fructomaltose** 4.4%
pH 4.3 (Dus/72)
Amylase 41 (Dus/69).   **Sucrase** 35.5 (Dus/72a).   **Peroxide number** (μg/g/h)
624.0 (Dus/71); 624.5 (308.5 after 10 minutes in sunlight, Dus/72)

Honey: physical and other properties
Colour greenish yellow (Dus/72); dark yellowish amber (Ord/83).
Electrical conductivity 0.000310/ohm cm (Vor/64)
Flavour of almond, may be rather bitter (Fed/55); strong, may be bitter
(Ord/83)

084 Centaurea solstitialis L.; Compositae

cockspur, St Barnaby's thistle, yellow Jack (En/AUS); Barnaby's thistle,
star-thistle, yellow star thistle (En/USA)
Herb, 30-100 cm, annual/biennial; fls pale yellow
Distribution temperate Europe, Oceania; subtropical N America; native to
Europe.   **Habitat** cultivated or waste ground; troublesome weed in grain
fields USA/CA
Soil dry.   **Temperature** – plant cut back by frost.   **Rainfall** drought
tolerant (Pel/76)

Economic and other uses
Food – as vegetable.   **Fodder** – hay, which needs moistening before use.
Other uses medicinal

Warning
Troublesome weed in grain fields (USA/CA)

Nectar rating; blooms, nectar flow; composition
N1 USA/CA(Jay/54; Pel/76; Van/41)
N2 ?ARG(Kat/68); AUS/NSW(Goo/47); USA/CA(Lov/56)
N3 AUS/VIC(Gum/73)
Blooms late spring to summer (AUS/NSW); vi-ix (EUR, Maz/82); vii – frost
(USA/CA).   **Nectar flow** stops in drought but restarts after rain
(Pel/76).   **Nectar secretion** 0.123 mg/fl/day (Sim/80); slow but continuous
(Pel/76).   **Sugar concentration** [medium] 51.2% (Sim/80); 38% (Van/41).
Sugar value (mg/fl/day) [medium] 0.107 (Sim/75); 0.164 (Sim/80)

Honey flow
Honey yield [moderate] <27 kg/colony/season (USA/CA, Van/41)

Pollen
P1 AUS/VIC; USA/CA. **P2** AUS/NSW. **Pollen value** important source in
Sacramento Valley (USA/CA) vii-x, pollen produced throughout the day (Van/-
41). **Chemical composition** (Maz/82; Shp/80). **Colour** yellow (Van/41);
dull purple (Wal/78). **Reference slide**

Honey: physical and other properties
Colour greenish (Lov/56; Van/41); greenish with yellow tinge like olive
oil (Pel/76). **Pfund** white or extra light amber (Lov/56; Van/41); white
(Pel/76; Wal/78)
Viscosity "heavy body" (Pel/76)
Granulation rapid (Van/41); v fine grained (Goo/47)
Flavour v sweet, almost cloying (Pel/76); delicate (Wal/78)

085 Cercidium floridum Benth.; Leguminosae
syn Cercidium torreyanum (S. Wats.) Sarg.

green bark acacia (En/USA); palo brea, palo verde (Es/MEX); palo verde
(USA)
Tree, <9 m, deciduous, lvs borne for short period only, bright green bark
(Pel/76); fls yellow
Distribution subtropical N America, C America. **Habitat** desert areas of
USA/CA, AZ and MEX
Rainfall drought tolerant (Pel/76)

Economic and other uses
Fuel. **Land use** shade

Nectar rating; blooms, nectar flow
N1 MEX(Ord/83)
N2 MEX(Ord/72); USA/AZ(Lov/56; Pel/76)
Blooms spring (tropical America, Ord/83). **Nectar flow** more reliable on
low ground with higher soil moisture (Pel/76)

Honey flow
Honey yield [moderate] 9-13 kg/colony/season (USA, Lov/56)

Pollen
P MEX

Honey: physical and other properties
Colour light yellow (Pel/76). **Pfund** light amber (Ord/83)
Viscosity "good body" (Pel/76)
Flavour distinctive, like bark of this tree (Ord/83; Pel/76)

086 Cicer arietinum L.; Leguminosae

chich pea, Bengal gram gram; harbara (INI)
Herb, 50-60 cm, annual, shrubby

Distribution temperate (Med) Europe; tropical Asia, Africa, C America, S
America; native to W Asia. **Habitat** cultivated crop plant especially in
dry regions; basins/river banks (SUD)
Soil heavy but not waterlogged. **Temperature** moderate. **Rainfall** needs
little rain and is not much affected by drought (Why/53); sometimes grown
under irrigation

Economic and other uses

Food - pods, young shoots, seeds used in dahl (INI). **Fodder** - hay. **Soil
benefit** soil renovation, green manure. **Other uses** - liquid from glandular
hairs (94% malic acid, 6% oxalic acid) used medicinally and as vinegar
(Pus/68)

Nectar rating + honeybee species; blooms, nectar flow

N1 INI/UTT(Cht/69)
N3 INI/MAH[ac](Chu/80)
  Blooms xii-i (INI/UTT). **Nectar flow** not reliable (Sig/62)

Honey flow

Honey yield [moderate] 2.0-2.5 kg/colony/season, migration to crop recommended (INI/UTT, Cht/69)

Pollen

P3 INI/MAH. **Reference slide**

Honey no data

087 Citrus aurantifolia (Christm.) Swingle; Rutaceae

lime; limero (Es/ARG); limón (Es/ELS); limbu (INI); kaghzi nimbu (PAK)
Tree, <4 m, evergreen, spiny; fls white
Distribution tropical S America, C America, Caribbean, Asia, Africa;
subtropical Middle East, Asia, N America; native to Asia
Soil poor soil tolerated. **Temperature** one of the least hardy Citrus spp;
frost not tolerated

Economic and other uses

Food - fruit for drinks/flavouring; preserves

Nectar rating + honeybee species; blooms, nectar flow; composition

N1 ELS(Woy/81); PAK,ac(Pak/77)
N2 INI/MAH[ac](Chu/80)
N ARG(Per/80); BRA,tm(Caa/7?); INI/HAR,ac(Vas/67); USA/FL(Mot/64)
Blooms vii-viii (BRA/SP, SC); i-xii (INI/MAH). **Sugar concentration**
[medium] 34-38% (Caa/72; Sao/54); 28% (nectar from honey sacs of bees,
humid area, Mof/74); 50% (Zma/80)

Pollen

P2 INI/MAH. **P PAK**

Honey no data

088 Citrus aurantium L.; Rutaceae

bitter orange, Seville orange, sour orange; naranjo agrio (Es); naranjo
(Es/CUB, VEN); naranja acida (Es/NIA); oranger (Fr/CEN, SEN); laranjeira
(Pt/BRA); arancio amaro (It)
Tree, 6-9 m, evergreen, spiny; fls white, fragrant. Names of C.
aurantium and C. sinensis often confused in literature; only reliably
authenticated data entered here
Distribution tropical Africa, C America, Caribbean, S America; subtropical
N America, Oceania; temperate (Med) Europe; native to Asia. **Habitat**
cultivated crop plant; grows wild on mountains in CUB; altitudes <1500 m
(INI/TAM)
Soil medium consistency, light and porous; not chalk or clay.
Temperature frost tender. **Rainfall** winter rainfall areas preferred; also
drier areas if irrigated

Economic and other uses
Food - fruit for marmalade. **Other uses** fls for perfumery; as stock plant
for grafting sweet orange

Nectar rating + honeybee species; blooms, nectar flow; composition
N1 ?CEN[tm](Dou/79); PUE(Ord/44); VEN(Cra/73)
N2 BRA/SC[tm](Wie/80); CUB/(Ord/44); ITA(Ric/78); NIA(Ord/63a)
N SEN[tm](Dou/70)
Blooms viii-ix (BRA/SC); iii-iv, viii-ix (INI/TAM, Chn/74); ii-iv(VEN,
Ste/71); winter-spring (CUB). **Nectar flow** affected by pH of soil
(Ord/83). **Nectar secretion** abundant (Ord/83). **Sugar concentration**
[medium] 25.8% (Fah/49). **Sugar analysis** (Vah/72)

Pollen
P BRA/SC; ITA. **Reference slide**

Honey: physical properties
Pfund-Lovibond grade (Aub/83)

089 Citrus bergamia Risso & Poiteau; Rutaceae

bergamot; bergamoteira, mexiriqueira (Pt/BRA); bergamotto (It)
Tree, <3 m, evergreen; fls white, fragrant
Distribution temperate (Med) Europe; subtropical S America; native to
China. **Habitat** cultivated crop plant
Temperature minimum 2°; frost tender

Economic and other uses
Food - fruit peel yields essential oil for liqueurs. **Other uses** oil for
perfumery

Nectar rating + honeybee species; blooms, nectar flow; composition
N1 ITA/Sicily(Ric/78)
N BRA/RG[tm](Jul/70)
Blooms viii-x (BRA/RG). **Sugar concentration** [medium] 35% (Fah/49); 20-
30% (Jul/70)

Pollen
P BRA; ITA

Honey no data

090 Citrus deliciosa Ten.; Rutaceae

Mediterranean mandarin, tangerine; mandarino (It)
Tree, 3-4 m, spiny; fls white, small, fragrant.   Names of C. deliciosa
and C. reticulata often confused in literature; only reliably authenti-
cated data entered here
Distribution tropical Africa; temperate (Med) Europe, Asia; native to SE
Asia.   **Habitat** cultivated crop plant
Soil fertile and well drained.   **Temperature** lower levels tolerated than by
the orange (C. sinensis)

Economic and other uses
Food - fruit; flavouring.   **Land use** amenity.   **Other uses** essential oil
from peel for perfumery/pharmacy

Nectar rating; composition
N1 ISR(Chi/65; Moa/55)
N2 ITA(Ric/78)
N JAP(Waa/61); MAQ(Far/79)
Sugar concentration [medium] 22% (Frj/70)

Honey flow
Honey yield (kg/colony/season) [high] 30-60, ISR; [moderate] 15-20, mixed
with other Citrus spp (ISR, Chi/65)

Pollen
P ITA

Honey: chemical composition
Water [high] 18.72, 26.27% (Waa/61)
Glucose [medium] 36.46, 38.62%.   **Fructose** [medium] 34.00, 37.60%.
Sucrose [medium] 0.95, 1.60%.   **Reducing sugars** 72.77, 79.02%

Honey: physical and other properties
Colour pale yellow
Relative density 1.359, 1.394.   **Optical rotation** -12.29, -13.14 deg

091 Citrus grandis (L.) Osbeck; Rutaceae

pomelo, pummelo, shaddock; toronja (Es); chakotra (PAK)
Tree, 9-15 m, evergreen, spines slender/absent on mature trees; fls white
Distribution tropical Asia, C America; subtropical N America; temperate
(Med) Europe; native to Asia.   **Habitat** cultivated crop plant; lowland
areas, brackish marshes, deltas (THA)

Soil light, well drained preferred; some salt tolerance. **Temperature**
uniformly warm climate preferred; not frost hardy

Economic and other uses
Food - fruit. **Other uses** essential oil from peel

Nectar rating + honeybee species; blooms, nectar flow; composition
Nl PAK,ac(Pak/77)
Blooms ii-iii (PAK). **Sugar concentration** in nectar [medium] 50% (Zma/80)

Pollen
P PAK

Honey no data

092 Citrus limetta Risso; Rutaceae

sweet lemon; metha (PAK)
Distribution subtropical, tropical Asia; native to tropical Asia

Economic and other uses
Food - fruit

Nectar rating + honeybee species; blooms, nectar flow; composition
Nl PAK,ac(Pak/77)
Blooms ii-iii (PAK). **Sugar concentration** in nectar [medium] 16-40% (Jul/72)

Pollen
P PAK

Honey no data

093 Citrus limon (L.) Burm. f.; Rutaceae

lemon; limón (Es/VEN); citronnier (Fr)
Tree, 3-6 m, evergreen, stout stiff thorns; fls white, petals pinkish
outside, v fragrant
Distribution temperate (Med) Europe, Oceania; tropical Africa;
subtropical Asia, Africa, N America; native to Asia. **Habitat** cultivated
crop plant; open forests in high rainfall areas (ZIM, Wil/72); semi-arid
areas
Temperature mild/warm; tree damaged by frost

Economic and other uses
Food - fruit; flavourings, juice, essential oil, liqueurs. **Land use**
hedges. **Other uses** as rootstock for grafting other Citrus spp; oil for
perfumery, cosmetics

Nectar rating + honeybee species; blooms, nectar flow; composition

N1 ISR(Chi/65; Moa/55);   PAK,ac(Pak/77);   VEN(Cra/73)

N2 AUS/VIC(Gom/73);   USA/CA(Pel/76)

N3 USA(Ord/83)

Blooms viii-ix (BRA/SC, SP, Caa/72);   ii-iii (PAK);   ii-iv (VEN, Ske/71).
Nectar secretion considered to be the lowest of Citrus spp; trees near
coast of USA secreted more than those inland (Ord/83).   **Sugar concen-**
tration [medium] 24.9-28.7% (Fah/49);   also [low] 15-18% (Frj/70);   nectar
from honey sacs of bees: 63.0% in dry area, 28.2% in humid area (Mof/74)

Honey flow

Honey yield (kg/colony/season) [high or moderate] 30-60 or 15-20, mixed
with other Citrus spp (ISR, Chi/65)

Pollen

P2 AUS/VIC.   **P** PAK.   **Chemical analysis** (Gil/80)

Honey: physical and other properties

Colour clear (Erb/83).   **Pfund** light amber (Ord/83)

Flavour delicate, "aromatic" (Erb/83);   strong, sour (Ord/83).   **Aroma**
delicate, "aromatic" (Erb/83);   characteristic, and like the plant (Ord/83)

094 Citrus medica L.; Rutaceae

citron; katta (PAK)

Tree/shrub, 2-3 m, deciduous, short stiff spines; fls white inside,
purplish outside, fragrant

Distribution subtropical S America, N America; temperate (Med) Europe;
tropical C America; native to Asia (E Himalayas).   **Habitat** cultivated
crop plant

Temperature warm, not humid; frost tender

Economic and other uses

Food - fruit for candied peel and juice.   **Other uses** medicinal

Nectar rating + honeybee species; blooms, nectar flow; composition

N1 PAK,ac(Pak/77)

N ELS(Woy/81)

Blooms ii-iii (PAK).   **Sugar concentration** in nectar [medium] 35.0-50.0%
(Shr/58)

Pollen

P PAK.   **Reference slide**

Honey no data

095 Citrus paradisi Macfad.; Rutaceae

grapefruit

Tree, 6-12 m, evergreen, spiny; fls white, fragrant
Distribution temperate (Med) Europe; subtropical N America, Africa;
tropical Africa, C America, Caribbean; native to ?Caribbean. **Habitat**
cultivated crop plant
Temperature tree hardier than C. grandis but not frost tolerant

Economic and other uses
Food - fruit, fruit juice, jam

Nectar rating + honeybee species; blooms, nectar flow; composition
Nl ISR(Chi/65; Moa/54); PAK,ac(Pak/77)
N USA/CA(Lov/56)
Blooms ii-iii (PAK); iii-iv (USA/FL). **Sugar concentration** [medium] 25.4%
(Fah/49); 21 or 50% (Zma/80); [low] 16% (Frj/70); nectar from honey sacs
of bees: 61.5% in dry area, 18.6% in humid area (Mof/74). **Amino acid
analysis** (Gil/80)

Honey flow
Honey yield "adds materially" to Citrus yield (USA/FL, Mot/64)

Pollen
P PAK. **Chemical analysis** (Gil/80)

Honey no data

096 Citrus reticulata Blanco; Rutaceae

mandarin orange; mandarino (Es/ARG); santara (PAK)
Tree, 3-4 m, evergreen, spiny; fls white, small, fragrant. Names of C.
reticulata and C. deliciosa often confused in literature; only reliably
authenticated data entered here
Distribution temperate Asia, (Med) Europe, N Africa; subtropical S
America, Africa, N America, Asia; tropical Africa; native to SE Asia.
Habitat cultivated crop plant
Soil fertile, well drained. **Temperature** frost tender

Economic and other uses
Food - fruit; flavouring. **Land use** amenity. **Other uses** essential oil
from peel for perfumery/pharmacy

Nectar rating + honeybee species; blooms, nectar flow; composition
Nl CHN(Mad/81); PAK,ac(Pak/77); URS(Glu/55)
N ARG[tm](Per/80)
Blooms ii-iii (PAK). **Sugar concentration** [medium] 42.5% (Fah/49); 22%
(Frj/70)

Pollen
P PAK; URS. **Reference slide**

Honey no data

097 Citrus sinensis (L.) Osb.; Rutaceae

sweet orange; orange (En/AUS); naranja de China (Es/DOR); naranja dulce
(Es/ELS, HOD, MEX); oranger doux (Fr); laranjeira-doce (Pt/BRA); arancio
dolce (It)
Tree, <8 m, evergreen, spines few or absent; fls white, fragrant.   Names
of C. sinensis and C. aurantium often confused in literature; only
reliably authenticated data entered here
Distribution temperate (Med) Europe, Asia; subtropical Africa, N America,
S America, Oceania; tropical Africa, C America; native to China.
Habitat extensively cultivated in areas with Med climate
Soil light loam.   **Temperature** not frost tolerant; tree less hardy than C.
aurantium.   **Rainfall** where low/irregular, irrigation required

Economic and other uses
Food - fruit, fruit juice.   **Fodder** - stock feed, pulp from fruit-
processing.   **Other uses** fragrant fls for decoration

Alert to beekeepers
Flow is early in season so beekeepers must make colonies strong to harvest
honey (SOU, Fal/74; USA/CA, FL, Pel/76).   Pesticide applications often
make citrus a dangerous crop for bees (Cra/83)

Nectar rating + honeybee species; blooms, nectar flow; composition
N1 ?ALG(Ske/72); CYP(Adm/54); DOR(Ord/64); ELS(Woy/81); HOD(Ord/63);
ISR(Kal/77); ITA(Ric/78): MEX(Ord/72); MOR(Cra/73); PAK,ac(Pak/77);
POR(Wat/73); SWA,tm(Ehr/77); USA/CA(Jay/54; Lov/56; Van/41)
N2 SOU,tm(Cri/57); USA(Pel/76)
N BRA/SP(Fle/63; Jul/70); MAQ(Far/79)
Blooms viii-ix (BRA/SP); vi-x (BRA/SC); viii-ix (SOU).   **Nectar flow** iii-
? (ISR, Kal/77); 3 or more wks (USA/CA, Van/42); 4-5 wks where Navels and
Valencias are grown together (SOU, Cri/57).   **Alert to beekeepers** flow is
early in season so beekeepers must make colonies strong to harvest honey
(SOU, Fal/74; USA/CA, FL, Pel/76).   **Nectar secretion** 20 μl/fl secreted
before petals unfold (Van/42); more reliable in interior valleys USA/CA,
coastal fogs unfavourable (Pel/76); heaviest with early rains, hot days,
cool nights, sufficient irrigation (Fal/74).   **Sugar concentration** [medium]
35% (Caa/72); 11.6-39.0% (Fah/49); 50% (Zma/80); [low] 11-18% (Frj/70);
18-20% (Jul/70).   Also 13-17% from bud, 20-50% from open fl (Van/42);
nectar from honey sacs of bees: 62.5-63.0% in dry area 19.8-22.5% in humid
area (Mof/74).   **Sugar analysis** (Kal/77; Van/41; Van/42).   **Amino acid
analysis** (Gil/80)

Honey flow
Honey yield (kg/colony/season) [high] 40-50, 140 (MOR, Cra/73); 34-79
(USA/FL, Pel/76); [medium] 10-20 (SOU, Fal/74); 10-40 (SOU, Guy/72a); 27
(USA/CA, Pel/76); also 10 kg/colony/day when sugar concentration of nectar
was 50% (USA/CA, Van/42); 5.3 kg/colony/day (10 days, USA, Pel/76)

Pollen
P1 ITA.   **P** ALG; BRA; DOR; PAK; ?USA/CA.   **Yield** Valencia oranges yield
viable pollen, 24 000 grains/ml nectar (Van/42); yield light; cv
Washington navel has no viable pollen.   **Chemical analysis** (Gil/80).
Colour of load yellow, load size small to medium (Van/41).   **Pollen grain**
described (Bah/73); illustrated and described (Sao/61; Smt/56a); under-
represented in honey (Bah/73); 1200-1500 grains in 10 g honey (Ric/75).
Reference slide

Honey: chemical composition
Water [medium] 15.4-17.5% (11 samples, The/77); 17.1% (1 sample, age 19 mths, Whi/62)
Glucose [low] 28.2-30.9% (The/77); 31.49% (Whi/62). **Fructose** [medium] 35.1-39.5% (The/77); 38.23% (Whi/62). **Sucrose** [high] 9.1-10.3%, unusually high (but 1.6-4.2% after 7 mths at room temp, Kal/77); 3.63-8.44% (1.80-6.51% after 1 mth at 27°, 0.63-6.1% after 2.5 mths, The/77); 2.68% (Whi/62). **Maltose** 2.6-3.5% (The/77); 7.41% (Whi/62). **Higher sugars** 1.47%
Ash [low] 0.084%
pH 3.60. **Total acid** 35.89 meq/kg. **Free acid** [medium] 22.77 meq/kg. **Lactone** 13.12 meq/kg
Amylase low (Ske/72). **HMF** 1-5 ppm (12-16 ppm after 7 mths at room temp, Kal/77)
Nitrogen 0.029% (Whi/62)
Other constituents - methyl anthranilate 0.084-3.90 μg/g (6 samples, Whi/66)

Honey: physical and other properties
Colour transparent and bright, light brown (Bah/73); light yellow, almost white (Ske/72). **Pfund** water white to v pale amber (Cri/57); white (Kal/77); white to extra light amber (Lov/56); clear white to dark amber, varies with season, darkens with age (Mot/64); white (Pel/76; Van/41)
Viscosity "heavy body" (Mot/64; Pel/76)
Granulation slow (Joh/75); few months (Mot/64)
Flavour aromatic (Bah/73); characteristic (Ske/72). **Aroma** characteristic of orange blossom (Joh/75; Lov/56; Mot/64)

098 Citrus unshiu (Mak.) Marc.; Rutaceae

orange, satsuma
Distribution temperate Asia

Economic and other uses
Food - fruit

Alert to beekeepers
Little pollen collected (Mcg/76)

Nectar rating; blooms, nectar flow; composition
N1 JAP(Sak/82)
Blooms early summer (JAP). **Sugar concentration** [medium] 29% (Ech/77).
Sugar analysis (Ech/77)

Pollen
Alert to beekeepers little pollen collected (Mcg/76)

Honey: chemical composition
Glucose [low] 30.49% (Ech/77). **Fructose** [medium] 38.20%. **Sucrose** [low] 0.91%. **Maltose** 5.19%
Other constituents - compounds probably contributing to flavour (Wab/80)

099 Clethra alnifolia L.;   Clethraceae

clethra; pepper bush, sweet pepperbush, white alder (En/USA)
Shrub, 1-3 m; fls white, fragrant
Distribution temperate and subtropical N America.   **Habitat** in large thickets along coast of USA/MA; also coasts of other southern states (USA); swampy stream banks
Soil sandy; damp

Economic and other uses
Land use amenity

Nectar rating; blooms, nectar flow; composition
N1 USA/MA(Lov/77);   USA/NC(Lor/79);   USA/RI(Lov/56c)
N2 USA/CT(Pel/76)
N USA/MA(Shw/50);   USA/VA(Gra/50)
Blooms vi-ix (USA, Lov/77).   **Sugar concentration** in nectar [medium] 19.5-34.0% (Shw/53)

Honey flow
Honey yield (kg/colony/season) [high] 68-77 (exceptional yr, USA/CT, Pel/76); 34 (USA/RI, Lov/56c)

Pollen
P USA/MA

Honey: chemical composition
Water [medium] 17.8% (1 sample, age 12 mths, Whi/62)
Glucose [medium] 31.30%.   **Fructose** [medium] 36.30%.   **Sucrose** [low] 0.81%.   **Maltose** 7.11%.   **Higher sugars** 1.63%
Ash [medium] 0.235%
pH 4.18.   **Total acid** 32.03 meq/kg.   **Free acid** [medium] 21.85 meq/kg.
Lactone 10.18 meq/kg
Amylase 12.0
Nitrogen 0.053%

Honey: physical and other properties
Colour tinged with yellow (Lov/56).   **Pfund** white (Lov/56; Pel/76);   50-70 mm, light amber (Whi/62)
Viscosity "thick" (Pel/76)
Flavour mild (Lov/56).   **Aroma** like the flower

100 Coccoloba belizensis Standl.;   Polygonaceae

Distribution tropical C America

Nectar rating; blooms, nectar flow
N1 BEL(Mul/79)
Blooms iii (BEL)

Honey no data

101 Coccoloba uvifera L.; Polygonaceae

sea grape; seaside plum (En/USA); uva caleta, uvero (Es/CUB); uva de mar
(Es/DOR); uvero de playa (Es/DOR, NIA); uva (Es/HOD)
Shrub/tree, evergreen; fls greenish yellow, small
Distribution subtropical N America, tropical C America, Caribbean, S
America. **Habitat** coastal regions; also inland CUB but trees much smaller
and more crooked; sandy slopes behind coastal vegetation (USA)
Soil sandy; salt tolerant

Economic and other uses
Food - fruit for jelly

Nectar rating; blooms, nectar flow
N1 CUB(Ord/44); HAI(Mul/78); JAM(Met/66)
N2 DOR(Ord/64); HOD(Ord/63); NIA(Ord/63a)
N3 USA/FL(Lov/56; Pel/76)
Blooms all yr (tropical America, Pel/76); iv-v and again later (CUB).
Nectar flow prolonged but less intense than Avicennia germinans (Ord/44);)
wind can cause first fls to fall (Ord/83). **Nectar secretion** copious until
well past noon, often till 17.00 h

Honey: chemical composition
Water high (Mot/64)

Honey: physical and other properties
Pfund amber (Lov/56; Pel/76); v light amber (Mot/64); light amber (Ord/83)
Flavour spicy (Mot/64); sharp (Ord/83)

102 Cochlospermum insigne St. Hil.; Cochlospermaceae

algodonillo (Es/BOL)
Tree, small, deciduous
Distribution tropical S America

Nectar rating + honeybee species; blooms, nectar flow
N1 BOL[tm](Kem/80)
Blooms vi-vii (BOL)

Honey no data

103 Cochlospermum vitifolium (Willd.) Spreng.; Cochlospermaceae

bototo (Es/VEN)
Tree/shrub, <20 m; fls yellow, resembling roses; end of dry season
Distribution tropical C America, Caribbean, S America. **Habitat** hill
slopes/plains of Pacific coast; deciduous forests
Soil rocky/v poor. **Temperature** v hot areas (tierra caliente)

Economic and other uses
Land use hedges

Nectar rating; blooms, nectar flow
N1 VEN(Cra/73)
Blooms i-iv (VEN, Ste/71)

Pollen
P tropical America (Ord/83)

Honey no data

104 Cocos nucifera L.; Palmae

coconut, coconut palm; cocotero (Es/DOR); cocotier (Fr); coqueiro
(Pt/MOZ); thengu (INI)
Tree, 30-40 m, evergreen; fls cream, small, monoecious, both male and
female fls have nectaries, male fls fragrant
Distribution tropical Asia, Oceania, Africa, C America, Caribbean;
subtropical N America; native to ?Indo-Malaysian region. **Habitat**
cultivated crop plant; irrigated plains, OMA; sea shores, inland lowlands
Soil salty, sandy soil tolerated; must be well drained. **Temperature**
light frost tolerated

Economic and other uses
Food - fruit, shoots, toddy from fermented sap. **Fodder** - pressed cake.
Timber. **Land use** amenity. **Other uses** oils/fats from copra; lvs for
thatching; fibres for ropes/mats

Alert to beekeepers
Where sap is tapped from unopened inflorescences for toddy-making, many
bees drown in the collecting pots (Kan/40)

Nectar rating + honeybee species; blooms, nectar flow; composition
N1 CHG/Diego Garcia(Sil/69); MEX(Brs/82); OMA(Bea/79); TAN/ZAN[tm](Cra/73)
N2 DOR(Ord/64; Ord/83); JAM(Ord/83); PUE(Phl/14); USA/FL(Ord/83)
N3 SRI,ac(Kud/81); USA/FL(Pel/76)
N BUM(Zma/80); INI/KAR,KER[ac](Kha/59); MOZ[tm](Cra/73); SEN[tm]-
(Dou/70); SEY(Sil/70); THA(Smt/83)
Blooms i-iii (BUM); most of yr (tropical America, Ord/83); peak in early
spring (MEX). **Nectar flow** heaviest before rainy season in May (MEX,
Brs/82); major source on Samui island (THA, Smt/83); production falls as
distance from coast and as altitude increase (Ord/83); unreliable (SEY,
Sil/70). **Sugar concentration** [medium] 24% (Zma/80). **Sugar analysis**
(Row/76; AA657/70). Bees also forage on young (2.5-cm) coconuts which
are coated with nectar for about a week (Mot/64); where sap is tapped
from unopened inflorescences for toddy-making, bees collect sap (SRI,
Kud/81). **Alert to beekeepers** many bees drown in the collecting pots
(Kan/40)

Honey flow
Honey yield (kg/colony/season) [high] 70-80 (MEX, Brs/82); [moderate]
estimated 1-3 (SRI, Kud/81)

Pollen

Pollen

Pl DOR; INI/KAR, KER; JAM; OMA; USA/FL. **P** CHG; SEY; USA/FL. **Yield**
6.1 g/inflorescence (Mcg/76). **Pollen value** "a useful perpetual source"
(SEY, Sil/70); major source (Kha/59). **Colour** white/yellowish-white
(Ord/83). **Pollen grain** illustrated (Bls/80)

Honey: physical and other properties

Colour may be greenish-yellow like motor oil (Mot/64); crystal clear if
monofloral (Ord/83). **Pfund** amber, but ?water white if monofloral
(Cra/75; Mot/64)
Granulation [medium] 3 mths (Mot/64)

105 Coffea arabica L.; Rubiaceae

Arabian coffee, coffee; café (Es/BOL, CUB, ELS); cafeto (Es/DOR, MEX,
PUE, VEN); cafeeiro (Pt/BRA)
Tree/shrub, evergreen, 2-9 m; fls white, fragrant
Distribution tropical C America, Caribbean, S America, Africa, Asia,
Oceania; native to Africa. **Habitat** cultivated crop plant; cool parts of
tropics; prospers in shade of taller trees; an understorey tree in high
altitude forests in ETH where it is native
Soil deep, slightly acid, well drained fertile loam preferred.
Temperature optimum 16-24°, damaged by frost or intense heat. **Rainfall**
optimum 190 mm with 2-3 mths dry period to initiate flowering; high
humidity; not drought tolerant

Economic and other uses

Food - toasted beans or dried lvs for beverage. **Fodder** for cattle. **Soil**
enrichment

Nectar rating + honeybee species; blooms, nectar flow; composition

N1 BOL(Mun/53); BRA(Mun/54; Smt/60); COL(Cor/76; Ken/76); DOR
(Ord/64); ELS(Woy/81); INI/TAM,ac(Chn/74); MEX(Ord/72); PUE(Ord/44);
TAN,tm(Smt/56); VEN(Cra/73; Ste/71)
N2 CUB(Ord/44; Ord/56); JAM(Ord/83); PUE(Ord/83); TAN(Smt/57)
N BRA[?tm](Fle/63)
Blooms all yr (BRA); peak in iii (CUB); iv-v (ELS); ii-iii (INI/TAM); ii-
iv (VEN). **Nectar flow** intense, and last in main flow in ELS (Woy/81);
dependent on rains (Chn/74). **Nectar secretion** reduced by drought
(Ord/44). **Sugar concentration** [medium] 30-37% (Caa/72); 38% (Frj/70); 32-
34% (Kem/80); 32-40% (cv semperflorens, Sao/54)

Pollen

P COL; DOR; INI/TAM. **Pollen grain** heavy and sticky (Frj/70); dry not
sticky (Mcg/76); illustrated and described (Smt/56a)

Honey: physical and other properties

Colour of whisky (Cor/76); light (Cra/75; Ord/83); clear (Ord/72;
Ste/71). **Pfund** white (Ord/44; Woy/81); amber (Ord/72; Ste/71); light
amber (Ord/83)
Flavour characteristic (Cra/75; Ord/83); characteristic of coffee
(Woy/81). **Aroma** delicate (Kem/80)

106 Combretum celastroides Laws.; Combretaceae
syn Combretum trothae Engl. & Diets

savanna bushwillow (En/SOU); Jesse-bush combretum (En/ZIM)
Shrub/tree, 4 m, often forms impenetrable thickets, "Jesse bush"; fls
greenish to yellow
Distribution tropical Africa. **Habitat** deciduous thickets in Itigi and
Manyoni areas (TAN, Smi/57); dry woodland on hillsides
Soil Kalahari sand; also rocky soil. **Rainfall** drought resistant

Nectar rating + honeybee species; blooms, nectar flow
N1 ?KEN,tm(Smt/57); TAN,tm(Smt/57)
Blooms "throughout the season" (TAN). **Nectar flow** xii-iii with peak in ii
(TAN)

Pollen
Pollen grain illustrated and described (Smt/54a)

Honey: physical and other properties
Pfund extra light to light amber (Smt/57)

107 Combretum fruticosum (Loefl.) Stuntz; Combretaceae

chapamiel, papamiel (Es)
Shrub, climber; fls red-orange, brush-like, fragrant
Distribution tropical C America, Caribbean, S America. **Habitat** dense rain
forests, dry forests and scrub on Pacific coast

Nectar rating + honeybee species; blooms, nectar flow; composition
N1 ELS(Ord/83)
Blooms autumn, sometimes spring and summer (tropical America, Ord/83).
Nectar secretion copious. **Sugar concentration** often low, sometimes
ignored by bees (Ord/83)

Honey flow
Honey yield "main honey crop" in Fonseca Gulf area (ELS, Ord/83)

Honey no data

108 Combretum imberbe Wawra; Combretaceae

leadwood; motswere (BOT)
Tree/shrub, 7-15 m; fls cream to yellow, fragrant
Distribution tropical Africa; native to Africa. **Habitat** medium to low
altitudes in mixed woodland, often along rivers, dry watercourses
Soil alluvial preferred

Economic and other uses
Fuel. **Timber** termite proof. **Other uses** whitewash/toothpaste from wood
ash; medicinal; dye from roots

Nectar rating + honeybee species; blooms, nectar flow
N1 BOT,tm(Cra/73)
Blooms xi-iii (southern Africa, Pag/77)

Honey no data

109 Combretum zeyheri Sond.; Combretaceae

large-fruited bushwillow (En/SOU); modubana (BOT)
Tree/shrub, >10 m, deciduous; fls greenish-yellow to yellow with orange
anthers, showy, fragrant
Distribution tropical Africa; native to southern Africa. **Habitat** medium
to low altitudes, in open woodland, on rocky hillsides; may grow by rivers
Soil wide range including v mineral-rich; usually acidic

Economic and other uses
Timber termite- and borer-proof. **Other uses** medicinal; fibrous roots for
weaving; tannin from fruit; dye from roots

Nectar rating + honeybee species; blooms, nectar flow
N1 BOT,tm(Cra/73)
Blooms ix-x (southern Africa, Pag/77). **Nectar secretion** abundant (ZAM,
Sto/82)

Honey no data

110 Cordia alba (Jacq.) Roemer & Schultes; Boraginaceae
syn Cordia dentata Poir.

chachalaco (Es/HOD); tiguilote (Es/HOD, NIA); zazamil (Es/MEX)
Tree, small, twisted; fls yellowish bells in panicles
Distribution tropical C America, Caribbean, S America; native to
?Caribbean. **Habitat** low deciduous woodland especially on Pacific coast of
C America
Soil calcareous and rocky types

Nectar rating; blooms, nectar flow
N1 MEX(Ord/72); NIA(Ord/63a)
N HOD(Ord/63)
Blooms vii-x and sometimes in spring (tropical America, Ord/83). **Nectar
flow** is during dearth period (CUB, Ord/44). **Nectar secretion** copious
(Ord/44)

Honey: physical and other properties
Pfund light amber (Ord/63a; Ord/72)

111 Cordia alliodora (Ruiz & Pavon) Cham.; Boraginaceae

capá prieto, muneco (Es/DOR); suchah (BEL); cypre (TRI)
Tree, <25 m, lvs smell of garlic; fls white, small, fragrant
Distribution tropical C America, Caribbean, S America; native to
Caribbean. **Habitat** low altitudes; tropical hardwood forests (BEL)
Rainfall dry zones of C America

Nectar rating; blooms, nectar flow; composition
N1 BEL(Mul/79)
N2 DOR(Ord/64)
N TRI(Lau/76)
Blooms iii (BEL); early spring (tropical America, Ord/83). **Sugar
concentration** in nectar [medium] 25% in fls 10 m from ground, 15% at 5 m
(AA336/79)

Honey flow
Honey yield [high] 75 kg/colony/season in region where it is a main honey
plant (BEL, Mul/79)

Pollen
P DOR

Honey no data

112 Cordia gerascanthus L.; Boraginaceae

prince-wood; Spanish elm (En/JAM); baría (Es/CUB); barillo, bojón (Es/MEX)
Tree, <25 m; fls grey-white, attractive
Distribution tropical C America, Caribbean (excluding PUE), S America
Soil calcareous preferred

Economic and other uses
Timber. **Land use** shade, amenity

Nectar rating; blooms, nectar flow
N1 CUB(Ord/44); MEX(Ord/72)
N2 CUB(Ord/56)
N JAM(Ord/83)
Blooms i-iii (tropical America). **Nectar flow** 2 wks, one of the earliest
major honey sources (Ord/83). **Nectar secretion** copious (Ord/44)

Recommended for planting to increase honey production
CUB(Ord/44). Useful for roadside planting

Honey: physical and other properties
Colour v light (Ord/44; Ord/83)
Flavour and aroma delicate (Ord/72; Ord/83)

113 Coreopsis borianiana Schultz-B.p.; Compositae
syn Bidens chaetodonta Sherff; Coreopsis abyssinica Schultz Dip. ex Walp

adey-abeba (ETH)
Herb
Distribution tropical Africa

Nectar rating + honeybee species
N1 ETH,tm(Bor/80)

Honey no data

114 Coriandrum sativum L.; Umbelliferae

coriander; dhania, kothmir (INI)
Herb, 30-90 cm, annual/perennial, aromatic; fls white/pink
Distribution subtropical Asia, temperate Europe; native to ?E Med.
Habitat widespread as cultivated crop plant and a weed of waste places
Soil central chernozem area (URS/Povoljie, N Caucasus, Ukraine).
Temperature warmth required

Economic and other uses
Food – seed as condiment and for oil; also in alcoholic beverages. **Other uses** medicinal

Nectar rating + honeybee species; blooms, nectar flow
N1 IRN(Cra/73); URS(Ave/78; Glu/55)
N2 INI/MAH[ac](Chu/80); INI/PUN,ac(Chd/77); ROM(Cir/77)
N3 INO,tm(Bee/77)
N ROM(Int/65); ZIM,tm(Pap/73)
Blooms vi-vii (ROM, Cir/80); spring-sown crop vii, winter-sown crop vi, both lasting for 1 mth (URS). **Nectar secretion** 0.09 mg/fl (Cir/77).
Sugar value [low] 0.022-0.045 (Bac/60)

Honey flow
Honey potential (kg/ha) [moderate] 200 (BUL, Iva/78; URS, Ave/78); 25-100 (GDR, Bec/67); 34 (ROM, Bac/60); 100-500 (ROM, Cir/77)

Pollen
P2 INI/MAH,PUN. P3 INO. P ROM. **Colour** of load red (Sha/76)

Honey: chemical composition
Water [medium] 16.80-18.48% (6 samples, Iva/78)
Sugars, total 72.25-73.70% (Iva/78). **Sucrose** [low] 0.00-2.18% (Iva/78). As % of total sugars (Maz/64): **glucose** 35.2%; **fructose** 48.1%; **maltose** 14.1%; **oligosaccharides** 1.1%
Ash [medium] 0.10-0.25% (Iva/78)
Total acid 21.0-33.0 meq/kg
Amylase 9.7-16.4. **HMF** 0.96-3.84 ppm (Iva/78)
Protein 0.0044, 0.0065% (Gen/67)

Honey: physical and other properties
Colour dark (Dem/64)
Optical rotation −1.75 to −2.40 deg (Iva/78). **Electrical conductivity** (per ohm cm) 0.000349-0.000520 (Iva/78); 0.000510 (Vor/64)

115 Croton floribundus Spreng.; Euphorbiaceae

capixingui (Pt/BRA)
Tree
Distribution subtropical S America. **Habitat** upland forest, latitudes 19-23 degrees S (BRA, Smt/60)

Nectar rating + honeybee species; blooms, nectar flow; composition
N1 BRA/SP[?tm](Mun/54; Smt/60)
N3 BRA[tm](Ord/83)
Blooms xi-ii (BRA/SP, Caa/72); xi-xii (S America, Ord/83). **Sugar concentration** [medium] 40-50% (Caa/72)

Pollen
P BRA

Honey: physical properties
Pfund white (Ord/83)

116 Cryptosepalum pseudotaxus Bak. f.; Leguminosae

mucube (ANA)
Tree/shrub
Distribution tropical Africa. **Habitat** often growing with Julbernardia and Brachystegia spp (ANA)
Rainfall 610-1400 mm (Ros/60)

Economic and other uses
Bark hives (ANA, Ros/60)

Nectar rating + honeybee species; blooms, nectar flow
N1 ANA,tm(Ros/60); ZAM[tm](Smt/59)
Blooms vii-viii (ANA). **Nectar flow** heaviest when rainfall >900 mm (Ros/60)

Honey no data

117 Cucumis melo L.; Cucurbitaceae

melon, muskmelon; melon (Fr); blewek (In)
Herb, trailing, annual, polymorphic; fls yellow, monoecious

Distribution tropical, subtropical, warm temperate zones; native to
?tropical Africa.   **Habitat** widely cultivated crop plant, also under glass
in cool temperate regions
Temperature frost tender

Economic and other uses
Food - fruit

Nectar rating + honeybee species; nectar composition
N1 MOZ,tm(Cra/73)
N2 INO[ac](Bee/77);   URS(Glu/55);   USA/CA,CO,MO,NM(Pel/76)
N ALG(Ske/72);   ZIM,tm(Pap/73)
Nectar secretion moderate (Mot/64);   see also Frj/70.   **Sugar concentration**
[medium] male fl 56%, female fl 27% (Frj/70);   31.6-32.7% (Mog/58).
Potassium content and **fluorescence** (AA491/80)

Honey flow
Honey yield "good surplus" in large fields (USA/FL, Mot/64).   **Honey
potential** [moderate] 10-40 kg/ha (ROM, Cir/80)

Pollen
P1 INO;   URS.   **P** ALG;   ZIM.   **Pollen value** good (Lov/77).   **Reference slide**

Honey: chemical composition
Water [low] 15.4% (1 sample, age 11 mths, Whi/62)
Glucose [medium] 34.51%.   **Fructose** [medium] 37.00%.   **Sucrose** [medium]
2.85%.   **Maltose** 5.41%.   **Higher sugars** 1.10%
Ash [medium] 0.203%
pH 3.80.   **Total acid** 41.57 meq/kg.   **Free acid** [medium] 31.28 meq/kg.
Lactone 10.20 meq/kg
Amylase 8.1
Nitrogen 0.021%

Honey: physical properties
Pfund 42-50 mm, light amber (Whi/62;   also Lov/56)

118 Cucurbita pepo L.; Cucurbitaceae

field pumpkin, pumpkin;   courgette (Fr)
Herb, trailing, annual;   fls yellow, large, monoecious
Distribution tropical, subtropical, temperate.   **Habitat** widely cultivated
crop plant
Temperature frost tender

Economic and other uses
Food - fruit

Nectar rating; composition
N1 CAE(Cra/73)
N2 URS(Fed/55)
N3 USA(Pel/76)
N ALG(Ske/72)

Sugar concentration in nectar [medium] 30-40% (Cir/80); 28.0% (Fah/49)

Honey flow
Honey potential (kg/ha) [moderate] 40-50 (ROM, Apc/68); 30 (URS, Fed/55)

Pollen
P ALG; URS; USA. **Yield** abundant (Pel/76). **Chemical analysis**
(Cir/80); protein content 26.4% (AA 1244/78). **Pollen grain** illustrated
and described (Sao/61); described (Sta/74). **Reference slide**

Honey: physical and other properties
Colour yellow (Fed/55)
Granulation rapid

119 Curatella alata Vent.; Dilleniaceae

bejuco chachaco, chachaco bejuco (Es/BOL)
Shrub, much branched; fls yellowish
Distribution tropical S America

Nectar rating + honeybee species; blooms, nectar flow
N1 BOL[tm](Kem/80)
Blooms iii-v (BOL)

Pollen
P BOL

Honey no data

120 Cynara cardunculus L.; Compositae

cardoon; artichoke, artichoke thistle (En/AUS); cardo de Castilla (Es/ARG)
Herb, perennial; fls violet-blue, large, thistle-like
Distribution temperate and subtropical S America, Oceania; temperate (Med)
Europe; native to (Med) Europe. **Habitat** cultivated crop plant; humid
pampas region of ARG
Rainfall 600-1000 mm (ARG)

Economic and other uses
Food - leaf-stalks, roots

Nectar rating + honeybee species; blooms, nectar flow
N1 ARG[?tm](Kat/68; Lut/63; Per/80)
N2 AUS/SA(Pur/68)
N3 AUS/VIC(Gom/73)
Blooms xii-ii or longer if summer rains are good (AUS/VIC). **Nectar flow**
uncertain, usually rather poor in AUS/VIC (Gom/73)

Pollen
P2 AUS/SA, VIC.   P ?ARG.   **Yield** good (Gom/73)

Honey: physical and other properties
Colour light (Cra/75); v light (Lut/63).   **Pfund** medium amber (Gom/73);
light amber (Pur/68)
Granulation coarse (Gom/73)
Flavour mild (Cra/75)

121 Cynometra alexandri C.H. Wright; Leguminosae

tembu (ZAI)
Tree
Distribution tropical Africa.   **Habitat** climax forest to N of Ituri River

Nectar rating + honeybee species; blooms, nectar flow
N1 ZAI,tm(Ich/81)
Blooms iii, during rainy season (ZAI)

Honey no data

122 Dalbergia sissoo DC.; Leguminosae

sissoo; shisham (INI)
Tree, 18-21 m, deciduous; fls yellowish, small
Distribution tropical, subtropical Asia; native to foothills of Himalayas.   **Habitat** tropical highlands; dry savannah woodlands; coastal sand
dunes and wasteland; canal banks in plains and lower hills
Soil salt tolerant.   **Temperature** <0 to 50° (Usa/79).   **Rainfall** 700-2000
mm with 3-4 mths drought (Usa/79)

Economic and other uses
Fodder - young branches/lvs.   **Fuel.**   **Timber.**   **Land use** windbreak, shade,
afforestation, amenity.   **Soil benefit** erosion control

Warning
Even light winds blow fls from branches reducing the nectar available in
windy seasons/areas (Sig/48)

Nectar rating + honeybee species; blooms, nectar flow
N1 INI/HIM[ac](Rah/41);   INI/PUN[ac](Rah/41);   INI/UTT,ad(Rae/80);
PAK,ac,ad(Pak/77; Shi/77; Shr/48)
N INI/HIM,ac(Sig/48)
Blooms iii-iv (INI);   iv-v (PAK).   **Nectar flow** 2 wks (INI, Sig/62)

Honey flow
Honey yield (kg/colony/season) [moderate] 4-9 (INI, Sig/62);   27, mixed
with honey from clovers (PAK, Shr/48)

Pollen
P PAK

Recommended for planting to increase honey production
INI/MAH (Sub/79). Propagate by suckers, root and shoot cuttings; grows
rapidly; suitable for dry zones. See **Warning**

Honey: chemical composition
Water [medium] 18.75% (Sig/48)
Glucose [medium] 34.6%. **Fructose** [medium] 39.1%. **Sucrose** [medium] 1.04%
Ash [medium] 0.18%

Honey: physical and other properties
Pfund amber to dark amber (Sig/62)
Flavour strong

123 Dalea revoluta S. Watson; Leguminosae

popote chiquito (Es/MEX)
<50 cm; fls white
Distribution subtropical C America; native to C America. **Habitat** ravines
and hills (MEX)

Nectar rating; blooms, nectar flow
N1 MEX(Ord/83)
Blooms autumn and spring (MEX). **Nectar flow** "one of the most valuable
honey plants of Sonora" (MEX, Ord/83)

Honey no data

124 Daniellia oliveri (Rolfe) Hutch & Dalz.; Leguminosae

santan (Fr/SEN)
Tree, 15-20 m; fls white, terminal panicles
Distribution tropical Africa

Economic and other uses
Bark used for hives in north IVO (Dou/80)

Nectar rating + honeybee species; blooms, nectar flow
N1 SEN,tm(Ndi/74)
N CEN,ac(Dou/79); IVO,ac(Dou/80)
Blooms xii (IVO)

Honey: physical and other properties
Colour black (Dou/80); dark (Ndi/74)
Granulation slow (Ndi/74)
Flavour acrid, unpleasant (Dou/79); strong (Ndi/74)

125 Daucus carota L.; Umbelliferae

carrot; cultivated carrot, wild carrot (En/USA); zanahoria (Es); carotte (Fr); carrotte sauvage (Fr/ALG)
Herb, biennial; fls white, small, in umbels
Distribution temperate N America, Europe, Asia; native to N America, cultivated form from AFG. **Habitat** cultivated crop plant grown world-wide; wild carrot grows in grassy places especially near the coast
Soil average neutral or weakly alkaline. **Temperature** warmer areas preferred. **Rainfall** drier areas preferred but some moisture required at start of growing season

Economic and other uses
Food - valuable root crop. **Fodder** - roots

Nectar rating; blooms, nectar flow; composition
N1 USA/CA(Pel/76)
N3 INI/MAH(Chu/80);   ITA(Ric/78)
N URS(Glu/55); USA/MA(Shw/50a)
Blooms vi-ix (Maz/82); for 1 mth (Mcg/76). **Sugar analysis** (Maz/82)

Honey flow
Honey yield "excellent honey plant" in Sacramento River Valley (USA/CA, Ord/83)

Pollen
P3 ITA. **P** USA/MA. **Yield** low (Maz/82)

Honey: physical and other properties
Colour dark yellow (Glu/55); dark (Ord/83; Van/41); grey (Ric/77).
Pfund light amber (Cha/48; Cra/75); white (Lov/56); medium amber (Lov/63a)
Flavour mild (Lov/56); distinctive (Lov/63a); strong (Van/41). **Aroma** fragrant (Glu/55); like the plant (Ord/83)

126 Dialium engleranum Henriques; Leguminosae
syn Dialium simii Phillips

Kalahari podberry (En/SOU, ZAM); mussala (ANA)
Tree, 7-20 m, much branched; fls small, petals absent, sepals creamy-white, golden brown and hairy outside
Distribution tropical Africa; native to BOT, NAM, ZAM. **Habitat** mixed forest/woodland; savannah
Soil deep white sand; Kalahari sand. **Rainfall** 600-800 mm (ZAM, Smt/59); drought tolerant

Economic and other uses
Food - fruit. **Other uses** medicinal

Nectar rating + honeybee species; blooms, nectar flow
N1 ANA,tm(Ros/60); ZAM,tm(Smt/59)
N ZAI,tm(Dul/50)
Blooms iv-v (ANA); vii-viii (southern Africa, PAG/77). **Nectar secretion** "one of the best nectar producers - south of Lunda" (ANA, Ros/60)

Honey: physical and other properties
Colour light (Dou/50)

127 Dillenia pentagyna Roxb.; Dilleniaceae

toothed dillenia (En/INI); muchilu (INI)
Tree; fls yellow
Distribution tropical Asia

Nectar rating + honeybee species; blooms, nectar flow
N1 INI/KAR,KER[ac](Kha/59)
Blooms iii-iv (INI/KAR, KER)

Pollen
P INI/KAR, KER

Honey no data

128 Dimocarpus longan Lour.; Sapindaceae
syn Nephelium longana (Lam.) Cambess.; Nephelium long-yan Bl.

longan
Tree, 9-12 m; fls yellowish-white, small
Distribution tropical and subtropical Asia; native to Asia. **Habitat**
cultivated crop plant; fruit-growing centres of Kuantung and Fujian (CHN,
Bia/79)

Economic and other uses
Food - fruit

Nectar rating; blooms, nectar flow
N1 ?CHN(Bia/79); TAI(Fan/52);
Blooms ii-iii, iv, v (south to north, TAI). **Nectar flow** abundant every 2
yrs (Fan/52)

Honey flow
Honey yield [high] 91 kg/colony/season, used for migratory beekeeping (TAI,
Fan/52)

Honey no data

129 Diospyros batocana Hiern; Ebenaceae

sand jackal-berry (En/SOU); batoka diospyros (En/ZAM)
Tree, <7 m; fls creamy-white tinged violet, fragrant
Distribution tropical Africa. **Habitat** low altitudes in hot woodland (S
Africa)

Economic and other uses
Food - fruit.   **Timber.**   **Other uses** bark and lvs for dyes;   medicinal

Nectar rating + honeybee species
N1 ZAM,tm(Smt/59)

Honey no data

130 Diospyros virginiana L.;   Ebenaceae

ebony, American persimmon;   possum-wood (En/USA)
Tree, <15 m, deciduous;   fls white with yellow tinge, bell-shaped
Distribution temperate (Med) Europe, N America;   subtropical N America;
native to SE states of N America.   **Habitat** cultivated crop plant;   also
wild in old pastures, wasteland;   hammocks, pinelands (USA/FL)

Economic and other uses
Food - fruit.   **Timber.**   **Other uses** rootstock for grafting D. kaki

Alert to beekeepers
Blooms early in season so beekeepers must make colonies strong to harvest
honey (USA, Pel/76)

Nectar rating; blooms, nectar flow
N1 URS(Glu/55);   USA/NC(Stp/54)
N2 USA/KS,NC(Pel/76);   USA/TN(Lit/54)
Blooms v in southern USA;   v-vi (USA/FL).   **Nectar flow** irregular and
brief, 2 wks (Pel/76).   **Alert to beekeepers** blooms early in season so
beekeepers must make colonies strong to harvest honey (USA, Pel/76).
Nectar secretion "plentiful" (Mot/64)

Honey flow
Honey yield (kg/colony/season) [high] 32 (Pel/76);   [moderate] 14-18 (Lov/77)

Honey: physical properties
Pfund light amber (Lov/56;   Mot/64);   amber (Pel/76)

131 Diplotaxis erucoides (L.) DC.:   Cruciferae

white wall-rocket;   rucola selvatica (It)
Herb, 5-50 cm, annual/biennial;   fls white/violet
Distribution temperate (and Med) Europe;   native to SW Europe.   **Habitat**
cultivated land/fields, also vineyards where it is a noxious weed

Nectar rating; blooms, nectar flow; composition
N1 FRA(Lou/81);   ITA(Ric/78)
Blooms iv-vi (Europe).   **Sugar analysis** of nectar (Bat/73a)

Honey flow
Honey yield "may be substantial in favourable yrs" (ITA/Marche, Ric/78)

Pollen
Pl FRA; ITA. **Pollen value** "may represent 100% of harvest ii-x"
(ITA/Marche, Ric/78)

Honey: physical and other properties
Colour greyish white (Ric/78)
Granulation fine, homogeneous
Aroma strong odour of rape when honey is dissolved in water

132 Dipsacus fullonum L.; Dipsacaceae
syn Dipsacus sylvestris Huds.

teasel; fuller's teasel (En/USA); cardère sylvestre (Fr/ALG)
Herb, 50-200 cm, biennial; fls pinkish-purple
Distribution temperate (and Med) Europe, N Africa, N America; native to
Europe. **Habitat** fields/roadsides/woods/streamsides; sometimes grown as a
crop

Economic and other uses
Dried fl heads used in processing woollen cloth

Nectar rating; blooms, nectar flow; composition
N1 IRN(Cra/73)
N3 USA/UT(Nye/71)
N ALG(Ske/72)
Blooms early summer (CAF; USA); vii-viii (Europe). **Sugar concentration**
[medium] 26% (Pec/65). **Sugar analysis** (Pec/61)

Pollen
P USA/UT. **Colour** white. **Pollen grain** illustrated and described
(Heu/71); described (Nye/71)

Honey: physical properties
Colour transparent (Pel/76). **Pfund** white (Lov/56; Pel/76)

133 Dipsacus pilosus L.; Dipsacaceae

shepherd's rod, small teasel
Herb, 30-120 cm, biennial; fls whitish
Distribution temperate Europe (not N), Asia; native to Europe. **Habitat**
damp/shady places; woods/hedges

Nectar rating; blooms, nectar flow
N1 IRN(Cra/73)
Blooms vi-ix (Europe)

Honey no data

134 Dombeya rotundifolia (Hochst.) Planch.;   Sterculiaceae

wild pear (En/SOU);   gewone drolpeer, wilde peer (Af)
Tree, 4-5 m, exceptionally 8 m, deciduous;   fls white rarely pink,
fragrant, appear before lvs
Distribution subtropical and tropical Africa.   **Habitat** wooded grassland/-
woodland;   often on termite mounds;   wide range of altitudes
Temperature several degrees of frost tolerated;   fire-resistant

Economic and other uses
Timber.   **Land use** amenity.   **Other uses** medicinal

Nectar rating + honeybee species;   blooms, nectar flow
N1 BOT,tm(Cra/73);   TAN,tm(Smt/57)
N2 ?ZAM,tm(Sto/82)
N3 SOU/NATAL,TVL,tm(And/73)
Blooms vii-x (southern Africa, Pag/77).   **Nectar flow** brief (SOU)

Pollen
P2 SOU/NATAL, TVL.   **Pollen grain** illustrated and described (Smt/56a)

Honey: physical properties
Colour light (And/73;   Cri/57).   **Pfund** extra light amber (Cra/75;
Smt/57);   light amber (Sto/82)

135 Dracocephalum moldavica L.;   Labiatae

dragonhead, Moldavian balm
Herb, <60 cm, annual;   fls violet/white
Distribution temperate Europe, Asia.   **Habitat** naturalized as weed and
ruderal, east and east-central Europe;   occasional casual elsewhere;   open
places
Soil dry

Economic and other uses
Land use amenity.   **Other uses** aromatic lvs were rubbed inside hives to
attract swarms (How/79)

Nectar rating + honeybee species;   blooms, nectar flow;   composition
N1 IRN(Cra/73);   URS(Fed/55)
N2 POL(Dem/64a)
Blooms vii-viii (ROM, Apc/68;   Cir/80).   **Nectar secretion** - influence of
RH (AA586/63);   of potassium fertilizer (AA719/72).   **Sugar concentration**
[medium] 30-48% (Cir/80);   47.87% (mean, 3 yrs, AA194/76).   **Sugar analysis**
(Jua/64)

Honey flow
Honey potential (kg/ha) [moderate] 133-362 (BUL, AA194/76);   200-400
(Dem/64a);   123-651 (POL, AA718/72);   300-400 (ROM, Cir/80);   >250 (URS,
Fed/55)

Honey: chemical composition
Hungary (as % of total, Maz/64);   **glucose** 29.8%;   **fructose** 53.5%;   **sucrose**
2.8%;   **maltose** 8.9%;   **fructomaltose** 5.0%

Honey: physical and other properties
Colour light (Fed/55)
Electrical conductivity 0.000120/ohm cm (Vor/64)
Flavour slightly of lemon (Fed/55)

136 Dryandra sessilis (Knight) Domin; Proteaceae

parrot bush (En/AUS)
Shrub, tall, prickly lvs; fls white to yellow
Distribution subtropical Oceania; native to W AUS
Soil well drained, cool

Nectar rating; blooms, nectar flow
N1 AUS/WA(Col/62)
Blooms vii-x (AUS/WA).   **Nectar flow** heaviest on coastal limestone hills
(Col/62)

Honey flow
Honey yield annual, and "possibly the greatest honey source" in AUS/WA
(Col/62)

Pollen
P2 AUS/WA.   **Pollen value** good (Col/62)

Honey: chemical composition
Sucrose [medium] 1.1-4.5% (12 samples, Smt/65)
HMF content increased by about 30 ppm on storage when Pfund increased by 10
mm (Smt/67)

Honey: physical and other properties
Pfund light to extra light amber (Cra/75); darkened about twice as rapidly
as other AUS/WA honeys; 55 mm, light amber (60 mm after 24 h at 50° or 1.5
h at 70°, 65 mm after 3.5 h at 70°, Smt/67)
Flavour characteristic (Cra/75)

137 Durio zibethinus Murr.; Bombacaceae

durian
Tree, 20-40 m, evergreen; fls pink/white
Distribution tropical Asia, C America; native to tropical Asia.   **Habitat**
haphazardly cultivated/semi-wild; humid lowland areas <800 m
Rainfall - trees do not thrive where there is a distinct dry season

Economic and other uses
Food - fruit and seeds

Warning
The large (20 kg) falling fruits are a hazard; fruit also attracts tigers,
elephants and monkeys

Nectar rating; blooms, nectar flow
N1 SIN(Kia/54; Smt/60)
Blooms ?viii-ix (SIN)

Pollen
P SIN

Honey no data

138 Dysophylla stellata Benth.; Labiatae

gomani (INI)
Herb, annual, erect; fls pink
Distribution tropical Asia. **Habitat** common in rice fields (INI/KAR);
roadsides, coastal marshes

Nectar rating + honeybee species; blooms, nectar flow
N1 INI/KAR,ac,ad,Trigona(Diw/64)
Blooms i-ii (INI/KAR)

Pollen
P INI/KAR

Honey no data

139 Echium lycopsis L.; Boraginaceae
syn Echium plantagineum L.

blue weed, Patterson's curse, salvation Jane, viper's bugloss (En/AUS);
flor morada, flor morena (Es/ARG)
Herb, 20-60 cm, annual/biennial; fls blue, becoming pink to purple
Distribution temperate Europe, Oceania; subtropical Oceania, S America;
native to S and W Europe. **Habitat** roadsides, fields, sandy areas near the
sea; humid pampas of ARG; noxious introduced weed in some wetter parts of
AUS where it is widespread and often dominant in pastures
Soil siliceous preferred. **Rainfall** drought tolerant

Economic and other uses
Fodder (Pig/77)

Warning; alert to beekeepers
Warning can be highly invasive weed in wetter parts of AUS (Cra/81).
Alert to beekeepers appearance of "bee paralysis" may be associated with
this pollen (AUS, Dol/61)

Nectar rating; blooms, nectar flow
N1 ARG(Cos/63; Per/80; Vit/65); AUS/SA(Pur/68)
N2 AUS/VIC(Com/73); AUS/WA(Col/62)
N3 AUS/NSW(Goo/47)

Blooms viii-x (AUS/WA); viii extending to xii (AUS/VIC)

Pollen
P2 AUS/SA, VIC, WA.   **Yield** heavy (Pur/68).   **Pollen value** good quality
(Col/62); doubtful quality (Pur/68); appearance of bee "paralysis" may be
associated with this pollen (AUS, Dol/61)

Honey: chemical composition
Sucrose [high] 9% (Smt/65)
pH 3.1 (Woo/78)
Other components several pyrrolizidine alkaloids identified (Cul/81)

Honey: physical and other properties
Colour white to light golden (Cra/75); light (Pur/68).   **Pfund** light amber
(Gom/73)
Granulation rapid
Flavour delicate (Cra/75); peculiar but not objectionable (Gom/73)

140 Echium vulgare L.;   Boraginaceae

viper's bugloss; blue thistle, blueweed (En/USA); viborera (Es/CHL);
vipérine (Fr); Natternkopf (De); erba viperina (It)
Herb, 20-90 cm, biennial/perennial; fls blue/bluish-violet
Distribution temperate Europe, N America, S America, Oceania; native to
Europe.   **Habitat** dry, open places: sand-dunes, roadsides, cultivated
fields, dry pastures, chalky hills, sea cliffs; altitudes <2000 m

Economic and other uses
Medicinal

Warning
Invasive weed in Canada (How/79)

Nectar rating; blooms, nectar flow; composition
N1 CAF/ONT(Tow/76);   CHL(Kot/38a; Vaq/39);   POL(Dem/64a);   URS(Fed/55)
N2 FRA(Lou/81);   GFR(Gle/77);   USA/MD(Die/71)
N3 ITA(Ric/78);   ROM(Cir/77);   USA/MD,NY,VA(Lov/56; Pel/76);
N ALG(Ske/72);   EUR(Maz/82);   NEZ(Wal/78);   USA/WV(Pel/76)
Blooms vi-viii (ROM); vi-vii (UK, How/79).   **Nectar secretion** 0.5-8.8
mg/fl/day, max at 15.00 h (Han/80; Maz/82); diurnal variations studied
(AA162,163/79).   **Sugar concentration** [medium] 17-43% (Han/80);   35.7%
(Pek/80); 20-35% (AA162,163/79); 23-41% (AA526/82).   **Sugar value**
(mg/fl/day) [medium] 1.64 (Cra/75); 0.09-1.3 (Cra/75);   0.23-2.56
(Han/80; Maz/82).   **Sugar analysis** (Maz/82; Pec/61; AA526/82)

Honey flow
Honey potential (kg/ha) [moderate] 128.8 (BUL, Pek/80);   400 (GDR,
Bek/67);   182-429 (EUR, Maz/82); >400 (POL, Dem/64a);   380-400 (ROM,
Apc/68);   180-430 (ROM, Cir/80);   300 (URS, Glu/55)

Pollen
P2 FRA; GFR; ITA. **P3** ROM. **Colour** of load light to dark blue
(Han/80); dark bluish-grey (Hod/74). **Pollen grain** illustrated and
described (Hod/74; Saw/81). **Reference slide**

Recommended for planting to increase honey production
UK (How/79). Propagate by sowing in wasteland; also on embankments of
roads/railways. See **Warning**

Honey: chemical composition
Water [medium] 16.4% (1 sample, age 17 mths, Whi/62)
Glucose [medium] 31.27%. **Fructose** [medium] 37.30%. **Sucrose** [medium]
1.28%. **Maltose** 8.43%. **Higher sugars** 2.53%. Contents of individual
sugars expressed as % of total sugars (Maz/64)
Ash [low] 0.039%
pH 3.8 (Lan/66); 3.88 (Whi/62). **Total acid** 16.50 meq/kg (Whi/62). **Free
acid** [low] 11.81 meq/kg. **Lactone** 4.69 meq/kg
Nitrogen 0.033% (Whi/62)

Honey: physical and other properties
Colour white to light golden (Cra/75); dull white (Mao/82). **Pfund** amber
(Cha/48); 24.2 mm, white (bulk honey, Lan/66); water white (Lov/56); 8-
12 mm, extra white (Whi/62)
Electrical conductivity 0.000111/ohm cm (Vor/64)
Granulation slow (Mao/82); fine, colour then dull (Wal/78)
Flavour delicate (Cra/75); flat (Wal/78)

141 Ehretia acuminata R. Br.; Boraginaceae

puna (INI)
Tree, medium size, deciduous; fls white, fragrant, sessile
Distribution subtropical and tropical Asia. **Habitat** margins of brush
forests/creek banks (AUS/NSW); altitudes <1500 m
Soil good, deep

Economic and other uses
Fodder - lvs. **Timber**. **Land use** amenity

Nectar rating + honeybee species; blooms, nectar flow
N1 INI,ac(Sig/62); ?INI/HIM[ac](Rah/40)
N2 INI/PUN[ac](Chd/77)
Blooms early iv. **Nectar flow** prolonged (INI, Sig/48)

Honey flow
Honey yield [moderate] 4.5 kg/colony/season (INI, Sig/62)

Pollen
P3 INI/PUN

Recommended for planting to increase honey production
INI (Sig/62). Propagate by root suckers or seed

Honey: physical and other properties
Pfund amber (Sig/62)
Flavour characteristic

142 Elaeagnus angustifolia L.; Elaeagnaceae

oleaster, Russian oleaster, Russian olive, wild olive
Tree/shrub, <7 m, spiny, deciduous; fls yellowish, fragrant
Distribution temperate S Europe, Asia; native to Asia. **Habitat** widely
naturalized in S Europe; bushy places by streams; damp places; by sea

Economic and other uses
Food - dried fruit, "Trebizond" grapes for cakes. **Timber.** **Land use**
amenity

Nectar rating
N1 IRN(Cra/73); URS(Fed/55)
Nectar secretion 0.405 mg/fl/day (Sim/80). **Sugar concentration** [medium]
34.96-38.10% (3 yrs, Pet/72); 42.1% (Sim/80). **Sugar value** [medium] 0.936
mg/fl/day (Sim/80)

Honey flow
Honey potential (kg/ha) [moderate] 100 (GDR, Bec/67; ROM, Cir/80)

Pollen
Yield 0.002 mg/10 fls (Sim/85)

Honey no data

143 Elaeis guineensis Jacq.; Palmae

oil palm; corozo (Es/DOR); palmera africana (Es/HOD, NIA)
Tree, palm, <9 m, erect, monoecious
Distribution tropical Africa, C America, Asia; native to NW Africa.
Habitat cultivated crop plant; coastal regions preferred
Soil rather poor soil tolerated. **Rainfall** - high rainfall preferred

Economic and other uses
Food - nuts; oil from nuts for margarine; sap for toddy making
Fodder - pressed cake. **Fuel** - outer nut shell. **Other uses** oil for soap

Nectar rating + honeybee species; blooms, nectar flow
N1 CAE[tm](Cra/73)
N2 DOR(Ord/64); NIA(Ord/63a)
N HOD(Ord/63); SEN(Dou/70)
Juice of fermenting fruit also collected by bees (Gor/64)

Pollen
P1 CAE.　**P DOR.**　**Yield** 85.2 g, male inflorescence (Mcg/76);　high
(Sil/76a).　**Pollen value** high, can be collected by hand, smells strongly
of aniseed and is most attractive to bees (Sil/76a).　**Chemical composition**
(Sta/74).　**Pollen grain** illustrated and described (Smt/56a)

Honey: physical and other properties
Colour v dark (Ord/83).　**Pfund** dark amber (Smt/56a)
Flavour strong, astringent (Hol/74;　Ord/83;　Smt/56a)

144 Epilobium angustifolium L.;　Onagraceae
syn Chamaenerion angustifolium (L.) Scop.

fireweed, rosebay willow-herb;　épilobe à feuilles étroites, herbe de Saint
Antoine, neritte (Fr);　schmalblättriges Weidenröschen, Waldweidenröschen
(De);　geitrans (No)
Herb, <2.5 m, perennial;　fls rose-pink
Distribution temperate Europe, N America, Asia.　**Habitat** woods and forests
especially where trees have been felled and the area burnt; gradually
crowded out by other growth;　wasteland;　taiga zone of URS
Soil burned-over forest with no chernozem (URS, Siberia, Altai)

Warning
Can be troublesome weed spreading rapidly by seed and underground stems

Nectar rating;　blooms, nectar flow;　composition
N1 CAF/BC(Dav/69);　CAF/ONT(Tow/64;　Tow/76);　CAF/QUE(Cha/48);
CAF/SASK(Mcc/58);　IRN(Cra/73);　ROM(Int/65);　URS(Ave/78; Fed/55;
Kov/55);　USA/AK(Liv/81);　USA/WI(Pel/76)
N2 CAF/ALTA(Wes/49);　FIN(Kay/79);　GFR(Gle/77);　KOS(Bek/65);　POL
(Dem/64a);　ROM(Cir/77);　UK(How/79);　USA/ID,MI,MN,MT,OR,WA,WI(Pel/76)
N3 ITA(Ric/78)
N　CAF/ALTA,MAN,ONT,QUE(Pel/76);　CAF/QUE(Cou/59);　FIN(Koc/74);
NOW(Lun/71);　USA/MA(Shw/50a);　?USA/NH(Pel/76)
Blooms vii-frost (CAF, USA, Pel/76);　vi-ix (URS/central, Ave/78).
Nectar flow vii to 1st half of viii (Ave/78);　throughout fl period unless
season is too dry;　heaviest during hot weather when air is clear, humid
and still (Pel/76).　**Nectar secretion** (mg/fl/day) 1.06-2.90 (Han/80;
Maz/82);　optimum temp 23-25° (Ave/78);　20-26°, optimum RH 60-70%
(Cir/80);　max secretion from 18.00-06.00 h, but highest sugar content from
10.00-14.00 h (Maz/82);　effect of minerals and soil moisture (AA1238/78).
Sugar concentration [medium] 44-60% (Cir/80);　44-63%, but only 13% at
06.00 h (Han/80;　Maz/82);　35.5% (AA450/61).　**Sugar value** (mg/fl/day)
[medium] 0.6-1.6 (Han/80;　Maz/82);　0.723 (AA450/61).　**Sugar analysis**
(Bat/73a;　Kay/78;　Pec/61;　Wyk/52;　AA538/67)

Honey flow
Honey yield (kg/colony/season) [high] 23-57 (USA/MI, Pel/76);　[moderate] 23-
30, ?together with Rubus idaeus honey (ROM, Int/65);　12 kg/colony/day
(URS/Siberia, Altai, Ave/78).　**Honey potential** (kg/ha) [high] 600 (GDR,
Dec/67);　1000 (exceptional, ROM, Cir/80);　[moderate] 50-600 (ROM,
Cir/77);　200-600 (ROM, Cir/80);　40 (ROM, AA450/61);　300-350 (URS,
Ave/78);　300-500 (URS, Fed/55);　140-240 (Maz/82)

Pollen
P1 GFR.   P2 ITA.   P3 USA/CA.   P FIN;   NOW;   ROM;   UK.   **Yield** 220-305
mg/100 fls (AA925/81).   **Chemical analysis** (Cir/80;   Gob/81;   Shp/79).
Colour of load dull green (Cir/80);   deep green-blue (Han/80);   bluish
(Hod/74;   How/79);   load large (Maz/82).   **Pollen grain** illustrated
(Hod/74;   Saw/81);   described (Saw/81);   under-represented in honey
(Ric/78).   **Reference slide**

Honey: chemical composition
Water [medium] 16.6% (1 sample, age 12 mths, Whi/62)
Glucose [low] 28.82%.   **Fructose** [medium] 40.00%.   **Sucrose** [low] 0.82%.
Maltose 8.45%.   **Higher sugars** 2.62%.   Contents of individual sugars
expressed as % of total sugars (Maz/64)
Ash [medium] 0.110%
pH 4.10.   **Total acid** 19.68 meq/kg.   **Free acid** [medium] 16.28 meq/kg.
Lactone 3.40 meq/kg
Amylase 17.6
Nitrogen 0.027%

Honey: physical and other properties
Colour greenish (Fed/55);   v light (Pel/76).   **Pfund** water white (Con/81;
How/79;   Lov/54c);   white (Cra/75);   27-34 mm, white (Whi/62)
Electrical conductivity 0.000128/ohm cm (Vor/64)
Granulation medium (Dem/64);   rapid (Glu/55);   rapid, fine (How/79)
Flavour rich (Fed/55);   almost none but v sweet (How/79);   mild (Lov/54c)

145 Erica arborea L.;   Ericaceae

tree heath;   bruyère arborescente (Fr/ALG);   erica, scopa (It)
Shrub/tree, 1-4 m, evergreen;   fls white/pale pink;   fragrant
Distribution temperate SW and Med Europe extending into Africa.   **Habitat**
woods, evergreen scrub and by streams;   dominates large areas of the maquis
in W Med
Soil sandy;   acid;   light and deep.   **Rainfall** good rainfall preferred

Economic and other uses
Timber - briar pipes and dice.   **Land use** amenity.   **Other uses** branches
and twigs for brooms and for screens against sun/wind

Nectar rating; blooms, nectar flow; composition
N1 ?ALG(Ske/72);   ITA(Ric/78);
N2 FRA(Lou/81);   GRC(Nic/55)
Blooms 11-v (EUR/Med, Lou/77).   **Sugar analysis** (Bat/72;   Bat/73a)

Pollen
P1 ITA.   P2 FRA.   **Colour** of load pink (Ric/78).   **Reference slide**

Honey: chemical composition
Sucrose [low] 0.30% (Spe/82).   **Reducing sugars** approx 75%.   As % of total
sugars (7 samples, Bat/73): **glucose** 51.03%;   **fructose** 43.76%;   **sucrose**
0.20%;   **maltose** 2.03%;   also **isomaltose, trehalose, gentiobiose**
Nitrogen 0.04% dry wt (Bos/78).   **Amino acids,** free 0.141%, protein 0.190%

Honey: physical and other properties
Colour golden (Bab/61); orange (Erb/83; Ric/78)
Optical rotation -32.65 deg (Bat/73)
Aroma intense (Erb/83); strong (Ric/78)

146 Erica cinerea L.; Ericaceae

bell heather, twisted heath; bruyère cendrée (Fr/ALG); purpurlyng (No)
Shrub, 15-75 cm, evergreen; fls reddish purple
Distribution temperate Europe; **Habitat** heaths, moorland, open woods, dry banks
Soil acid; rocky ground; dry moorland

Economic and other uses
Land use amenity

Nectar rating; blooms, nectar flow
N1 ?ALG(Ske/72); UK(How/79; Wht/54)
N2 FRA(Lou/81)
N3 IRR(Hil/68)
N NOW(Lun/71)
Blooms vi, vii-ix (UK, Wht/54)

Pollen
P1 UK. **P2** FRA. **P** NOW. **Colour** of load white to silver grey; dark grey/brownish if mixed with Cladosporium (Hod/74; Maz/82). **Pollen grain** illustrated (Hod/74). **Reference slide**

Honey: chemical composition
Water [medium] 17.8% (5 samples, Gon/65)
Total acid 61.5-63.0 meq/kg. **Free acid** [medium] 37.0 meq/kg (42.50 after 2 yrs at 20°). **Lactone** 25.3 meq/kg
Inhibine 187.2 µg hydrogen peroxide/g/h (Dus/71). **HMF** 18.2-25.0 ppm

Honey: physical and other properties
Colour brownish "port-wine" (Cra/75); reddish "port-wine" (How/79)
Granulation rapid (How/79)
Flavour characteristic (Cra/75); distinctive or pronounced (How/79)

147 Erica herbacea L.; Ericaceae
syn Erica carnea L.

spring heath; Schneeheide (De)
Shrub, evergreen
Distribution temperate Europe. **Habitat** mountain regions of Europe; open woods; rocky places
Soil wide range, including nutrient-poor soil

Economic and other uses
Land use amenity

Nectar rating; blooms, nectar flow; composition
N1 GFR(Gle/77)
Blooms early spring (EUR, Maz/82).   **Sugar analysis** (Maz/82; Pec/61)

Pollen
P3 GFR.   **Yield** pollen available throughout day; max foraging 11.00-13.00
h (Maz/82).   **Pollen value** important for bees in mountain areas (EUR,
Maz/82).   **Colour** of load white to silver-grey; dark grey/brownish if
mixed with Cladosporium; load 4.2 mg (Hod/74; Maz/82).   **Pollen grain**
illustrated (Hod/74; Saw/81).   **Reference slide**

Honey: chemical composition
Fructose:glucose ratio 1.55 (Maz/82)
Can be v rich in enzymes

Honey: physical and other properties
Colour light to dark yellow
Flavour v sharp.   **Aroma** strong

148 Erica manipuliflora Salisb.; Ericaceae
syn Erica verticillata Forssk.

heather
Shrub, <50 cm, evergreen; fls pink
Distribution temperate (Med) Europe into Africa.   **Habitat** rocky hillsides,
dry woods and evergreen thickets
Soil limestone and sand; stony ground

Nectar rating; blooms, nectar flow
N1 GRC(Adm/64)
N2 GRC(Mai/52; Nic/55)
Blooms viii-x (EUR/Med)

Honey flow
Honey yield "important secondary source" on Aegean/Ionian Islands and Crete
(GRC, Nic/55)

Pollen
Reference slide

Honey no data

149 Eriobotrya japonica (Thunb.) Lindl.; Rosaceae

loquat; néflier du Japon (Fr)
Tree, <10 m, evergreen; fls white, woolly

Distribution temperate (Med) Europe, N Africa, Asia; subtropical Asia; native to China. **Habitat** cultivated crop plant; altitudes <1800 m, optimum 900-1200 m
Soil well drained, light loam. **Temperature** – tree hardy but fruit damaged by frost. **Rainfall** drought tolerant

Economic and other uses
Food – fruit; liqueur from fruit in BER. **Land use** amenity

Nectar rating + honeybee species; blooms, nectar flow; composition
N1 AFG(Cra/73; Hof/72); CHN/N,central(Tse/54); LEB(Yaz/53); PAK[ac]-(Cra/73); PAK,ac(Pak/77); PAK/NWFP(Shi/77)
N2 JAP(Sak/82)
N3 INO[ac](Bee/77); JAP(Inu/57); LEB(Fli/62)
N INI/UTT,ac(Koh/58)
Blooms viii-i in 3 flushes (INI/north, Koh/58); autumn (tropical America, Ord/83). **Nectar flow** during dearth period (INI/north); reduced by covering of dust during droughts (Pak/77). **Sugar concentration** [high] 30.5-65.0% (Shr/58)

Honey flow
Honey yield (kg/colony/season) [moderate] 3.6 (PAK, Pak/77); >1 (ac, PAK/77); 20 (LEB, Yaz/53); every 2 yrs (JAP, Inu/57)

Pollen
P1 INO. **P** PAK. **Pollen grain** illustrated and described (Nak/65).
Reference slide

Honey: physical properties
Pfund amber (Cra/75; Mot/64)

150 Eucalyptus accedens W. Fitzg.; Myrtaceae

paper-barked wandoo, powder bark (En/AUS)
Tree, <20 m
Distribution subtropical Oceania; native to Australia. **Habitat** very restricted area of western AUS/WA (Pen/61); high ground in wandoo zone
Soil rocky laterite

Economic and other uses
Timber. **Other uses** tannin

Nectar rating; blooms, nectar flow
N1 AUS/WA(Col/62)
Blooms xii-iv, 3-4 yr cycle, buds 3 yrs before flowering (AUS, Pen/61).
Nectar flow limited (Smt/69)

Honey flow
Honey yield [high] 25-35, occasionally 45 kg/colony/season (AUS, Pen/61)

Pollen
P1 AUS/WA. **Yield** heavy (Pen/61). **Pollen value** good (Pen/61; Smt/69)

Honey no data

151 Eucalyptus alba Reinw. ex Blume; Myrtaceae

poplar gum (En/AUS); eucalipto alba (Pt/BRA)
Tree, <18 m, deciduous; fls cream
Distribution subtropical S America; temperate Africa; tropical Asia, C
America, Oceania; native to AUS. **Habitat** usually flats; sometimes
hillsides (AUS)
Soil granite slopes and rocky hillsides (AUS)

Economic and other uses
Timber. **Land use** windbreak, amenity. **Other uses** tannin from bark

Nectar rating + honeybee species; blooms, nectar flow; composition
Nl BRA/SC[tm](Wie/80); GUY(Smt/60)
Blooms v-viii (AUS, Pen/61); ii-iv (BRA). **Sugar concentration** [medium]
30% (Caa/72); 22-24% (Jul/72); 35-45% (Wie/80)

Pollen
P BRA

Honey: physical properties
Colour dark (Wie/80). **Pfund** medium amber (Bla/72)

152 Eucalyptus albens Benth.; Myrtaceae

white box (En/AUS, SOU)
Tree, <24 m; fl stamens cream
Distribution subtropical Oceania, Africa; native to AUS. **Habitat** drier
areas in SOU
Soil basaltic/granitic/alluvial; black soil (AUS/QD). **Temperature** -
resistant to cold but not to heat. **Rainfall** medium to high (AUS); not
drought resistant

Economic and other uses
Fodder for cattle. **Timber.** **Land use** windbreak, shade, amenity

Alert to beekeepers
Pollen inadequate for brood rearing, especially in winter/early spring
(AUS, Pen/61; AUS/VIC, Gom/73); heavy losses of bees can occur (AUS/QD,
Bla/72)

Nectar rating + honeybee species; blooms, nectar flow; composition
Nl AUS/NSW(Goo/47); AUS/QD(Bla/72); SOU/CAPE,OFS,TVL,tm(Bey/68)
N2 SOU/CAPE,OFS,TVL,tm(And/73)
N AUS/VIC(Gom/73)
Blooms late winter to spring, iii-v (AUS, Pen/61); iii-vii (AUS/QD); all
yr but best v-viii (SOU); xii-iii (SOU/Cape); buds 9-12 mths before
flowering (AUS, Gom/73). **Nectar secretion** profuse (Gom/73)

Pollen
P2 AUS/QD.   P3 AUS/VIC;   SOU/CAPE, OFS, TVL.   P SOU/CAPE, OFS, TVL.
Alert to beekeepers pollen inadequate for brood rearing, especially in winter/early spring (AUS, Pen/61; Gom/73); heavy losses of bees can occur (AUS/QD, Bla/72).   **Chemical analysis** 20.6%, 24.3% crude protein (AA1244/78)

Recommended for planting to increase honey production
AUS (Aus/83); recommended for Gippsland only, not central VIC.   See **Alert to beekeepers**

Honey: chemical composition
Water [medium] 17.8% (Woo/76)
Sugars, total 77.5% (75.1% after 44 days at 50°, Woo/76a).   **Glucose** [low] 29.6% (28.7%).   **Fructose** [medium] 36.4% (37.0%).   **Sucrose** [medium] 1.4% (0.8%).   **Maltose** 2.9% (3.1%).   **Melezitose** 1.1% (0.9%).   **Turanose** 5.8% (6.3%)
pH 4.22 (4.25 after 44 days at 50°, Woo/76); 3.9 (Woo/78).   **Total acid** 35.6 (34.0) meq/kg.   **Free acid** [medium] 28.9 (28.0) meq/kg.   **Lactone** 6.5 (6.0) meq/kg
Amylase 17.8 (Edw/75)
Nitrogen 0.033% (0.034% after 44 days at 50°, Woo/76).   **Amino acids**, free 1688.1 (297.3) μM/100 g, also contents of individual acids (proline 80% of total) (Woo/76a)
Volatile compounds 43 present, 12 named (Woo/78a)

Honey: physical and other properties
Colour light (Gom/73); dark (Wie/80); slightly cloudy (Bla/72).   **Pfund** white to light amber (And/73); 19-46 mm, white to extra light amber (Bla/72; Lei/72; Roc/68); almost water white (Cra/75); medium amber (Pur/68); 47.3 mm, extra light amber (123.5 mm, dark amber, after 44 days at 50°, Woo/76)
Granulation rapid (And/73; Gom/73); rapid, smooth, creamy (Bla/72); rapid fine (Cra/75)
Flavour mild (Gom/73)

153 Eucalyptus anceps (Maiden) Blakely; Myrtaceae

Kangaroo Island mallee, peaked crown top mallee, sand mallee, white mallee
Tree, 6 m, small mallee
Distribution subtropical Oceania; native to Australia
Soil sandy; sandy-loam.   **Rainfall** areas with <1000 mm AUS/SA; semi-arid/arid AUS

Economic and other uses
Fuel.   **Timber**

Nectar rating; blooms, nectar flow
N1 AUS/SA(Boo/72);   AUS/WA(Lei/72)
N2 AUS/SA(Lei/72);   AUS/WA(Col/62)
Blooms i-ii AUS (Pen/61);   i-iv, buds 2-3 yrs before flowering (AUS/WA, Lei/72)

Honey flow
Honey yield good every 5 yrs, otherwise poor to medium (AUS/SA, Boo/72); occasional moderate yields (AUS/SA)

Pollen
P2 AUS/WA.   P3 AUS/SA.   **Yield** high every 2 yrs (Boo/72);   poor (Lei/72).   **Pollen value** good (Boo/72);   average (Lei/72)

Honey: physical properties
Pfund light amber (Boo/72);   medium/amber (Lei/72;   Pur/68)

154 Eucalyptus caleyi Maiden; Myrtaceae

Caley's ironbark, drooping ironbark (En/AUS)
Tree, <12 m; fl stamens creamy-white/pinkish
Distribution subtropical and temperate Oceania;   native to AUS.   **Habitat** hilly country above 450 m (AUS)
Soil sandy/stony. **Rainfall** arid, semi-arid (AUS/QD)

Economic and other uses
Timber.   **Land use** amenity

Alert to beekeepers
No pollen (AUS, Lei/72);   pollen inadequate for brood rearing (AUS, Pen/61)

Nectar rating; blooms, nectar flow
N1 AUS/NSW(Cok/63);   AUS/QD(Bla/72)
N3 AUS/NSW(Goo/47)
Blooms iv-viii (AUS, Pen/61);   vi-x (AUS/QD)

Pollen
Alert to beekeepers no pollen (Lei/72);   pollen inadequate for brood rearing (Pen/61)

Honey: physical and other properties
Colour bright, exceptionally clear (Bla/72).   **Pfund** 2.0-15.5 mm, extra white (Bla/72;   Lei/72;   Pur/68;   Roc/68)
Granulation rapid, smooth, transparent grain (Bla/72)
Flavour sweet

155 Eucalyptus calophylla R.Br.; Myrtaceae

marri, red gum (En/AUS);   white flowering gum (En/SOU)
Tree/shrub, <45 m, much smaller in coastal areas;   fl filaments pink/white
Distribution subtropical Oceania, Africa;   tropical Africa;   native to AUS.   **Habitat** coastal plain and jarrah/karri forest AUS/WA;   sub-humid SOU
Soil alluvial/lighter sandy soils preferred.   **Temperature** - frost can damage young plants.   **Rainfall** "good" rainfall preferred

Economic and other uses
Timber. **Land use** shade, amenity.   **Other uses** tannin

Nectar rating + honeybee species; blooms, nectar flow
N1 AUS/WA(Col/62)
N2 AUS/WA(Smt/69)
N3 SOU/CAPE,NATAL,TVL,tm(And/73)
Blooms ii-iii, buds 2 mths before flowering (AUS, Pen/61); fls sensitive
to humidity (Pen/61).   **Nectar flow** reduced by hot dry winds (Smt/69);
also by high soil moisture (Col/62)

Honey flow
Honey yield fairly dependable annual source (Smt/69); major yield every 3-
7 yrs (Pen/61)

Pollen
P1 AUS/WA.   **P3** SOU/CAPE, NATAL, TVL.   **Pollen value** high (Smt/69)

Recommended for planting to increase honey production
SOU (Dai/70); sub-humid zone

Honey: chemical composition
Sucrose [medium] <2% (Smt/65)

Honey: physical and other properties
Pfund light amber (Cra/75; Smt/69); 46-64 mm (Smt/65) and 40 mm (Smt/67),
both light amber
Granulation fine (Cra/75; Smt/69)

156 Eucalyptus camaldulensis Dehnh.; Myrtaceae
syn Eucalyptus rostrata Schlechtd.

Murray red gum, river red gum (En/AUS)
Tree, <40 m depending on soil moisture
Distribution temperate (Med) Europe, Africa; subtropical S America,
Africa, Oceania, Asia; tropical Africa, Asia; native to AUS.   **Habitat**
most widely planted eucalypt; banks of inland rivers/alluvial flats
subject to periodic flooding; cold tablelands; sub-humid SOU; some
provenances will grow at >1200 m
Soil deep silt with clay subsoil preferred; salt tolerance varies with
provenance; free lime not usually tolerated.   **Temperature** high temperatures and hot dry winds tolerated; some provenances frost hardy.
Rainfall 200-1250 mm, commercial plantations >400 mm; arid and semi-arid
areas, some provenances more drought tolerant than others

Economic and other uses
Fodder - lvs, but not always favoured.   **Fuel** - charcoal particularly
important.   **Timber** termite resistant.   **Land use** windbreak, shade,
afforestation, amenity.   **Other uses** medicinal; paper-pulp; rayon;
tannin from bark

Warning; alert to beekeepers
Warning young trees susceptible to fire (Usa/80).   Other plants will not grow around this sp (Usa/80).   **Alert to beekeepers** when the bug Nysius vinitor Berg. infests fls in some areas, there is no flow (AUS, Pen/61)

Nectar rating + honeybee species;   blooms, nectar flow;   composition
N1 AUS/NSW(Goo/47);   AUS/QD(Bla/72);   AUS/SA(Boo/72);   AUS/VIC(Gom/73); BRA/RG[tm](Cor/70);   MOR(Cra/73);   SOU,tm(Mou/72);   URS(Glu/55)
N2 AUS/WA(Col/62);   ITA(Ric/78);   PAK,ac(PAK/77);   SOU,tm(And/73)
N ALG(Ske/72);   ZIM,tm(Pap/69)
Blooms all yr, heavily every 2 yrs (AUS, Pen/61);   buds 9-12 mths before flowering (AUS, Gom/73);   vii-ix (BRA).   **Nectar flow** ix-ii (ISR, Eis/80); too short for appreciable colony gains (SOU, And/73).   **Alert to beekeepers** when Nysius vinitor infests fls in some areas, there is no flow (AUS, Pen/61).   **Nectar secretion** 4.1-15.4 mg/fl/day (Eis/80);   very profuse, one of the heaviest yielders (Gom/73).   **Sugar concentration** [high] 61-81% (10 fls, Sao/54;   Wie/80);   [medium] 16.5-24.8% (various dates, Eis/80);   30% (Pel/76);   >50% (Zma/80).   **Sugar value** [medium] 0.56-2.90 mg/fl/day (Eis/80)

Honey flow
Honey yield (kg/colony/season) [high] 60 (AUS, Pen/61);   55 (AUS/QD, Bla/72);   100-120, mixed with honey from E. cladocalyx (MOR, Cra/73)

Pollen
P1 AUS/QD, SA, VIC;   ITA.   **P2** AUS/WA;   SOU.   **P** PAK;   ZIM.   **Yield** prolific (Gom/73);   high in alternate yrs (Boo/72).   **Pollen value** good (Boo/72;   Pen/61).   **Chemical analysis** 25.8% crude protein (AA1244/78). **Colour** of load greyish-brown (Ric/78).   **Pollen grain** illustrated and described (Smt/56a)

Recommended for planting to increase honey production
AUS (Aus/83);   BRA (Wie/80);   FRA (Sab/82);   SOU, sub-humid zone (Loo/83).   Propagate by seed which is freely produced;   choice of provenance v important (Usa/80).   V valuable for both brood rearing and honey production, and especially to beekeepers wanting colonies to store pollen before working pollen-deficient flow (Gom/73).   See **Warning;   alert to beekeepers**

Honey: chemical composition
Water - refractive index 1.4935 (Moh/82)
Glucose [medium] 32.70%.   **Fructose** [medium] 38.20%.   **Sucrose** [medium] 1.79-2.30%.   **Reducing sugars** 68.82%.   **Maltose** 6.60%.   **Raffinose** 1.60%. Contents of sugars as % of total sugars (Peo/72)
Ash [medium] 0.12% (Moh/82);   K 0.148%, Na 0.0079%, Ca 0.001%
pH 4.2 (Lan/66);   5.3 (Moh/82).   **Free acid** [medium] 20.70 meq/kg (Moh/82)
Amylase 29.4 (Lan/66)
Nitrogen 0.035% dry wt (Bos/78).   **Amino acids**, free 0.157, protein 0.120% dry wt (Bos/78);   0.68%, also contents of 15 individual acids (Peo/72a; Peo/74)

Honey: physical and other properties
Colour clear golden (And/73;   Sou/63);   grey with chestnut tint (Erb/83); straw coloured (Gom/73);   v variable, generally light grey, may be darker (Ric/78);   v clear (Ske/72).   **Pfund** 32-63 mm, white to light amber (Bla/58;   Lei/72;   Roc/68);   light amber (Boo/72;   Pur/68);   medium amber (Cra/75);   34.1 mm (also 3 "bulk honeys" 37.7-51.3 mm), extra light amber (Lan/66);   51 mm, light amber (Peo/72)

Viscosity 103.20 poise (Moh/82).   **Optical rotation** -8.55 deg.   **Other physical properties** - may froth on extraction (Bla/58)
Granulation rapid (And/73; Sou/63);   slow, large brown crystals (Bla/58; Cra/75);   rapid, hard (Gom/73);   medium (Moh/82);   fine-grained and compact (Ric/78)
Flavour mild, woody (Bla/58; Cra/75; Lei/72).   **Aroma** characteristic (Ric/78)

157 Eucalyptus citriodora Hook.; Myrtaceae

lemon-scented gum (En/SOU);   eucalyptus odorant (Fr/MAY);   eucalipto limão (Pt/BRA)
Tree, 25-40 m;   fl filaments cream
Distribution subtropical Africa, Asia, S America, Oceania;   tropical Africa, Asia, S and C America;   temperate (Med) Europe;   native to AUS/QD.   **Habitat** widely cultivated;   sea level to 600 m, and (rarely) at 2000 m (SRI)
Soil rather heavy soils in AUS but also deep sandy loam and clay;   must be well drained.   **Temperature** 29-35° mean monthly max;   subtropical Med type;   frost tender when young.   **Rainfall** >900 mm preferred;   moderate with max in summer (AUS);   also in areas of uniform rainfall

Economic and other uses
Fuel.   Timber.   Land use windbreak, shade, afforestation, amenity.   **Other uses** oil from lvs for perfume

Warning
Brittle branches;   unsuitable for urban/domestic sites (Usa/80)

Nectar rating + honeybee species; blooms, nectar flow; composition
N1 BRA(Smt/60);   BRA/SP(Mun/54);   MAE(Chl/75);   MAY(Bro/82; Cra/73);   URS(Glu/55)
N2 SOU/NATAL,TVL,tm(And/73)
Blooms vi-vii (AUS, Key/77);   ix-ii (BRA, Wie/80).   **Sugar concentration** [medium] 38-45% (Wie/80).   **Sugar value** [low] 0.0253 mg/fl/day (Jua/64)

Honey flow
Honey potential [moderate] 113 kg/ha (ROM, Jua/64)

Pollen
P3 SOU/NATAL, TVL.   **Pollen grain** illustrated and described (San/61)

Recommended for planting to increase honey production
BRA (Wie/80).   Propagate by seed, freely produced;   grows rapidly (Usa/80).   See **Warning**

Honey: physical and other properties
Pfund light amber (And/73);   medium amber (Cra/75)
Granulation v firm (And/73)

158 Eucalyptus cladocalyx F. Muell.; Myrtaceae

sugar gum (En/AUS)
Tree, <30 m (6-12 m in poor conditions); fls v fragrant
Distribution subtropical Africa, Oceania; temperate (Med) Africa; native
to AUS/SA, VIC. **Habitat** mixed woodland AUS; sub-humid Cape coastal belt
(SOU) but not where there are salt-laden sea-breezes
Soil wide range especially quartzite ridges, acid soil (AUS); poor soil,
infertile wasteland (SOU). **Temperature** hot dry conditions (AUS);
susceptible to frost damage only when young. **Rainfall** 500 mm but higher
rainfall areas preferred (AUS); winter rainfall zone, but not heavy rains,
drier districts of western SOU

Economic and other uses
Fodder – lvs, but can be toxic to sheep/cattle/horses (Pen/61). **Fuel.**
Timber. **Land use** windbreak, shade, amenity

Warning
Lvs can be toxic to animals (Pen/61; Boo/72); other plants will not grow
around this sp (Anr/74)

Nectar rating + honeybee species; blooms, nectar flow
N1 MOR(Cra/73); SOU/CAPE,tm(And/73; Mou/72)
N2 AUS/NSW(Goo/47); AUS/SA(Boo/72)
N3 AUS/VIC(Gom/73)
Blooms i-ii, may be every 2 yrs; buds 13 mths before flowering (AUS,
Pen/61). **Nectar flow** annual (SOU, Anr/74); lengthy (SOU, Loo/82); few
days (AUS/VIC). **Nectar secretion** heavy, especially on warm slightly humid
days; hot dry winds reduce flow

Honey flow
Honey yield (kg/colony/season) [high] 100-120, often with E. camaldulensis
(MOR, Cra/73); 15-25, max 90 (SOU/Cape, Anr/74); "indifferent yields"
outside winter rainfall area (SOU, And/73)

Pollen
P2 AUS/WA. **P3** AUS/SA, WA; SOU/CAPE. **Yield** moderate, every 2 yrs
(Boo/72); small (Anr/74). **Pollen value** disagreement as to value
(Pen/61; Sou/65); bees prefer pollen from other sources (Gom/73); poor
quality (Boo/72); little collected (And/73)

Recommended for planting to increase honey production
AUS (Aus/83); SOU (Anr/74; Loo/82; Sou/65); sub-humid zone. Propagate
by seed; grows rapidly; first fls at 4-5 yrs but initial nectar yield low
(Anr/74). See **Warning**

Honey: chemical composition
Water [low] 14.6% (by gravimetry), 15.4% (by refractometry) (Anr/74)
Glucose [low] 25.2%. **Fructose** [medium] 41.9%. **Maltose** 12.2%
Ash [medium] 0.3%
Nitrogen 0.02%

Honey: physical and other properties
Colour pale straw (And/73 Gom/73); light (Mou/72). **Pfund** 41 mm, extra
light amber (Anr/74); light amber (Boo/72; Pur/68)

Relative density 1.438 (Anr/74).   **Viscosity** 21.36 poise.   **Other**
properties froths on heating (Sou/65)
Granulation slow, none if monofloral (And/73;   Mou/72)

159 Eucalyptus cornuta Labill.;   Myrtaceae

yate (En/AUS, SOU)
Tree, <21 m but can form stunted thickets
Distribution subtropical Oceania, Africa;   native to AUS.   **Habitat** <300 m
altitude (Pen/61)
Soil moist gravelly loam;   also alkaline and saline soil.   **Temperature**
thrives under hot wet conditions;   frost resistant.   **Rainfall** low rainfall
areas

Economic and other uses
Timber.   **Land use** windbreak, shade, amenity

Nectar rating + honeybee species;   blooms, nectar flow
N1 AUS/WA(Col/62)
N2 SOU/CAPE,tm(And/73;   Mou/72)
Blooms i-ii (AUS, Pen/61);   xii-i (SOU/CAPE);   usually every 2 yrs
(Pen/61).   **Nectar flow** fair (Pen/61)

Pollen
P2 AUS/WA.   **P3** SOU/CAPE.   **Yield** fair (Pen/61).   **Pollen value** good (Pen/61)

Honey: physical and other properties
Colour light medium (And/73)
Granulation rapid, fine
Flavour fairly strong

160 Eucalyptus crebra F. Muell.;   Myrtaceae

narrowed-leaved red ironbark (En/AUS)
Tree, <30 m
Distribution subtropical Oceania, Africa;   native to AUS.   **Habitat**
undulating/hilly country <600 m (AUS)
Soil deep, moderately good acid soil;   also sandy soil with hard clay
subsoil.   **Temperature** seedlings frost hardy.   **Rainfall** minimum 25 mm in
driest mths;   seedlings drought tolerant (AUS);   summer rainfall area, sub-
humid interior (SOU)

Economic and other uses
Timber

Nectar rating + honeybee species;   blooms, nectar flow
N1 AUS/QD(Bla/72)
N3 AUS/NSW(Goo/47);   SOU,tm(And/73)
Blooms v-i (AUS, Pen/61);   vii-xii, main period ix-xi (SOU).   **Nectar flow**
"reasonable", enhanced by heavy rains prior to blooming (Pen/61)

Honey flow
Honey yield [high] 82 kg/colony/season; good yield once every 3 yrs, when many trees flower together (AUS/QD, Bla/72)

Pollen
P1 AUS/QD.   **P3** AUS/NSW;   SOU.   **Yield** minor to medium (Bla/72);   deficient (Cok/63);   fair (Goo/47);   major producer (Pen/61)

Recommended for planting to increase honey production
SOU, sub-humid zone (And/73; Dai/70).   Propagate by seed; grows rapidly at first, slowing down at pole-size (AUS, Pen/61)

Honey: physical and other properties
Pfund light amber (And/73);   extra white to extra light amber (Bla/72);   14-35 mm, white (Lei/72; Roc/68)
Viscosity "heavy body" (Bla/72)
Granulation slow, coarse whitish grain
Flavour mild, sweet

161 Eucalyptus diversicolor F. Muell.;   Myrtaceae

karri (En/AUS, SOU)
Tree, 45-70 m; forms up to 50% of karri forest in AUS/WA
Distribution subtropical Africa, Oceania, S America;   native to AUS/WA.
Habitat undulating hilly country;   altitude <300 m
Soil deep loam;   light soils derived from granite/gneiss.   **Rainfall** high, 1150-1500 mm (AUS);   higher rainfall areas of W Cape and constant rainfall areas of Knysna and George (SOU)

Economic and other uses
Timber.   **Land use** firebreaks.   **Other uses** tannin from bark

Alert to beekeepers
Karri forest in AUS/WA has many shrubs whose spring flowering can lead to uncontrollable swarming (Smt/63);   pollen may be inadequate for brood rearing (AUS/WA, Smt/63)

Nectar rating + honeybee species;   blooms, nectar flow
N1 AUS/WA(Col/62)
N2 SOU/CAPE,tm(And/73)
Blooms v-xii, buds 18-24 mths before flowering (AUS, Pen/61);   ii-iii (SOU).   **Nectar flow** i-iii (AUS/WA, Smt/63), one of the longest and heaviest in the world (Col/62);   but heavy flow infrequent (4, 5, 8, 12 yr intervals recorded);   2 wet winters needed for maximum bloom and a dry autumn for nectar collection (Pen/61);   sensitive to hot summers (Col/62).   **Alert to beekeepers** karri forest in AUS/WA has many shrubs whose spring flowering can lead to uncontrollable swarming (Smt/63)

Honey flow
Honey yield [high] 150-200 kg/colony/season (AUS/WA, Col/62);   135-225 (Smt/63);   major yield every 4-12 yrs (Col/62; Pen/61)

Pollen
P3 SOU/CAPE. **Pollen value** adequate for summer maintenance (AUS/WA, Smt/63); poor (Col/62). **Alert to beekeepers** pollen may be inadequate for brood rearing (AUS/WA, Smt/63)

Recommended for planting to increase honey production
SOU(Dai/70). Grows rapidly (Pen/61). See **Alert to beekeepers**

Honey: physical and other properties
Colour light straw (And/73). **Pfund** extra light amber (Cra/75; Smt/69); 22 mm, white (Smt/67)
Granulation slow (And/73); rapid, fairly coarse (Cra/75)
Flavour mild, characteristic (Cra/75)

162 Eucalyptus drepanophylla F. Muell. ex Benth.; Myrtaceae

grey ironbark, Queensland grey ironbark, white ironbark (En/AUS)
Tree, <40 m
Distribution subtropical Oceania; native to AUS. **Habitat** coastal; tops of ridges, slopes and valleys (AUS)
Soil sandy loam but also poor stony ridges (AUS). **Rainfall** growth is stunted in low rainfall areas

Economic and other uses
Fuel. Timber

Alert to beekeepers
Pollen "unsatisfactory" (AUS, Pen/61)

Nectar rating; blooms, nectar flow
N1 AUS/QD(Bla/72)
Blooms viii-x (AUS, Pen/61); vii-xii (AUS/QD)

Honey flow
Honey yield "one of the valuable trees of coastal QD"; high yield every 3 yrs (AUS/QD, Bla/72)

Pollen
P3 AUS/QD. **Alert to beekeepers** pollen "unsatisfactory" (Pen/61)

Honey: physical and other properties
Pfund 20-40 mm, white to extra light amber (Bla/72; Roc/68)
Flavour - this sp usually contributes to mixed honeys, in which its flavour predominates (AUS/QD, Bla/72)

163 Eucalyptus falcata Turcz.; Myrtaceae
syn Eucalyptus falcata var. ecostata Maiden

smooth-fruited mallet, white mallet (En/AUS)
Tree, <10 m

Distribution subtropical Oceania; native to AUS. **Habitat** coastal; ridges and hills (AUS)
Soil ironstone

Economic and other uses
Timber. **Other uses** tannin from bark

Nectar rating; blooms, nectar flow
N1 AUS/WA(Col/62)
Blooms xi-xii (AUS, Pen/61)

Pollen
P2 AUS/WA

Honey no data

164 Eucalyptus fasciculosa F. Muell.; Myrtaceae

pink gum (En/AUS)
Tree, 5-11 m
Distribution subtropical Oceania; native to AUS/SA, VIC. **Habitat** dry sterile hilly sites; woodlands/open forests
Soil poor, dry, sandy/stony soil. **Temperature** cool winters, hot summers. **Rainfall** wet winters, dry summers

Economic and other uses
Fuel. **Timber.** **Land use** windbreak, shade

Alert to beekeepers
Pollen not collected by bees (AUS/SA, Boo/72)

Nectar rating; blooms, nectar flow
N1 AUS/SA(Boo/72)
Blooms xii-v (AUS, Pen/61). **Nectar flow** every 2-4 (or more) yrs (AUS/NSW, Boo/72)

Pollen
Alert to beekeepers pollen not collected by bees (AUS/SA, Boo/72)

Honey: chemical composition
Water [low] 15.8% (Che/74)
Glucose [low] 25.5%. **Fructose** [high] 45.1%. **Sucrose** [low] 0.6%.
Reducing sugars 70.6%
Ash [low] 0.04%
pH 3.88. **Total acid** 11.8 meq/kg. **Free acid** [low] 8.6 meq/kg. **Lactone** 3.2 meq/kg
Amylase 18. **HMF** 2.0 ppm

Honey: physical properties
Pfund white (Boo/72); 16 mm, extra white (Che/74); light amber (Pur/68)

165 Eucalyptus ficifolia F. Muell.; Myrtaceae

scarlet gum (En/NEZ); red-flowering gum (En/SOU); scarlet bloom (En/USA)
Tree, 9–12 m; fls scarlet, some cvs pink
Distribution subtropical Africa, S America, N America, Oceania; temperate
(Med) Europe; native to AUS. **Habitat** widely planted throughout the
world; best in coastal areas
Soil poor gravelly/sandy soil; tree adapts to other types. **Temperature**
frost tender; hot east winds scorch trees. **Rainfall** >750 mm preferred;
winter rainfall areas

Economic and other uses
Timber. **Land use** shade, amenity – probably the best known ornamental
eucalypt

Alert to beekeepers
Honey may be "stringy/ropy" (And/73; Pry/52); viscosity increases to
sudden stiffness on rapid stirring (dilatancy) (Pry/50; Pry/52)

Nectar rating + honeybee species; blooms, nectar flow
N1 USA/CA(Pel/76)
N2 SOU/CAPE,NATAL,TVL,tm(And/73);
N NEZ(Wal/78)
Blooms xii–ii (AUS, Key/77); x–iii, best xii–ii (SOU). **Nectar flow** 1 mth
(USA/CA, Pel/76). **Nectar secretion** v heavy; nectar drips in long threads
from fls (And/73)

Pollen
P3 SOU/CAPE, NATAL, TVL. **P** NEZ. **Yield** small (And/73)

Honey: chemical composition
Dextrin 7.2% "dextran" [?dextrin] (Pry/52)

Honey: physical and other properties
Colour light straw (And/73); rather dark (Wal/78). **Pfund** water white
(Pel/76)
Viscosity (alert to beekeepers) may be "stringy/ropy" (And/73; Pry/52);
increases to sudden stiffness on rapid stirring (dilatancy) (Pry/50;
Pry/52)
Granulation slow (And/73)
Flavour pleasant, but feels slimy in the mouth (And/73); pronounced,
characteristic (Wal/78)

166 Eucalyptus globulus Labill.; Myrtaceae

fever tree, southern blue gum, Tasmanian blue gum (En/AUS); blue gum
(En/USA)
Tree, 55–70 m, evergreen
Distribution tropical Africa, Asia, C America; subtropical S America, N
America, Africa; temperate Oceania, Europe (Med); native to E Tasmania.
Habitat the most widely cultivated eucalypt; altitudes <3000 m· one of
the best adapted spp in south BRA, MEX plateau and C America

Soil loam with adequate moisture;   shallowness, poor drainage and salinity
limit cultivation.   **Temperature** mild temperate;   cool tropical highlands
without extremes of heat/cold preferred;   damaged by frost.   **Rainfall**
native range 800 to >1500 m, well distributed year-round;   not drought
tolerant

Economic and other uses
Fuel.   Timber termite resistant.   **Land use** windbreak, shade, amenity.
Soil benefit erosion control.   **Other uses** oil;   pulp for paper-making, rayon

Warning
Dead lvs and bark are fire hazard

Nectar rating + honeybee species;   blooms, nectar flow;   composition
N1 AUS/VIC(Gom/73)
N2 PAK,ac(Pak/77);   USA/CA(Pel/76)
N3 SPA(Sau/82)
N ALG(Ske/72);   INI/TAM[ac](Cha/74);   NEZ(Wal/78);   URS(Glu/55)
Blooms vi-xi (AUS, Pen/61);   buds 15-18 mths before flowering (Gom/73);
spring (NEZ);   x-xi (SPA);   v-viii (BRA/SC);   late winter-early spring but
not every yr (USA/CA).   **Nectar flow** adversely affected by dry windy
conditions (Wal/78).   **Sugar concentration** [medium] 35-40% (Wie/80);   [low]
17% (Pel/76).   **Sugar analysis** (Maz/59)

Honey flow
Honey yield not reliable in NEZ (Wal/78);   yield every 4 yrs (USA/CA,
Lov/77);   this sp produces most of the eucalypt honey in USA/CA (Ord/83)

Pollen
P1 INI/TAM.   **P2** AUS/VIC.   **P** NEZ;   PAK.   **Pollen value** useful for building
up colonies after winter (Gom/73);   fair (NEZ).   **Colour** pale yellow or
cream to white (Wal/78)

Recommended for planting to increase honey production
BRA (Wie/80).   Propagate by seed;   easy to establish, grows rapidly.   See
Warning

Honey: physical and other properties
Colour dark (Glu/55;   Ske/72);   fairly dark (Ord/83);   cloudy (Wal/78).
Pfund light amber (Gom/73);   70 mm, medium amber (Lov/61c;   Roc/68;
Wal/78);   amber (Pel/76)
Viscosity "heavy body" (Pel/76);   "light body" (Wal/78)
Granulation medium (Pel/76);   slow (Wal/78)
Flavour distinctive (Glu/55;   Lov/61c);   acid (Pel/76);   like muscatel
grapes (Wal/78)

167 Eucalyptus gomphocephala DC.;   Myrtaceae

tuart (En/AUS, SOU)
Tree, <42 m, occurs as forest in AUS/WA
Distribution subtropical Oceania, Africa, Asia;   temperate (Med) Europe;
native to AUS/WA.   **Habitat** coastal (AUS;   SOU), semi-arid/sub-humid zones
(Usa/80)

Soil sandy loam overlying limestone (AUS); limestone areas (AUS/WA); calcareous sand; slightly saline soil tolerated (Usa/80); waterlogging not tolerated. **Temperature** absolute minimum -4°; poor frost tolerance. **Rainfall** absolute minimum 300 mm; range in AUS/WA 700-1000 mm with 6 dry summer mths (Usa/80)

Economic and other uses
Fuel. **Timber.** **Land use** windbreak, shade, amenity. **Soil benefit** stabilizes dunes; soil protection

Warning; alert to beekeepers
Warning young plantations susceptible to fire (Usa/80). **Alert to beekeepers** flow reduced if the weevil Haplonyx tibialis has caused severe bud drop (AUS, Pen/61)

Nectar rating + honeybee species; blooms, nectar flow
N1 AUS/WA(Col/62); URS(Glu/55)
N3 SOU/CAPE,NATAL,tm(And/73)
N AUS/WA(Smt/69)
Blooms every 4-7 yrs; i-iii, buds 2 yrs before flowering (AUS, Pen/61); autumn (SOU, Dai/70). **Nectar flow** iii-iv (AUS/WA). **Alert to beekeepers** flow reduced if the weevil Haplonyx tibialis has caused severe bud drop (AUS, Pen/61). **Nectar secretion** rain during blooming stops flow (Pen/61), but nectar may be produced again later (Col/62)

Pollen
P2 SOU/CAPE, NATAL. **P3** AUS/WA. **Yield** abundant (Pen/61). **Pollen value** poor (Col/62; Pen/61)

Recommended for planting to increase honey production
SOU (Dai/70); good for calcareous areas in sub-humid zones. Propagate by seeds, which are produced infrequently (Pen/61). See **Warning; alert to beekeepers**

Honey: physical and other properties
Colour light (And/73; Smt/69)
Granulation rapid, fine (Smt/69)
Flavour of caramel (And/73; Dai/70). **Aroma** strong (Dai/70)

168 Eucalyptus gracilis F. Muell.; Myrtaceae

small-budded mallee, snap and rattle, white mallee, yorrell (En/AUS)
Shrub/tree, <18 m, mallee-like, forms clumps/thickets
Distribution subtropical Oceania; native to AUS. **Habitat** widespread in southern AUS
Soil sandy soil (AUS). **Rainfall** dry areas (AUS)

Economic and other uses
Fuel. **Timber**

Alert to beekeepers
Heavy bee losses have occurred, pollen probably inadequate for brood rearing (AUS/VIC, Gom/73); little/no pollen collected (AUS/SA, Boo/72)

Nectar rating; blooms, nectar flow; composition
N1 AUS/VIC(Gom/73)
N2 AUS/SA(Boo/72); AUS/WA(Col/62; Lei/72)
Blooms iv-viii (AUS, Pen/61); iii-xi, best viii-x, rarely iv-v (AUS/WA);
main flowering ix-x, buds 3-6 mths before flowering (AUS/VIC, Gom/73).
Nectar flow good every 4-15 yrs (AUS/NSW, Boo/72). **Sugar concentration**
probably high, since a colony working this flow uses up to 1 litre of water
per day (AUS, Col/62)

Honey flow
Honey yield (kg/colony/season) [high] mean 27-36, max 113 (AUS/VIC, Gom/73)

Pollen
P3 AUS/VIC, WA. **Alert to beekeepers** heavy bee losses have occurred,
pollen probably inadequate for brood rearing (AUS/VIC, Gom/73); little/no
pollen collected (AUS/SA, Boo/72). **Colour** grey to dirty white (Gom/73)

Honey: chemical composition
Amino acids 373.23 µg/g (contents of 13 individual acids given, Peo/72a;
Peo/74)

Honey: physical and other properties
Pfund light amber (Boo/72; Gom/73; Lei/72; Pur/68); 55.4 mm, light
amber ("bulk honey", Lan/66)
Flavour mild (Gom/73; Lei/72)

169 Eucalyptus grandis W. Hill ex Maiden; Myrtaceae

flooded gum, rose gum (En/AUS); saligna gum (En/SOU)
Tree, 40-60 m; fls creamy-white. Names of E. grandis and E. saligna
often confused in literature (Guy/74); data for SOU and AUS only entered
here
Distribution subtropical Africa, Oceania, S America, Asia; tropical
Africa, Oceania, Asia, S America; temperate (Med) Africa; native to
AUS. **Habitat** widely cultivated; most important commercial timber crop in
SOU; altitudes from sea level to 800 m in AUS but up to 2700 m elsewhere
in tropics; wetter coastal belt/tablelands/lower slopes and flats in
native AUS
Soil moist but well drained; boron-deficiency not tolerated; rather
infertile soil with low phosphorus content (AUS). **Temperature** native
range, -1 to -3° winter min and <40° summer max; sudden freezing v
damaging. **Rainfall** mean 1000-1800 mm, with summer/autumn predominance,
dry spring (AUS); >2500 mm, or <600 mm if irrigated; >750 mm preferred
(SOU)

Economic and other uses
Fuel. Timber, and pulp for paper/rayon. **Land use** shade, afforestation,
amenity. **Other uses** tannin

Warning; alert to beekeepers
Warning no lignotubes, so vulnerable to fire. **Alert to beekeepers** larvae
of Drosophila flavohirta can consume all nectar (SOU, Her/83)

Nectar rating + honeybee species;   blooms, nectar flow;   composition
N1 SOU,tm(And/73;   Mou/72);   SWA[tm](Ehr/78)
N2 ZAM,tm(Sil/76a)
N3 AUS/QD(Bla/72)
N ZIM,tm(Pap/69)
Blooms vi-viii (AUS, Pen/61);   regularly, sometimes heavily, iii-v
(AUS/QD);   ii-v or vii, depending on altitude (SOU, Guy/74).   **Nectar flow**
difficult to predict (SOU, Joh/77);   iii-vi in low veld, ii-iv Zululand
(SOU, Mou/72).   **Alert to beekeepers** larvae of Drosophila flavohirta can
consume all nectar (SOU, Her/83).   **Nectar secretion** highest in cooler
weather (ZAM, Sil/76a);   not stopped by frosts;   enhanced by good rains 6
mths before flowering (SOU, Joh/77).   **Sugar concentration** [medium] 35-40%
(Wie/80)

Honey flow
Honey yield (kg/colony/season) [high] 15-45 (SOU/Natal, Guy/74);   mean 50,
max >100 (SOU/Zululand, Guy/74);   7-65 (SOU/TVL, AA902/78);   lower yields
in AUS

Pollen
P1 ZAM.   **P2** AUS/QD;   SOU.   **P** ZIM.   **Yield** good (Pen/61).   **Pollen value**
high, useful in autumn/early winter (Pen/61);   colonies build up rapidly
(Sil/76a).   **Colour** of load cream (Guy/74)

Recommended for planting to increase honey production
BRA (Wie/80);   SOU (Loo/83).   Propagate by seed, provenance v important
(Usa/80).   Grows v rapidly;   blooms in 2nd/3rd yr.   Vulnerable to sudden
freezing.   See **Warning; alert to beekeepers**

Honey: chemical composition
Water low (Guy/74);   high (Sil/76a)

Honey: physical and other properties
Colour pale to medium amber (Bla/58);   medium dark (Cra/75);   darkens on
heating (Guy/74;   Joh/75).   **Pfund** medium amber (And/73;   Her/83);   lighter
at higher altitudes, darker on Zululand coast and in lower Transvaal (SOU,
Guy/74)
Other physical properties - froths on heating (And/73;   Her/73)
Granulation medium, slaty, soft, smooth grain (Bla/58);   rapid, fine
(Cra/75;   Her/73);   coarse in warm regions, fine in colder areas where it
sets rock hard (Guy/74);   rapid, may set rock hard (Joh/75)
Flavour strong, slightly objectionable (Bla/58);   strong if dark, impaired
by heating (Guy/74);   **Aroma** when fresh, but fades (Guy/74)

170 Eucalyptus incrassata Labill.; Myrtaceae
syn Eucalyptus incrassata var. costata N.T. Burbridge

angulosa mallee, giant angular mallee, lerp mallee, ridge-fruited mallee,
yellow mallee (En/AUS)
Tree, 2-5 m
Distribution subtropical and temperate Oceania;   native to AUS
Soil sandy.   **Rainfall** low rainfall areas in AUS

Economic and other uses
Timber. **Land use** windbreak

Nectar rating; blooms, nectar flow
N1 AUS/VIC(Gom/73); AUS/WA(Col/62)
N3 AUS/SA(Boo/72)
Blooms iii-iv, buds 10-15 mths before flowering (AUS, Pen/61); conflicting reports but generally x-iv (AUS/VIC)

Honey flow
Honey yield "heavy" (AUS, Aus/83)

Pollen
P1 AUS/VIC, WA. **P2** AUS/SA, VIC, WA. **Yield** quite reasonable (Pen/61); average (Gom/73)

Recommended for planting to increase honey production
AUS (Aus/83)

Honey: chemical composition
Amino acids 507.99 µg/g, also contents of 14 individual acids (Peo/72a; Peo/74)

Honey: physical properties
Pfund medium amber (Boo/72; Gom/73; Lei/72; Pur/68)

171 Eucalyptus jacksonii Maiden; Myrtaceae

red tingle (En/AUS)
Tree, <60 m
Distribution subtropical Oceania; native to AUS. **Habitat** coastal hills/high forest and lower reaches of rivers/southern rainforest; only small areas now existing (AUS/WA)
Soil deep red loam. **Rainfall** 1300-1500 mm (AUS)

Economic and other uses
Timber

Nectar rating; blooms, nectar flow
N1 AUS/WA(Col/62)
Blooms i-ii, usually every 3-4 yrs (AUS, Pen/61). **Nectar secretion** higher in hot weather, sometimes reduced by thrip attack (Pen/61)

Pollen
P2 AUS/WA. **Pollen value** good (Pen/61). **Colour** bright yellow (Pen/61)

Honey no data

172 Eucalyptus leucoxylon F. Muell.; Myrtaceae

South Australian blue gum, yellow gum (En/AUS); leucoxylon gum (En/SOU); white ironbark (En/USA)
Tree, <27 m; fls with pink/crimson/white filaments
Distribution subtropical Africa, N America, Oceania; tropical Africa; temperate (Med) Africa, Oceania; native to AUS. **Habitat** undulating country/valleys; also open woodland/forests in moist valleys (AUS)
Soil heavy alluvium, stiff clay, sandy loam. **Temperature** hot dry winds tolerated; some frost tolerated but seedlings need protection. **Rainfall** 500 mm (AUS/SA); drought tolerant (AUS); drier areas more suitable (SOU)

Economic and other uses
Timber. **Land use** windbreak, shade, amenity. **Other uses** medicinal oil from lvs

Alert to beekeepers
Sometimes bees will not collect nectar (Pen/61); no pollen (AUS, Pen/61); pollen inadequate for brood rearing (AUS/VIC, Gom/73)

Nectar rating + honeybee species; blooms, nectar flow
N1 AUS/SA(Boo/72); AUS/VIC(Gom/73)
N2 SOU,tm(And/73); USA/CA(Pel/76)
N NEZ(Wal/76); URS(Glu/55)
Blooms iv-xii, best v-ix (SOU); v-xii, buds 6-10 mths before flowering (AUS, Pen/61). **Nectar flow** heavy every 2 yrs. **Alert to beekeepers** sometimes bees will not collect nectar (Pen/61). **Nectar secretion** generally yielded freely (Gom/73)

Pollen
P3 SOU. **Alert to beekeepers** no pollen (AUS, Pen/61); pollen inadequate for brood rearing (AUS/VIC, Gom/73)

Recommended for planting to increase honey production
AUS (Aus/83). Propagate by seed; regenerates fairly well (Pen/61). Useful shelter tree; cv Rosea for amenity planting. See **Alert to beekeepers**

Honey: chemical composition
Water [medium] 15.4, 16.8% (Che/74; also gives data for samples age 5-12 mths)
Glucose [medium] 27.7, 30.6%. **Fructose** [medium] 43.7, 40.7%. **Sucrose** [medium] 1.0, 4.8%. **Reducing sugars** 71.4, 71.3%
Ash [medium] 0.24, 0.11%
pH 5.19, 3.88 (Che/74); 4.3 (Lan/66). **Total acid** 10.6, 32.2 meq/kg (Che/74). **Free acid** [low] 8.6, 22.8 meq/kg
Amylase 28 (Che/74); 13.9 (Lan/66). **HMF** 2.0, 1.4 ppm

Honey: physical and other properties
Colour clear, pale straw (And/73; Cra/75; Gom/73). **Pfund** light amber (Boo/72); 20 mm, white; also 68 mm, light amber (Che/74); white (Glu/55); 62.4 mm (67.4 mm after 16 h at 66°), light amber (Lan/66; Lei/72; Pur/68); 23.3-55.0 mm, white to light amber (4 bulk honeys, Lan/66); medium amber (Lei/72)
Viscosity "good body" (Cra/75)

Granulation rapid (And/73; Cra/75); fine (Gom/73)
Flavour mild (And/73; Cra/75; Gom/73); like vanilla (Glu/55; Pel/76)

173 Eucalyptus loxophleba Benth.; Myrtaceae

York gum (En/AUS)
Tree, <13 m
Distribution subtropical Oceania; native to AUS. **Habitat** widespread in
AUS/WA
Soil granite with clay subsoil (AUS)

Economic and other uses
Timber

Alert to beekeepers
Pollen may be inadequate for brood rearing (AUS, Pen/61)

Nectar rating; blooms, nectar flow
N1 AUS/WA(Col/62)
N AUS/WA(Smt/69)
Blooms v-xii, buds 9-10 mths before flowering (AUS, Pen/61). **Nectar flow**
viii-xii; hot weather required for flow (Col/62)

Honey flow
Honey yield every 2 yrs (AUS, Col/62)

Pollen
P3 AUS/WA. **Alert to beekeepers** pollen may be inadequate for brood rearing
(AUS, Pen/61)

Honey: physical properties
Colour light to medium, varies with season (Cra/75). **Pfund** medium amber
(Col/62; Smt/69); light amber (Lei/72)

174 Eucalyptus macrorhyncha F. Muell. ex Benth.; Myrtaceae

red stringybark (En/AUS)
Tree, <35 m
Distribution subtropical and temperate Oceania, native to AUS. **Habitat**
foothills and lowland ranges (AUS); grows best as an isolated tree; up to
900 m altitude (AUS)
Soil wide range of well drained soils; sedimentary or clay soils preferred
to shale (AUS); poorer stony, clay and acid soils (AUS/VIC). **Rainfall**
high rainfall areas (AUS)

Economic and other uses
Timber. **Land use** windbreak, shade, amenity. **Other uses** medicinal,
natural source of the drug rutin

Warning
Caterpillars sometimes devastate large areas of forest (AUS/VIC, Gom/73)

Nectar rating; blooms, nectar flow
N1 AUS/NSW(Goo/47)
N3 AUS/SA(Boo/72)
Blooms ii, buds 15-18 mths before flowering (AUS, Pen/61).   **Nectar flow**
heavier in less dry areas (AUS/VIC, Gom/73);   good yield every 20-30 yrs
(AUS/SOU, Boo/72)

Honey flow
Honey yield "heavy" (Aus/83);   not very reliable (AUS/VIC, Gom/73)

Pollen
P2 AUS/NSW, SA.   **Yield** heavy (AUS, Aus/83);   moderate, annual (Boo/72).
Pollen value average quality (Boo/72);   useful (Pen/61)

Recommended for planting to increase honey production
AUS (Aus/83).   See **Warning**

Honey: chemical composition
Ash sulphated 0.337% (also ash analysis, Peo/70)
pH 6.0-6.4 (3 samples, Lan/66)
Amylase 23.8-38.5
Amino acids 16 acids identified (Peo/71)

Honey: physical and other properties
Colour clear, strong colour (Gom/73);   dark (Pur/68).   **Pfund** amber
(Boo/72);   47.4 mm, extra light amber, and 57.9 mm, light amber (also 62.8-
77.3 mm, light amber for 3 "bulk honeys", Lan/66);   72 mm, light amber
(Roc/68)
Granulation medium (Gom/73)

175 Eucalyptus maculata Hook.;   Myrtaceae

spotted gum (En/AUS, SOU)
Tree, <45 m;   fire resistant
Distribution subtropical Oceania, Africa;   tropical Africa;   temperate
(Med) Africa, Oceania, (Med) Europe, S America;   native to AUS/QD, VIC.
Habitat altitude <1000 m, coastal (AUS);   stony ridges in forest country
(AUS/QD);   now being replaced by other eucalypts (SOU)
Soil wide range but sandy shale preferred;   or deep moist soil (SOU).
Temperature light frost tolerated when tree well established.   **Rainfall**
summer rain required

Economic and other uses
Fodder - lvs for koala bears.   **Timber.**   **Land use** windbreak, amenity.
Other uses wood pulp

Alert to beekeepers
Bees "angry" during this flow (SOU, Guy/72)

Nectar rating + honeybee species; blooms, nectar flow
N1 AUS/NSW(Cok/63);   SOU/CAPE,NATAL,TVL,tm(Bey/68)
N2 AUS/NSW(Goo/47);   AUS/VIC(Gom/73);   SOU/CAPE,NATAL,TVL,tm(And/73)
N3 AUS/QD(Bla/72)
Blooms vii-viii, during dearth period (SOU/NATAL, Coe/67);   buds 12-18 mths
before flowering (AUS, Pen/61);   6 mths before (AUS/VIC);   unpredictable,
up to 18 mths before (SOU, Guy/72).   **Alert to beekeepers** bees "angry"
during this flow (SOU, Guy/72)

Honey flow
Honey yield [high] 60 kg/colony/season (AUS/NSW, Pen/61);   yields well
every 3-4 yrs (AUS/QD, Bla/72);   major flow every 3-4 yrs (SOU, Dai/70)

Pollen
P1 AUS/NSW, VIC.   **P2** AUS/QD.   **P3** or **P** SOU/CAPE, NATAL, TVL.   **Yield**
medium to abundant (Pen/61).   **Pollen value** good (Pen/61).   **Chemical
analysis** 33.3% crude protein (AA1244/78).   **Colour** cream (Gom/73)

Recommended for planting to increase honey production
SOU(Dai/70; Loo/82).   Grows fairly rapidly.   See **Alert to beekeepers**

Honey: chemical composition
Water [medium] 16.8% (Che/74)
Glucose [medium] 31.2%.   **Fructose** [high] 45.9%.   **Sucrose** [low] 0.3%.
Reducing sugars 77.1%
Ash [medium] 0.30%
pH 4.24 (Che/74);   4.7 (Woo/78).   **Total acid** 26.6 meq/kg (Che/74).   **Free
acid** [medium] 20.1 meq/kg.   **Lactone** 6.5 meq/kg
Amylase 22.   **Inhibine number** 2 (Woo/78).   **HMF** 3.1 ppm (Che/74)
Nitrogen 0.025-0.043%

Honey: physical and other properties
Colour orange in fresh combs (Guy/72).   **Pfund** amber (And/73);   49-65 mm,
extra light to light amber (Bla/72; Roc/68);   52 mm, light amber
(Che/74);   medium amber (Cra/75);   dark amber (Guy/72)
Other physical properties stringy/ropy (And/73; Guy/72);   froths very much
on heating (Bla/72; Guy/72)
Granulation slow, coarse (And/73; Dai/70; Guy/72);   brown grain (Bla/72)
Flavour strong (And/73; Cra/75; Guy/72)

176 Eucalyptus melliodora A. Cunn. ex Schauer; Myrtaceae

yellow box, yellow jacket (En/AUS, SOU)
Tree, <25 m, drooping habit;   fl stamens white, rarely pink
Distribution subtropical Oceania, Africa, Asia;   tropical Africa;   native
to AUS.   **Habitat** gentle slopes/foothills in well watered E AUS;   river
flats in drier W AUS;   always <900 m altitude;   high veld SOU
Soil wide range;   heavy alluvial soil preferred, but also sandy loam,
granites;   not poor sand.   **Temperature** fairly frost resistant.   **Rainfall**
mostly within 380-760 mm;   very drought resistant

Economic and other uses
Fodder. **Fuel.** **Timber.** **Land use** windbreak, shade, amenity

Alert to beekeepers
Pollen inadequate for brood rearing (Gom/73; Loo/83; Pen/61)

Nectar rating + honeybee species; blooms, nectar flow
N1 AUS/NSW(Cok/63; Goo/47); AUS/QD(Bla/72); AUS/VIC(Gom/73);
KEN[tm](Tow/69); SOU,tm(And/73; Bey/68; Loo/70; Mou/72)
N URS(Glu/55); ZIM[tm](Pap/69)
Blooms x–xii (SOU); ix–ii (sometimes vi), heavily every 2 yrs, buds 10–12
mths before flowering (AUS, Pen/61). **Nectar flow** maintained in dry
periods, often copious (Pen/61); heaviest in warm moist conditions, 6 wks
duration (SOU, Mou/75)

Honey flow
Honey yield [high] 25, max 75 kg/colony/season (AUS, Mou/75)

Pollen
Alert to beekeepers pollen inadequate for brood rearing (Gom/73; Loo/83;
Pen/61))

Recommended for planting to increase honey production
AUS (Aus/83); SOU (Mou/75). Propagate by seed; blooms when little other
nectar is available. See **Alert to beekeepers**

Honey: chemical composition
Water [low] 14.0% (Che/74; also gives data for 2 samples age 5–12 mths);
17.2% (Woo/76; Woo/76a)
Sugars, total 79.5% (77.6% after 44 days at 50°, Woo/76a). **Glucose**
[medium] 30.4% (Che/74); 33.3% (28.0%, Woo/76a). **Fructose** [medium] 42.9%
(Che/74); 36.1% (38.2%, Woo/76a). **Sucrose** [medium] 5.1% (Che/74); 1.2%
(0.7%, Woo/76a). **Reducing sugars** 73.3% (Che/74). **Maltose** 5.1% (4.2%,
Woo/76a). **Melezitose** 2.3% (2.0%). **Turanose** 0.0% (4.5%)
Ash [low] 0.06%
pH 4.10 (Che/74); 4.18 (4.05, Woo/76); 4.4 (Woo/78). **Total acid** meq/kg
17.2 (Che/74); 22.3 (21.5, Woo/76). **Free acid** (meq/kg) [medium] 12.2
(Che/74); 17.8 (16.5, Woo/76). **Lactone** (meq/kg) 5.0 (Che/74); 4.5 (5.0,
Woo/76)
Amylase 30 (Che/74); 26.1 (Edw/75). **HMF** 1.9 ppm
Nitrogen 0.025–0.043% (Che/74); 0.020% (Woo/76). **Amino acids,** free 579.8
uM/100g (237.9, Woo/76); contents of individual acids (proline 80% of
total, Woo/76a)
Volatile compounds, major: acetoin and ?hexenyl butyrate; also 6 other
compounds (Grd/79); 48 present, 13 named (Woo/78a)

Honey: physical and other properties
Colour pale straw if monofloral (And/73; Gom/73; Joh/75; Sou/63);
usually extra light (Cra/75). **Pfund** 12–45 mm, extra white to extra light
amber (Bla/72; Lei/72; Roc/68); 21 mm, white (26 mm in sample age 5–12
mths, Che/74); 26.9 mm, white (115.2 mm, dark amber after 44 days at 50°,
Woo/76)
Granulation slow, none if monofloral (Bla/72; Joh/75; Mou/75)
Flavour sweet, cloying, pronounced (And/73; Bla/72; Cra/75; Mou/75).
Aroma characteristic (And/73); v aromatic (Sou/63)

177 Eucalyptus moluccana Roxb.; Myrtaceae

brown box, grey box, grey iron-box, gum-topped box, white box (En/AUS)
Tree, <15 m; fl stamens white
Distribution subtropical Oceania; native to AUS. **Habitat** forest country;
coastal (AUS/QD)
Soil clay

Economic and other uses
Fuel.   Timber

Alert to beekeepers
Pollen often inadequate for brood rearing (Bla/72)

Nectar rating; blooms, nectar flow
N1 AUS/QD(Bla/72)
Blooms ii-iv (AUS/QD)

Pollen
Alert to beekeepers pollen often inadequate for brood rearing (Bla/72)

Honey: chemical composition
Fermentation likely, unless fully ripe before extraction (Bla/72)

Honey: physical and other properties
Colour darkens on storage (Bla/72).   **Pfund** 35-75 mm, extra light to medium
amber (Bla/72; Roc/68)
Granulation rapid, colour then greyish (Bla/72)

178 Eucalyptus oleosa F. Muell ex Miq.; Myrtaceae

acorn mallee, giant mallee, oil mallee, red mallee (En/AUS)
Tree/mallee, <10 m
Distribution subtropical Oceania; native to southern AUS
Soil grey and brown calcareous soils; sometimes on sand (AUS).
Temperature frost resistant.   **Rainfall** 300-500 mm average (AUS); v
drought tolerant

Economic and other uses
Fuel.   Timber.   Land use windbreaks in low rainfall areas; shade.
Other uses essential oils

Nectar rating; blooms, nectar flow
N1 AUS/VIC(Gom/73)
N2 AUS/NSW(Goo/47)
N3 AUS/SA(Boo/72)
Blooms v-vii every 2 yrs, buds 12 mths before flowering (AUS, Pen/61); xii-
v usually each year (AUS/VIC)

Honey flow
Honey yield ⎨high⎬ 54 kg/colony/season (AUS/VIC, Gom/73); usually good
every 5 or more yrs, smaller yields between (AUS/SA, Boo 72)

Pollen
P2 AUS/NSW.   **P3** AUS/SA, VIC.   **Yield** low to moderate, annual (Boo/72);
average (Gom/73).   **Pollen value** average (Boo/72);   ?poor (Gom/73).
Colour cream

Honey: physical and other properties
Pfund light amber (Boo/72; Pur/68);   medium amber, sometimes darker
(Gom/73; Lei/72)
Flavour mild (Gom/73; Lei/72)

179 Eucalyptus panda S.T. Blake; Myrtaceae

Brogan's ironbark, corky ironbark, tumble-down ironbark (En/AUS)
Tree, <18 m, often crooked and stunted; fl stamens white
Distribution subtropical Oceania; native to AUS.   **Habitat** sandstone areas
(AUS/QD)
Soil sandy

Nectar rating; blooms, nectar flow
N1 AUS/QD(Bla/72)
Blooms iv-x (AUS/QD)

Honey flow
Honey yield every 5 yrs (AUS/QD, Bla/72)

Pollen
P2 AUS/QD.   **Yield** high, useful to bees especially in yrs when tree blooms
early (Bla/72)

Honey: physical and other properties
Pfund extra light to light amber (Bla/72);   light amber (Lei/72)
Granulation slow (Bla/72)
Flavour mild, sweet

180 Eucalyptus paniculata Smith; Myrtaceae

grey ironbark (En/AUS)
Tree, <42 m; fl stamens white/creamy-yellow
Distribution subtropical Oceania, Africa, S America; tropical Africa;
native to AUS/NSW, QD.   **Habitat** moist valleys preferred, also ironstone
ridges (AUS); coastal areas and humid interior SOU
Soil good sandy loam; best in deep well drained soil but will adapt to
poor, dry sites (SOU); not poor sand; salt tolerant.   **Temperature** frost
tender.   **Rainfall** somewhat drought resistant but growth more rapid in rain-
fed areas; >750 mm preferred (SOU)

Economic and other uses
Timber termite resistant.   **Land use** windbreak, amenity

Alert to beekeepers
Pollen can be inadequate for brood rearing (AUS, Pen/61; SOU, Sou/63);
honey granulates in hive if cold (Coe/71)

Nectar rating + honeybee species; blooms, nectar flow; composition
N1 AUS/NSW(Cok/63; Goo/47); KEN,tm(Smt/60; Tow/69); SOU,tm(Mou/72)
N2 SOU/CAPE,NATAL,TVL,tm(And/73)
Blooms v-xii usually every 3 yrs (AUS, Pen/61); iii-iv (BRA/RS, Jul/72);
iv-vi, regular (SOU, Sou/63); during dearth period (SOU, Coe/67). **Nectar
flow** irregular and difficult to predict (SOU, Coe/67). **Nectar secretion**
most copious in dry periods after good rains during bud development (SOU,
Guy/71). **Sugar concentration** [medium] 28-30% (Jul/72)

Honey flow
Honey yield (kg/colony/season) [high] 100 (AUS, Pen/61); 50 (SOU, Guy/71)

Pollen
P3 SOU/CAPE, NATAL, TVL. **Alert to beekeepers** pollen can be inadequate for
brood rearing (AUS, Pen/61; SOU, Sou/63)

Recommended for planting to increase honey production
SOU (Dai/70). See **Alert to beekeepers**

Honey: physical and other properties
Colour light (And/73; Cra/75); pale straw (Coe/71; Guy/71)
Granulation slow or medium (And/73); slow, fine (Coe/71; Cra/75;
Guy/71); **(alert to beekeepers)** granulates in hive if cold (Coe/71)

181 Eucalyptus platypus Hook.; Myrtaceae

moort (En/AUS)
Tree, <9 m; fl stamens yellow
Distribution subtropical Oceania; native to AUS/WA. **Habitat** moist
depressions on low hills and flats (AUS); southern mallee areas
Soil sandy loam; heavy, grey, clayey soil; moist loam. **Temperature**
moderately frost resistant. **Rainfall** 350 mm, but 400-700 mm with
predominance in winter preferred; drought resistant

Economic and other uses
Timber. **Land use** windbreak (low shelter), amenity. **Other uses** tannin
from bark

Nectar rating; blooms, nectar flow
N1 AUS/WA(Col/62; Lei/72)
Blooms vi-x every 4 yrs, varies with climatic conditions; fl period rarely
>2 mths (Pen/61). **Nectar flow** starts and finishes abruptly (Col/62)

Pollen
P1 AUS/WA. **Yield** heavy (Pen/61). **Pollen value** excellent (Pen/61)

Recommended for planting to increase honey production
AUS (Aus/83). Excellent low shelter

Honey no data

182 Eucalyptus polyanthemos Schauer; Myrtaceae

red box (En/AUS, SOU)
Tree, <23-30 m; poor form
Distribution subtropical Africa; temperate Oceania, (Med) Africa; native
to AUS/NSW, VIC.   **Habitat** in mountainous areas, only in sheltered valleys
<600 m (AUS); woodland; drier areas of SOU
Soil poor, dry, stony/gravelly and poor-class heavy soils (AUS).
Temperature moderately frost tolerant.   **Rainfall** moderately drought tolerant

Economic and other uses
Fuel.   Timber.   Land use windbreak, shade

Alert to beekeepers
Pollen inadequate for brood rearing (Gom/73; Loo/83); honey "difficult to
extract" (Gom/73)

Nectar rating + honeybee species; blooms, nectar flow; composition
N1 SOU/TVL[tm](Bey/68)
N2 ?SOU/TVL[tm](Loo/70)
N3 AUS/NSW(Goo/47);   AUS/VIC(Gom/73);   SOU/TVL,tm(And/73)
Blooms ix-xii (AUS, Pen/61); usually every 2 yrs, buds 10-12 mths before
flowering (Gom/73).   **Nectar flow** xii, ii (ISR, Eis/80); unreliable (AUS,
Gom/73).   **Nectar secretion** 0.4 mg/fl/day (Eis/80).   **Sugar concentration**
[low] 18.7% (Eis/80).   **Sugar value** [low] 0.08 mg/fl/day (Eis/80)

Pollen
P3 SOU/TVL.   **Alert to beekeepers** pollen inadequate for brood rearing
(Gom/73;   Loo/83)

Recommended for planting to increase honey production
SOU (Loo/83;   Sou/63).   Grows rather slowly (Pen/61).   See **Alert to
beekeepers**

Honey: physical and other properties
Colour pale, dull (And/73;   Gom/73;   Lei/72)
Viscosity (alert to beekeepers) honey "difficult to extract" (Gom/73)
Granulation slow, none if monofloral (And/73;   Gom/73;   Sou/63)
Flavour usually slightly oily or like tallow, but this disappears after 12
mths (Gom/73);   oily (Lei/72;   Loo/70)

183 Eucalyptus propinqua Deane & Maiden; Myrtaceae

grey gum, grey irongum, small-fruited grey gum (En/AUS)
Tree, <45 m, never in pure stands
Distribution temperate and subtropical Oceania; native to AUS.   **Habitat**
coastal slopes <300 m altitude; forest/hill country (AUS/QD)

Soil wide range but moist, well drained better-class soils preferred; not heavy clay or poorest sandy soil

Economic and other uses
Timber

Nectar rating; blooms, nectar flow
N1 AUS/NSW(Goo/47)
N3 AUS/QD(Bla/72)
Blooms xii-iv every 3 yrs (AUS, Pen/61); i-iii (AUS/QD)

Pollen
P1 AUS/NSW.   **P2** AUS/QD.   **Yield** fair (Pen/61).   **Colour** ?cream (Bla/72)

Honey: physical properties
Colour reddish tint (Goo/47).   **Pfund** extra light to medium amber (Bla/72); medium amber (Pen/61)

184 Eucalyptus robusta Smith; Myrtaceae

swamp mahogany, swamp messmate (En/AUS); robusta gum (En/SOU);   eucalyptus rouge (Fr/MAE, MAY); eucalipto (Pt/BRA)
Tree, <27 m; fl stamens white
Distribution subtropical Africa, S America, Oceania, Asia; tropical Africa, Asia, S America; temperate Europe; native to coastal AUS/NSW, QD.   **Habitat** cultivated in warmer Americas; coast, saltwater flats/edges of saltwater lagoons, low-lying forest country (AUS); thrives in coastal/inland districts (SOU); mist belt of midlands to coast (SOU/NATAL)
Soil wide range; sandy soil; waterlogged soil; salt tolerant; swampy ground in subsaline areas (EUR); some flooding tolerated (AUS).
Temperature frost-tender.   **Rainfall** high rainfall areas (AUS); summer rainfall areas (SOU)

Economic and other uses
Timber.   **Land use** windbreak, shade, afforestation, amenity

Nectar rating + honeybee species; blooms, nectar flow; composition
N1 BRA[tm](Smt/60; Wie/80); MAE(Chl/75); MAY(Bro/82; Cra/73); URS(Glu/55)
N3 AUS/NSW(Goo/47); AUS/QD(Bla/72); SOU/CAPE,NATAL,TVL,tm(And/73)
Blooms ix-xi (subtropical AUS, v temperate AUS, Pen/61); vi-vii (AUS/QD); iii-vii (BRA).   **Nectar flow** reliable (SOU, Dai/70).   **Sugar concentration** [medium] 51% (Caa/72); 32-37% (Jul/72); 38-45% (Wie/80).   **Sugar analysis** (Maz/59)

Pollen
P3 AUS/QD; SOU/CAPE, NATAL, TVL.   **P** BRA.   **Yield** small, but reliable in winter (AUS, Pen/61); fair (SOU, Loo/83); good (Aus/83).   **Pollen grain** described (Bah/73).   **Reference slide**

Recommended for planting to increase honey production
AUS (Aus/83); BRA (Wie/80); SOU (Dai/70; Sou/63)

Honey: chemical composition
Water [medium] 17.0, 17.5% (Fle/63)
Ash [medium] 0.211, 0.201%
pH 4.2, 4.3

Honey: physical properties
Pfund medium amber (Bla/72); dark amber (Cra/75); 85.1, 80.1 mm, amber
(Fle/63)

185 Eucalyptus rubida Deane & Maiden; Myrtaceae

candle bark gum (En/AUS, SOU)
Tree, <30 m
Distribution subtropical Africa, Oceania; native to AUS/NSW, TAS.
Habitat <1200 m altitude (AUS); sheltered valleys or occasionally on
slopes/ridges
Soil moist alluvial flats; slate, igneous soils; sandy and black turf
soil (SOU); deeper dry soils. **Temperature** warm moist situations
preferred; frost and cold wind tolerated. **Rainfall** 750-1000 mm (AUS/SA);
summer rainfall area, sub-humid interior (SOU); only moderately drought
tolerant

Economic and other uses
Fuel. **Timber**, but not durable. **Land use** windbreaks (useful for cold
areas); amenity. **Other uses** pulp for paper

Warning
Frequently attacked by leaf-eating beetles (AUS, Key/77)

Nectar rating + honeybee species; blooms, nectar flow
N1 SOU/TVL[tm](Bey/68)
N2 AUS/SA(Boo/72)
N3 AUS/VIC(Gom/73); SOU/NATAL,OFS,TVL,tm(And/73)
Blooms i-ii, buds 12-15 mths before flowering (AUS, Pen/61); summer (SOU,
Dai/70). **Nectar flow** heavy every 2 yrs (AUS/SOU, Boo/72)

Pollen
P2 AUS/SA. **P3** AUS/VIC; SOU/NATAL, OFS, TVL. **P** SOU/TVL. **Yield**
moderate, annual (Boo/72). **Pollen value** fair (Pen/61); moderate (Boo/72)

Recommended for planting to increase honey production
SOU (Dai/70; Loo/82; Sou/63). See **Warning**

Honey: physical properties
Colour dark (And/73); clear (Gom/73). **Pfund** amber (Gom/73)

186 Eucalyptus saligna Smith; Myrtaceae

Sydney blue gum (En/AUS)

Tree, <45 m; often confused with E. saligna, only data for BRA and AUS included here, data for SOU entered under E. grandis (Sou/77)
Distribution subtropical S America, Oceania; tropical C America; temperate (Med) Africa; native to AUS. **Habitat** valleys <1200 m altitude; widely cultivated for timber
Soil deep, moist, rich, well drained; heavier shale preferred (AUS); poor soils (BRA). **Temperature** - moderately frost resistant. **Rainfall** high (AUS)

Economic and other uses
Fuel. Timber. Land use windbreak, shade, afforestation, amenity.
Other uses pulp for paper, rayon

Nectar rating + honeybee species; blooms, nectar flow; composition
N1 BRA/SP,tm(Mun/54; Smt/60; Wie/80)
N3 AUS/NSW(Goo/47)
Blooms i-iii every 2 yrs (AUS, Pen/61); iii-ix (BRA). **Sugar concentration** [medium] 48% (Caa/72); 35-40% (Wie/80)

Honey flow
Honey yield [high] 100 kg/colony/season (BRA, Wie/80)

Pollen
P BRA/SP. **Yield** good (Aus/83); "reasonable" (Pen/61)

Recommended for planting to increase honey production
AUS (Aus/83); BRA (Wie/80). Propagate by seed, which is freely produced; grows rapidly

Honey no data

187 Eucalyptus sideroxylon A. Cunn. ex Woolls; Myrtaceae

black ironbark (En/AUS, SOU); mugga, red ironbark (En/AUS)
Tree, <30 m, only 18 m in dry areas; fl stamens pink/white
Distribution subtropical Africa, Asia, Oceania, S America; temperate (Med) Africa, Oceania; tropical Africa; native to AUS/NSW, QD, VIC. **Habitat** high rainfall coastal areas; also on ridges, plains and undulating hill country (AUS); open forest hillsides (AUS/QD); sub-humid zone (SOU)
Soil poor shallow soils including clay, gravel and sand (AUS).
Temperature fairly frost tolerant; high summer max >100°. **Rainfall** low to medium (AUS); drought tolerant (SOU)

Economic and other uses
Fuel. Timber. Land use windbreak, shade, amenity. **Other uses** oil; tannin

Alert to beekeepers
Pollen inadequate for brood rearing; colony populations decrease especially after heavy flow (AUS, Pen/61)

Nectar rating + honeybee species; blooms, nectar flow
N1 AUS/NSW(Cok/63); AUS/QD(Bla/72); AUS/VIC(Gom/73); SOU/TVL,tm(And/73)
N2 AUS/NSW(Goo/47); PAK,ac(Pak/77)
Blooms v–vii (AUS, Pen/61); buds 5–6 mths before flowering; winter
(Gom/73); vii–x (AUS/QD); autumn–spring, profuse (SOU, Dai/70). **Nectar
secretion** plentiful (Gom/73); so profuse that nectar can be shaken out by
hand (Loo/83)

Pollen
P3 AUS/QD; PAK. **Alert to beekeepers** pollen inadequate for brood
rearing; colony populations decrease especially after heavy flow (AUS,
Pen/61)

Recommended for planting to increase honey production
AUS (Aus/83); SOU, sub-humid zone (Dai/70). Propagate by seed, which is
freely produced but often of low viability (Pen/61). Seed provenance
important (Pen/61). See **Alert to beekeepers**

Honey: chemical composition
pH 4.6 (Lan/66)

Honey: physical and other properties
Colour light straw (And/73); pale (Gom/73). **Pfund** 41 mm, extra light
amber (Bla/72; Lei/72; Roc/68); 47.4 mm, extra light amber (bulk honey,
Lan/66)
Viscosity "good body" (Bla/72)
Granulation rapid, fine (And/73; Bla/72; Gom/73)
Flavour mild (Gom/73)

188 Eucalyptus socialis F. Muell. ex Miq.; Myrtaceae

Christmas mallee (En/AUS)
Tree, 7–20 m, mallee
Distribution temperate and subtropical Oceania; native to AUS/VIC, SOU.
Habitat mallee plains and adjacent interior (AUS/SOU)
Soil red sandy loam (AUS). **Rainfall** dry areas best (AUS)

Economic and other uses
Fuel. Timber. Land use windbreak, shade

Nectar rating; blooms, nectar flow
N1 AUS/VIC(Gom/73)
Blooms ix–xii or a little later, reported to last 8 wks (AUS/VIC). **Nectar
flow** can be v heavy (Gom/73)

Honey flow
Honey yield [high] 27, max 54 kg/colony/season (AUS/VIC, Gom/73); good
every 4 yrs or so, smaller in other yrs (AUS/SOU, Boo /72)

Pollen
P2 AUS/VIC. **Yield** good (Gom/73); high every 4 yrs (Boo/72). **Pollen
value** good (Boo/72). **Colour** cream (Gom/73)

Recommended for planting to increase honey production
AUS (Aus/83)

Honey: physical and other properties
Pfund light amber (Boo/72); medium amber (Gom/73)
Flavour "not altogether objectionable" (Gom/73)

189 Eucalyptus tereticornis Smith; Myrtaceae

blue gum, forest red gum, Queensland blue gum, red gum, red iron gum
(En/AUS); slaty gum (En/SOU); forest gray gum (En/USA): eucalyptus blanc
(Fr/MAY)
Tree, <45 m; fl stamens white
Distribution tropical Africa, Oceania, S America; subtropical S America,
Africa, Asia, Oceania; temperate (Med) Europe, Oceania; native to eastern
AUS and Papua. **Habitat** coastal/subcoastal districts, alluvial flats,
stream banks (AUS)
Soil wide range if not too dry/shallow/acid (AUS); alluvium (AUS/QD);
deep moist soil, not acid (AUS/VIC); marshy ground (BRA); some
waterlogging tolerated. **Temperature** wide range

Economic and other uses
Timber. **Land use** windbreak, shade, afforestation. **Other uses** paper from
pulp

Nectar rating + honeybee species; blooms, nectar flow; composition
N1 ?BRA/RG[tm](Jul/72); BRA/SP(Mun/54; Smt/60); MAY(Cra/73)
N2 PAK,ac(Pak/77)
N3 AUS/QD(Bla/72); AUS/VIC(Gom/73); SOU,tm(And/73)
N BUM(Zma/80)
Blooms viii-x (AUS, Pen/61); spring-early summer (AUS/VIC); annually but
heavy flowering every 3-4 yrs (AUS/QD, Bla/72); iv-ix (BRA, Jul/72;
Sao/54; Wie/80); ix-x (BUM, Zma/80). **Sugar concentration** [high] 70-80%
(Sao/54); 30-40%, max 70-79% (Wie/80); [medium] 45% (Zma/80)

Honey flow
Honey yield [moderate] 18 kg/colony/season (AUS/QD, Bla/72)

Pollen
P1 AUS/QD. **P2** AUS/VIC; SOU. **P** PAK. **Yield** none (SOU, Orm/63).
Pollen value high in late winter/early spring, when it encourages brood
rearing (AUS, Pen/61)

Recommended for planting to increase honey production
BRA (Wie/80)

Honey: chemical composition
pH 4.2 (Lan/66)
Amylase 23.8

Honey: physical and other properties
Colour extra light straw (And/73);   dark (Cra/75).   **Pfund** light amber
(Bla/72);   44.9 mm, extra light amber (67.7 mm, light amber after 16 h at
77°, Lan/66)
Viscosity "good body" (And/73)
Granulation rapid, medium grain
Flavour characteristic, like caramel (Bla/72);   strong (Cra/75)

190 Eucalyptus viminalis Labill.; Myrtaceae

blue gum, manna gum, ribbon gum, rough barked ribbon gum, white gum (En/AUS)
Tree, <45 m; fl stamens white
Distribution subtropical Africa, S America, N America, Oceania;   temperate
(Med) Africa, Oceania;   native to AUS/SA, TAS.   **Habitat** widely distributed
in AUS;   valleys <1300 m;   watercourses, dry open country and mountain
forests (AUS/VIC);   cool sheltered valleys, protected slopes (AUS/SA)
Soil deep basaltic soil preferred, not poor sand (AUS);   deep sandy loam on
river banks preferred (SOU).   **Temperature** cooler parts of AUS/VIC;   frost-
hardy but scorched by severe cold (SOU);   v frost resistant (to -20°)
(BRA).   **Rainfall** 635-900 mm; summer rainfall area (SOU);   not drought
tolerant on shallow dry soil (SOU)

Economic and other uses
Fuel.   Timber, low grade.   **Land use** windbreak, shade, afforestation,
amenity.   **Other uses** paper from pulp

Warning
Susceptible to attack by snout beetles in v cold areas of SOU (Loo/82)

Nectar rating + honeybee species;   blooms, nectar flow
N1 SOU/TVL,tm(Bey/68)
N2 AUS/NSW(Goo/47);   AUS/QD(Bla/72);   AUS/SA(Boo/72);   AUS/VIC(Gom/73)
N NEZ(Wal/78)
Blooms ii-iv (AUS, Pen/61);   xii-iv with peak ii-iii (AUS/VIC);   buds hang
for considerable time before opening (AUS/QD);   i-ii (BRA/RG, Jul/72)

Honey flow
Honey yield annual, poor to good (AUS/SOU, Boo/72)

Pollen
P1 AUS/NSW.   **P2** AUS/QD, SA, VIC.   **P** NEZ;   SOU/TVL.   **Yield** - freely
produced (Pen/61);   medium to abundant (Gom/73);   low to moderate
(Boo/72).   **Pollen value** good, provides winter stores for stimulating brood
rearing (Pen/61);   good quality (Boo/72).   **Colour** whitish (Bla/72)

Recommended for planting to increase honey production
BRA (Wie/80);   SOU (Dai/70;   Loo/82;   Sou/63).   Grows rapidly;   v frost
resistant (to -20°).   See **Warning**

Honey: chemical composition
"Contains considerable air" (Wal/78)

Honey: physical and other properties
Colour clear (Gom/73; Sou/63). **Pfund** pale to medium amber (Bla/72);
light amber (Boo/72); medium amber (Lei/72; Pur/68)
Granulation slow, coarse brownish grain (Bla/72); rapid (Cra/75; Gom/73;
Pen/61)
Flavour distinctive, sweet (Gom/73; Sou/63)

191 Eucalyptus wandoo Blakely; Myrtaceae
syn Eucalyptus redunca Schauer var. elata Benth.

wandoo, white gum (En/AUS)
Tree, 20–30 m
Distribution subtropical Oceania; temperate (Med) Africa; native to
AUS/WA. **Habitat** <300 m altitude (AUS)
Soil granite soil with clay (AUS); brown/sandy loam, gravelly soil
(AUS). **Rainfall** drought tolerant

Economic and other uses
Fuel. Timber. Land use shade, amenity. **Other uses** tannin from bark
and wood

Nectar rating; blooms, nectar flow
N1 AUS/WA(Col/62)
N AUS/WA(Smt/69)
Blooms winter in N of range and summer in S, buds 2–3 yrs before flowering,
but flowers annually (AUS, Pen/61)

Honey flow
Honey yield [high] 90 kg/colony/season (AUS, Pen/61)

Pollen
P3 AUS/WA. **Yield** abundant (Pen/61). **Pollen value** poor (Pen/61)

Honey: physical and other properties
Pfund light amber to amber (Cra/75); extra light to light amber (Lei/72;
Smt/69); 36 mm, extra light amber (Smt/67)
Granulation medium grain, and colour then light cream (Cra/75; Pen/61)
Flavour mild (Cra/75; Lei/72; Smt/69)

192 Eucryphia cordifolia Cav.; Eucryphiaceae

ulmo (Es/CHL)
Tree, <24 m, evergreen; fls white, fragrant
Distribution temperate S America; native to and found only in CHL.
Habitat permanent rain forest (CHL)
Soil moist; some lime tolerated. **Temperature** damaged by frost, and
woodland protection is helpful

Economic and other uses
Timber.   **Land use** amenity.   **Other uses** tannin from bark

Nectar rating; blooms, nectar flow; composition
Nl CHL(Car/38; Fis/38; Kar/56; Kar/60)
N CHL(R.B/38)
Blooms ii-v.   **Nectar flow** used less in yrs with good flow from Weinmannia sp

Pollen
P CHL.   **Pollen grain** illustrated and described (Heu/71)

Honey: physical and other properties
Colour pale pink (Car/38)
Granulation, colour then almost white
Aroma characteristic

193 Eucryphia lucida (Labill.) Baill.; Eucryphiaceae

leatherwood (En/AUS)
Tree, 7-17 occasionally up to 30 m, evergreen, resinous shoots and young
lvs; fls white, fragrant
Distribution temperate Oceania; native to and found only in W half of
AUS/TAS.   **Habitat** rain forests; part of understorey in some eucalypt
forests of higher rainfall areas; sea level to 800 m
Soil some acidity preferred

Economic and other uses
Timber.   **Land use** amenity

Nectar rating; blooms, nectar flow
Nl AUS/TAS(Par/77; Rap/69)
Blooms summer (AUS/TAS).   **Nectar flow** 6 wks but can be as short as 10 days
(Par/77)

Honey flow
Honey yield [high] 36 kg/colony/season, Eucryphia spp (AUS/TAS, Rap/69);
main honey-producing tree of AUS/TAS (Par/77)

Recommended for planting to increase honey production
AUS/TAS (Par/77).   Propagate by seed; grows slowly

Honey: physical and other properties
Pfund light amber (Rap/69)
Flavour distinctive, pronounced when liquid, but less so when granulated.
Aroma quite strong

194 Eucryphia milligannii Hook.f.; Eucryphiaceae

leatherwood (En/AUS)
Shrub, <4 m; fls white

Distribution temperate Oceania; native to and found only in AUS/TAS.
Habitat rain forests at higher altitudes than other AUS/TAS Eucryphia spp
Soil moist, slightly acid; no free lime; poorer soils. **Temperature**
frost tender

Nectar rating; blooms, nectar flow
N1 AUS/TAS(Par/77)
Blooms summer, 3 wks later than E. lucida

Honey flow
Honey yield [high] 36 kg/colony/season, Eucryphia spp (AUS/TAS, Rap/69)

Recommended for planting to increase honey production
AUS/TAS (Par/77). Propagate by cuttings/layers; grows slowly

Honey: physical and other properties
Colour lighter than E. lucida honey (Par/77)
Flavour milder than E. lucida honey

195 Eugenia spicata Lamk.; Myrtaceae

bhedas (INI)
Tree, <5 m; fls white
Distribution tropical Asia. **Habitat** moist deciduous forest; open forest
and on outskirts of evergreen forests (INI/KAR)

Nectar rating + honeybee species; blooms, nectar flow; composition
N1 INI/KAR,ac,ad,Trigona(Diw/64)
Blooms ii-iv (INI/KAR)

Honey no data

196 Eugenia zeylanica Wight; Myrtaceae

gudda panneralu (INI)
Tree; fls white
Distribution tropical Asia

Nectar rating + honeybee species; blooms, nectar flow
N1 INI/KAR,KER[ac](Kha/59)
Blooms ii-iv (INI/KAR, KER)

Honey no data

197 Eupatorium odoratum L.;   Compositae

white snake root;   bizat (BUM);   sap-sua (THA)
Herb
Distribution tropical Asia, ?Caribbean;   native to Thailand.   **Habitat**
common weed;   forest areas of lower Burma

Alert to beekeepers
Bees do not collect pollen from this plant (Zma/80)

Nectar rating + honeybee species;   blooms, nectar flow;   composition
N1 BUM(Zma/80);   THA,ac(Bur/80)
Blooms xi-i (BUM);   xi-i (THA).   **Sugar concentration** in nectar [medium] up
to 50% (Cra/84)

Pollen
Alert to beekeepers   bees do not collect pollen from this plant (BUM, Zma/80)

Honey: chemical composition
Water [medium] 19-20% (Cra/84)

Honey: physical and other properties
Pfund amber (Cra/84);   light amber (Zma/80)
Granulation slow, fine (Cra/84;   Zma/80)
Flavour mild (Cra/84).   **Aroma** slight

198 Euphoria longan (Lour.) Steud.;   Sapindaceae

longan
Tree
Distribution tropical Asia.   **Habitat** cultivated crop plant

Economic and other uses
Food - fruit

Nectar rating + honeybee species;   blooms, nectar flow
N1 THA[ac](Bur/80)
Blooms ii-iii (THA).   **Nectar flow** 6 wks (THA, Cra/84)

Honey flow
Honey yield [high] mean 40, max 45-50 kg/colony/season (THA, Cra/84)

Honey: chemical composition
Water [medium] 18% (THA, Cra/84);   [high] 21.4-23.2% (4 samples, THA, Lin/77)
Glucose [medium] 29.9-35.2%.   **Fructose** [medium] 38.6-39.4%.   **Sucrose**
[low] 0.1-0.8%
Ash [medium] 0.09%-0.25%
pH 4.4-4.7.   **Free acid** [medium] 16.1-17.7 meq/kg (19.0-24.8 after 1 yr
storage)

Honey: physical properties
Colour of 4 samples: orange, deep brown, red brown, earth yellow
(Lin/77).   **Pfund** dark amber (Cra/84)

Relative density 1.38-1.40 (3 samples, Lin/77)
Granulation [slow] rare (Cra/84)
Flavour characteristic.   **Aroma** v fragrant

199 Fagopyrum esculentum Moench.;   Polygonaceae
syn Polygonum fagopyrum L.

buckwheat; blé noir, sarrasin (Fr); trigo mourisco, trigo sarraceno
(Pt/MOZ); grano saraceno (It)
Herb, 15-60 cm, annual; fl sepals white, fragrant; 2 fl forms: long
stamens and short style, short stamens and long style
Distribution temperate Europe, Asia, N America; subtropical Africa;
native to Asia.   **Habitat** cultivated crop plant in Europe, URS, USA;
elsewhere an occasional naturalized ruderal
Soil sandy or other light soil, well drained soil preferred; low-
fertility, abandoned land being brought back into production, also v acid
soil; not heavy clay.   **Temperature** cool, but damaged by frost; not hot
climate.   **Rainfall** moist climate preferred

Economic and other uses
Food - flour from grain; grain for breakfast-food.   **Fodder** - grain for
poultry and pigs.   **Soil benefit** green manure.   **Other uses** game coverts;
root protection for saplings

Nectar rating + honeybee species; blooms, nectar flow; composition
N1 ?AUT(Pla/52); CAF/MAN(Mit/49); CAF/ONT(Tow/76); CAF/QUE(Cha/48);
CHN(Mad/81); INI/KAS,ac(Sha/78); IRN(Cra/73); MOZ[tm](Cra/73); POL(Dem/-
64a); SOU/OFS,TVL,tm(And/73); URS(Ave/78; Fed/55; Kov/65); USA/?IA,MI,-
NY,OH,PA(Pel/76); USA/MD(Die/71)
N2 CAF/QUE(Cou/59); INI/KAS,ac(Sar/73); JAP(Inu/57; Sak/82);
SOU,tm(Mou/72); URS(Dan/75); USA/CT,KY,?OK,WI(Pel/76); USA/MD(Les/54);
YUG(Kul/59)
N3 ?CAF/NB(Pel/76); ITA(Ric/78); ?UK(How/79); USA/MA(Shw/50a)
N ALG(Ske/72); CZE(Svo/58); USA/WV(Pel/76); ZIM,tm(Pap/73)
Blooms vii-ix (INI/KAS); vi-vii (EUR, Sta/74).   **Nectar flow** copious, 2-3
wks (Cri/57); flow mainly in mornings, and bees become "angry" in
afternoons when it ceases (Lov/77).   **Nectar secretion** (mg/fl/day) 0.2-0.4
(Han/80); 0.87, 1.27 (5-yr means, spring crop, second crop, AA275/62);
secretion higher in plants dusted with boric acid (AA816/63); moisture,
and lime/phosphorus/nitrogen in soil improve sugar concentration by 20-
50%; on fertile soil with adequate moisture secretion maintained at low
temperatures; 70% of total sugar secreted during 1st half of flowering
(Frj/70); cool nights and mean temperature <70° best (How/79); 16-26°
optimum (Maz/82).   **Sugar concentration** [medium] 35-45% (Frj/70); 7-45%
(Han/80; Maz/82); 35.0% (Shr/58); 41.0-48.5% (Shw/53).   **Sugar value**
(mg/fl/day) [medium] 0.10-2.68 (Cra/73); 0.03-0.17 (Han/80; Maz/82);
0.138, 0.176 (2 yrs, AA342/64); 0.04-0.16(AA628/74).   **Sugar analysis**
(Ech/72; Ech/77; Maz/82; Wyk/52; Zbo/68)

Honey flow

Honey yield (kg/colony/day) up to 4-6 (ROM, AA765/65); 3-7 (URS, Ave/78); 1.1, 1.7 (5-yr means for spring crop, late crop URS, AA275/62); 10 (USA, Pel/76). 90, 292 (2 yrs, POL, AA342/64); **Honey potential** (kg/ha) [moderate] 90, 292 (2 yrs, POL, AA342/64); 60-70 (ROM, Cir/77); 70-90 (URS, Ave/78); 60-70 (URS/Ukraine, Fed/55); <500 (Cra/75)

Pollen

P2 JAP. P3 ITA; SOU/OFS, TVL. P CAF/QUE; URS; ZIM. **Yield** little if any (Cri/57); average 0.17 mg/fl (Maz/82). **Pollen value** high but dried pollen may be toxic to bees (Sta/74). **Chemical analysis** (Shp/79). **Colour** of load brown (Ric/78); light-greyish yellow; average wt of load 5.2 mg (Maz/82). **Pollen grain** larger in short-styled fls than in long-styled fls (Frj/70); illustrated and described (Ert/63); [over-represented in honey] >100 000 grains per 10 g (Fagopyrum sp, Sta/74). **Reference slide**

Recommended for planting to increase honey production

UK (How/79); URS (Ave/78). Propagate by seed, sow in spring after danger of frost is past. Attract bees initially to crop by sowing a small % of an early flowering cv or sow some of field earlier than the rest (Lov/77). Successional sowings from May onwards recommended for UK (How/79)

Honey: chemical composition

Water [high] 20.5% (Aso/60); sometimes 33% (Roo/74); 20.42-22.14% (Ryc/65); 18.5-20.5% (Sha/79); 16.2% (1 sample, age 12 mths, Whi/62) **Glucose** [medium] 33.40% (Aso/60); 33.38% (Whi/62). **Fructose** [medium] 33.35% (Aso/60); 37.05% (Whi/62). **Sucrose** [medium] 2.46% (Aso/60); 0.57% (Whi/62). **Reducing sugars** 66.75% (Aso/60). Contents of these sugars also given as % of total sugars (Ryc/65). **Maltose** 5.69% (Whi/62). **Higher sugars** 1.18%
Ash [medium] 0.118%
pH 3.62-4.19 (Ryc/65); 3.98 (Whi/62). **Total acid** 54.23 meq/kg (Whi/62). **Free acid** [high] 46.29 meq/kg. **Lactone** 7.94 meq/kg. **Alkalinity number** 6.81-14.08 (Ryc/65)
Amylase, sucrase values, but method not stated (Ryc/65)
Nitrogen 0.124% (Whi/62). **Protein** 0.0095-0.0167% (17 samples, Gen/67).
Lipids identified (Pop/79a). **Colloids** 3.40-4.58 mm (Lund, Ryc/65); data for 3 colloidal constituents (Hel/53)
Vitamins C 41-82 ppm (Ryc/65)
Other constituents - rutin in 6 of 10 samples, quercetin in others; after 4 days quercetin (only) present in all (Zbo/68)

Honey: physical and other properties

Colour light to dark brown (Maz/82); all other reports say dark, with further descriptions: brown (Aso/60; Fos/56; Joh/75); reddish (Fed/55); purple or black (Lov/56); purplish when extracted (Roo/74); yellow with reddish tinge, to brown (Sha/70). **Pfund** >114 mm, dark amber (Lov/58; Whi/62); Pfund-Lovibond grade (Aub/83)
Viscosity "thick", difficult to extract (How/79); "heavy body" (Lov/56)
Granulation slow, coarse (And/73); rapid, "flakey" (Fos/56)
Flavour sharp (Fed/55); burnt (Fos/56); all other reports say strong, with further description: characteristic (Cra/75; How/79; Lov/58); somewhat bitter (Maz/82); unusual, slightly nauseating to some (Pel/76).
Aroma unpleasant (Cri/57); "aromatic" (Fed/55; Fos/56); v strong (Ske/72); strong, characteristic (Sol/63)

200 Faurea saligna Harv.; Proteaceae

beechwood; bushveld boekenhout, Transvaal boekenhout (Af); monyena (BOT)
Tree, 5-10 m, deciduous; fls greenish-creamy white, honey-scented
Distribution tropical Africa; native to southern and central Africa.
Habitat common in mixed woodland and wooded grassland at medium/high
altitudes; hillsides/river banks; sourveld (SOU)
Soil poor sandy or rocky acid soil

Economic and other uses
Timber.   **Other uses** dye and tannin from wood

Nectar rating + honeybee species; blooms, nectar flow; composition
N1 BOT[tm](Cra/73); SOU,tm(Mou/72); SOU/NATAL,TVL,tm(Bey/68)
N2 SOU/NATAL,TVL,tm(And/73)
Blooms ix-xii (southern Africa, Pag/77).   **Nectar flow** heavy; blooms
annually but does not always produce nectar (Guy/71a).   **Nectar secretion**
abundant (Sto/82)

Honey flow
Honey yield not annual; poor yields are often ascribed to veld fires
(And/73); yield every 5 yrs (Guy/71a)

Pollen
P3 SOU/NATAL, TVL

Honey: physical and other properties
Colour dark red-brown (And/73; Guy/71a)
Viscosity "high" (And/73)
Granulation slow
Flavour strong, malty (And/73); strong (Guy/71a).   **Aroma** rather musty
(And/73)

201 Fuchsia excorticata (J. & G. Forst.) L.f.; Onagraceae

konini, kotukutuku (NEZ)
Shrub/tree, 5-15 m, deciduous; fls purple and red
Distribution temperate Oceania; native to NEZ.   **Habitat** grows profusely
on the fringes of the bush on N and S Islands, NEZ

Economic and other uses
Land use amenity.   **Soil benefit** conservation; river banks

Nectar rating + honeybee species; blooms, nectar flow; composition
N1 NEZ (God/52)
N NEZ (Wal/78)
Blooms viii to ix or xii (NEZ).   **Nectar secretion** profuse (Wal/78)

Pollen
P1 NEZ.   **Colour** deep blue, v sticky (Wal/78)

Recommended for planting to increase honey production
NEZ (Wal/78).   Can be transplanted

Honey: chemical composition
Water content may be high
Fermentation on storage likely (Wal/78)

Honey: physical and other properties
Colour light (Wal/78)
Granulation rapid, smooth
Flavour delicate, rather like overheated clover honey

202 Geranium pratense L.; Geraniaceae

meadow cranesbill
Herb, 30-80 cm, perennial; fls bright violet-blue
Distribution temperate Europe.   **Habitat** frequently cultivated ornamental
widely naturalized in meadows/ditches/streamsides

Economic and other uses
Land use amenity

Nectar rating; blooms, nectar flow; composition
N1 URS(Fed/55)
N3 POL(Dem/64a)
Blooms vi-ix (EUR).   **Nectar secretion** 1.3-1.5 mg/fl/day (Dem/63a;
Han/80).   **Sugar concentration** [high] 57-71% (Dem/63a; Han/80).   **Sugar
value** (mg/fl/day) [medium] 0.744-1.103 (Dem/63a); 0.31-0.36 (Han/80).
Sugar analysis (Pec/61)

Honey flow
Honey potential (kg/ha) [moderate] 28-80 (POL, Dem/63a); 50-100 (POL,
Dem/64a)

Pollen
P2 URS.   **Colour** of load blue (Han/80).   **Pollen grain** illustrated and
described (Saw/81).   **Reference slide**

Honey: chemical composition
Sugars (as % of total, Maz/64): **glucose** 34.9%; **fructose** 47.6%; **sucrose**
10.4%; **maltose** 5.5%; **fructomaltose** 1.6%

Honey: physical properties
Electrical conductivity 0.000337/ohm cm (Vor/64)

203 Gilibertia arborea (L.) March.; Araliaceae

Tree, <20 m; fls greenish-yellow, small, in umbels
Distribution tropical C America, Caribbean, S America.   **Habitat** rain
forests and river banks; altitudes up to 2400 m

Soil wide range but calcareous types preferred

Nectar rating; blooms, nectar flow
N1 ANT(Ord/66)
N3 ANT(Ord/83); MEX(Ord/72)
Blooms vii to ix or x depending on altitude (tropical America, Ord/83).
Nectar flow heavy; "one of the most valuable spp in Antilles as it blooms when there is no other flow" (Ord/83)

Honey: physical properties
Pfund amber (Ord/83)

204 Gilibertiodendron dewevreii (De Wild.) Leonard; Leguminosae

mbau (ZAI)
Tree
Distribution tropical Africa. **Habitat** Ituri forest (ZAI), one of the 3 characteristic types of climax forest

Nectar rating + honeybee species; blooms, nectar flow
N1 ZAI,tm(Ich/81)
Blooms v-vi (ZAI)

Honey no data

205 Gleditsia triacanthos L.; Leguminosae

honey locust, thorny locust (En/USA)
Tree, <45 m, deciduous, spreading, thorny; thornless cvs available; fls greenish, honey-scented
Distribution subtropical Oceania, Asia, Africa; temperate N America; tropical Africa; native to N America. **Habitat** tropical highlands; rich bottom land USA; steppes URS; veld SOU; outback AUS
Soil wide range; acid and alkaline; sand to clay; deep sandy loam best; slightly saline soil tolerated. **Temperature** v frost tolerant. **Rainfall** 500-2500 mm with 6-8 mths max dry period; drought tolerant where deep soil moisture available; semi-arid conditions

Economic and other uses
Food - pods. **Fodder** - pods for cattle/pigs; lvs. **Fuel. Timber.**
Land use hedges, windbreak, shade, amenity. **Soil benefit** conservation; erosion control

Warning
Forms dense thickets and has become a nuisance in AUS/QD (Usa/79)

Nectar rating + honeybee species; blooms, nectar flow
N1 PAK,ac(Pak/77)
N2 URS(Glu/55)

N3 SOU,tm(And/73);   USA/LA(Pel/76)
Blooms iii-iv (PAK);   xi-iii (SOU);   v-vi (USA).   **Nectar flow** too short
for large yield (Pel/76).   **Nectar secretion** 0.154 mg/fl/day (Sim/80).
Sugar concentration [medium] 45.5% (Sim/80).   **Sugar value** (mg/fl/day)
[medium] 0.156 (Sim/75);   0.189 (Sim/80)

Honey flow
Honey potential [moderate] 250 kg/ha (Rom, Apc/68;   Cir/80)

Pollen
P3 SOU.   P PAK.   **Yield** 0.004 mg/10 fls (Sim/75).   **Chemical analysis**
(Sta/74).   **Pollen grain** illustrated (Lie/72);   illustrated and described
(Ada/76).   **Reference slide**

Honey no data ("pure honey not known", And/73)

206 Gliricidia sepium (Jacq.) Walp.;   Leguminosae

nut pine;   madre de cacao (Es/ANT, HOD, MEX, NIA);   mata ratón (Es/COL,
MEX);   piñón amoroso (Es/CUB);   piñón de Cuba (Es/DOR);   palo de hierro
(Es/ELS);   madero (Es/NIA);   gliricidia (Fr/MAT)
Tree, 5-10 m, deciduous, thornless;   fls white/pink, after lvs have fallen
Distribution tropical C America, Caribbean, S America, Africa, Asia;
subtropical N America.   **Habitat** humid/sub-humid areas, altitudes <1600 m,
paddy bunds, edges of roads
Soil moist or dry, and even if v alkaline.   **Temperature** 22-30°.   **Rainfall**
1500-2300 mm with 2-3 mths max dry period

Economic and other uses
Food - roasted fls.   **Fodder** - lvs for cattle only;   toxic to many other
animals.   **Fuel.**   **Timber** termite resistant.   **Land use** hedges round
apiaries (CUB, Ord/44), windbreaks, shade for crops and hives, amenity,
living fence, firebelts.   **Soil benefit** conservation;   N-fixation;   organic
manure.   **Other uses** pesticides from seed/bark/roots;   weed control

Warning;   alert to beekeepers
Warning roots, bark, seed toxic to man;   lvs ?toxic to man if not cooked
(Usa/80).   Lvs toxic to many animals except cattle.   **Alert to beekeepers**
bees do not collect pollen (MAT, Bal/76)

Nectar rating + honeybee species;   blooms, nectar flow
N1 ANT(Ord/66);   ?BEL(Mul/79);   COL(Cor/76;   Ken/76);   DOR(Ord/64);
ELS(Woy/81);   HOD(Ord/63);   UGA[tm](Nsu/77)
N2 ?ANT(Ord/83);   CUB(Ord/44;   Ord/56);   MEX(Ord/72);   NIA(Ord/63a)
N3 MAT(Bal/76)
Blooms xii-ii (ELS);   i-ii (MAT).   **Nectar flow** v sensitive to climate,
intense flow after rain (Mot/64);   ceases during long drought periods;
bees continue to forage on fallen fls (Ord/83)

Pollen
Alert to beekeepers bees do not collect pollen (MAT, Bal/76)

Recommended for planting to increase honey production
ELS (Esp/64).   Propagate by seeds or large (2m) cuttings.   See **Warning;**
alert to beekeepers

Honey: chemical composition
Water [high] 23.57% (Sig/62)
Sucrose [high] ?7.91%
Ash [medium] 0.15%
Free acid [medium] 18.72 meq/kg

Honey: physical properties
Pfund amber (Ken/76; Ord/83; Woy/81); light amber (Mot/64)
Viscosity "very heavy" (Mot/64); "thick" (Woy/81)
Aroma none (Bal/76)

207 Glycine max (L.) Merr.; Leguminosae

soya bean; soja (Es)
Herb, annual, 45-120 cm depending on cv; fls white/purple
Distribution subtropical N and S America, Asia, Africa; temperate Asia, N
America; tropical S America, Caribbean, Asia, Africa; native to SE
Asia.   **Habitat** cultivated crop plant
Soil wide range, even poor; loose soil preferred, moist but not
waterlogged; some acidity tolerated, but plant may be v sensitive to
acidity.   **Temperature** optimum 20-25°; warm but not too hot in summer;
damaged by frost.   **Rainfall** drought tolerant after germination and early
development; wet seasons tolerated

Economic and other uses
Food - seed; oil and meal from seed; bean sprouts; green vegetable.
Fodder - hay/pasture/silage/concentrates; meal from oil extraction. **Soil
benefit** cover, green manure, N-fixation. **Other uses** linoleum, printing
ink, glycerine, insecticides, rubber substitutes, paints, soaps

Nectar rating; blooms, nectar flow; composition
N1 CHN(Tse/54); USA/?AR(War/65); USA/MS(Tat/56)
N2 ?USA/AL(Bas/67); USA/LA(Lie/72); USA/TN(Lit/54)
N3 USA/NC,TN(Pel/76)
Blooms xii-iii (BRA, Jul/72); late vi-early viii (USA/IL, Jay/70); mid vii-
viii (USA/WI, Eri/75). **Nectar flow** starts 4 days after first flowers
open, thereafter only at air temp above 21-24° (Eri/75). **Nectar secretion**
highest with hot days and warm nights, reduced by low humidity and cool
weather (Lov/77); v dependent on soil fertility (Frj/70); none on clay
soil (Lov/57b); nectar volume may be small and its attractiveness varies
greatly; sometimes bees ignore it (Frj/70). **Sugar concentration** [medium]
31-38% (Jul/72); the following results are for nectar from honey sacs of
bees: 33-36% (several cvs, Eri/75); 43% (mean, 55 samples, Jay/70);
28.5% (mean, 2 yrs, Lid/81); 18-55% (AA694/77); 33.0-39.7% (Hark cv,
AA1242/77)

Honey flow
Honey yield (kg/colony/season) [high] 14-45 (Frj/70); up to 45 (USA/AR,
IL, Jay/70); 36 (USA/AR, Lov/77)

Pollen
P ?USA/NC, TN.   **Colour** of load grey to brown, small (Eri/75;   Jay/70).
Pollen grain illustrated (Lie/72).   **Reference slide**

Honey: physical and other properties
Colour v light (Lov/58).   **Pfund** white to extra light amber (Eri/75);
light amber (Lov/56);   water white (Lov/57b);   47-49 mm, extra light amber
(Pel/76, Roc/68)
Viscosity "rather thin" (Cra/75;   Pel/76);   "medium body" (Lov/56)
Granulation rapid (Pel/76)
Flavour characteristic (Eri/75);   distinctive (Lov/56);   unusual (Pel/76)

208 Gmelina arborea Roxb.;   Verbenaceae

melina;   malayna (GAM);   kumil (INI)
Tree, 15-80 m, deciduous;   fls yellow and brown, fragrant, produced when
tree is leafless
Distribution tropical Africa, Asia;   native to Asia.   **Habitat** tropical
forests and plantations;   humid lowlands
Soil good, well drained but moist alluvium;   plant stunted by dry sand,
leached acid soil and v thin impermeable soil.   **Temperature** <52°;
severely damaged by frost.   **Rainfall** 750-4500 mm;   where <1000 mm, grows
along water courses or irrigated areas;   some provenances drought tolerant

Economic and other uses
Fuel.   Timber.   Other uses paper from pulp

Warning
Produces heavy shade limiting other growth;   dead lvs create a mild fire
hazard;   trees may die young (10 yrs);   cattle eat lvs/bark and may cause
damage

Nectar rating + honeybee species;   blooms, nectar flow
N1 GAM[tm](Sve/80);   INI/TAM,ac(Chn/74)
N GAM[tm](Mea/76)
Blooms ?ii (GAM)

Honey flow
Honey yield [high] mean 20, max <100 kg/colony/hive

Pollen
P1 INI/TAM

Honey no data

209 Gossypium barbadense L.;   Malvaceae

Egyptian cotton, long staple cotton, Pima cotton, sea island cotton
Herb, 1.0-1.2 m, annual, but perennial if ground does not freeze;   fls
yellow with maroon petal spot;   nectary in fl, also extrafloral nectaries

on inner circum-bracteal and outer sub-bracteal areas, lvs, young petioles,
fl peduncles (Mcg/76)
Distribution temperate (warm) Europe, Asia; subtropical N America, Asia;
tropical Africa, C and S America, Asia; native to W Indies. **Habitat**
cultivated crop plant
Soil wide range; fertile soil preferred. **Temperature** 25° mean for 150
days; damaged by frost, frost-free period >180 days required. **Rainfall**
900 mm of well distributed rain preferred; irrigation necessary in low
rainfall areas

Economic and other uses
Food - oil from seed. **Fodder** - seed cake and meal for ruminants, but can
be toxic to poultry/pigs. **Other uses** cloth from lint; oil for industry;
rayon production

Warning
Pesticides kill more bees on this crop than on any other (Cra/83). Seed
cake and meal can be toxic to poultry and pigs (Liz/76)

Nectar rating + honeybee species; blooms, nectar flow; composition
N1 URS(Ave/78); USA/AZ(Wae/82)
N SEN,tm(Ndi/74)
Nectar flow bees prefer extrafloral nectar (Mcg/76). **Nectar secretion** 0.54-
0.82 mg/plant[?]/day (AA1087L/76); higher than G. hirsutum (Rad/61);
secretion affected by cv, climate, soil fertility and moisture. **Nectar
yield** up to 3.4 litres/ha/day (But/72); 1.4 kg/ha/day (Wae/82). **Sugar
concentration** [medium] mean 21.5%, max 65% at 08.00 h (But/72); 25%
(Mcg/59); 34% (Mcg/76); 34.5% (AA1087L/76); [low] 12.3-15.9% (Ivn/50).
Also, 40% (leaf nectar), 17.6% (sub-bracteal nectar), 45.3% (from nectaries
between bracts and sepals) (But/72). **Sugar value** [high] 12.2 mg/fl/day
(irrigated, Rad/61). **Sugar analysis** (But/72; Ivn/50; Mcg/76)

Honey flow
Honey potential [moderate] 50 kg/ha (USA/AZ, Wae/82); higher on irrigated
land and higher than G. hirsutum (Rad/61)

Pollen
Yield 45 000 grains/fl (Mcg/76). **Pollen value** collected by bees only when
there is no more attractive pollen (Mcg/76); see also Wae/82. **Pollen
grain** illustrated and described (Smt/56a). **Reference slide**

Honeydew produced (Mcg/76)

Honey no data were found for this species, but there is considerable
information on cotton as a honey plant (and some on cotton honey), which
does not identify the species referred to (e.g. Fed/55, Pel/76)

210 Gossypium hirsutum L.; Malvaceae

upland cotton; algodon (Es/ELS)
Herb, 1.0-1.2 m, annual, but perennial if ground does not freeze; nectary
in fl, also extrafloral nectaries (But/72)
Distribution subtropical Asia; tropical C America, Africa; temperate (S)
Europe; native to C America. **Habitat** cultivated crop plant

Soil wide range.   **Temperature** damaged by frost.   **Rainfall** – irrigation necessary in low rainfall areas

Economic and other uses
Food – oil from seed.   **Fodder** – seed cake and meal for ruminants, but toxic to poultry/pigs.   **Other uses** lint for cloth; oil for industry; rayon production

Warning
Pesticides kill more bees on this crop than on any other (Cra/83).   Seed cake and meal can be toxic to poultry and pigs (Liz/76)

Nectar rating; blooms, nectar flow; composition
N1 URS(Ave/78)
N BUL(Rad/61); ELS(Woy/81)
Blooms ix-xi (ELS); vii-ix (USA/AZ, AA888/82).   **Nectar flow** vii-ix, highest viii-early ix (USA/AZ, AA1055/76).   **Nectar secretion** 18.13 mg/fl/day (irrigated, Rad/61); 0.46-61 mg/plant[?]/day (AA1087L/76); increases from 09.00 h to max at 17.00 h (AA1055/76).   **Sugar concentration** [medium] 36.4% (But/72); 26.3% (Stoneville 213, AA1055/76); [low] 9.3-11.9% (?irrigated, Ivn/50).   Also 49.1% (leaf nectar), 15.3% (sub-bracteal nectar) (But/72).   **Sugar value** [high] 5.32 mg/fl/day (irrigated, Rad/61).   **Sugar analysis** (But/72; Ivn/50).   **Amino acid analysis** (AA527/82)

Honey flow
Honey potential (kg/ha) [moderate] 90 (URS, Wae/82);   50-60, but includes also G. barbadense honey (URS, Ave/78); 29 (USA/AZ, Wae/82)

Pollen
P ELS.   **Yield** abundant but rarely collected (Wae/82).   **Pollen value** low, grains probably too large and spiny and too difficult for bees to pack, or lacking in attractive chemicals (Wae/82)

Honey: chemical composition
Water [low] 15.6, 16.2% (2 samples, age 24, 5 mths, Whi/62)
Glucose [medium] 33.40% (Moh/82); 33.39, 36.93% (Whi/62).   **Fructose** [medium] 39.70% (Moh/82); 36.97, 39.91% (Whi/62).   **Sucrose** [medium] 1.13-1.68% (Moh/82); 3.02, 2.32% (Whi/62).   **Reducing sugars** 71.43% (Moh/82).   **Maltose** 1.80% (Moh/82); 5.56, 4.59% (Whi/62).   **Melezitose** 0.00, 0.68% (Whi/62)
Ash [medium] 0.38 (Moh/82); 0.146, 0.402% (Whi/62).   K, Na, Ca contents (Moh/82)
pH 5.9 (Moh/82); 4.10, 4.20 (Whi/62).   **Total acid** 20.37, 35.52 meq/kg (Whi/62).   **Free acid** (meq/kg) [medium] 30.00 (Moh/82); 16.59, 28.69 (Whi/62).   **Lactone** 3.78, 6.63 meq/kg (Whi/62)
Nitrogen 0.030, 0.018% (Whi/62)

Honey: physical and other properties
Colour light, but may be dark on dry sandy soils (Cra/75);   fairly clear (Woy/81).   **Pfund** 17-42 mm, white to extra light amber (Whi/62)
Viscosity "may be thin bodied" (Cra/75).   **Optical rotation** -13.72 deg (Moh/82)
Granulation medium
Flavour of honey from plants on dry sandy soils may be strong (Cra/75)

211 Gouania lupuloides (L.) Urban;   Rhamnaceae

bejuco de indio (Es/DOR)
Shrub, climber
Distribution tropical C America, S America, Caribbean

Economic and other uses
Food - used instead of hops in brewing.   **Other uses** tooth-picks

Nectar rating;   blooms, nectar flow;   composition
N1 DOR(Ord/64)
N2 ?ANT(Ord/83);   DOR(Ord/83)
N JAM(Ord/83)
Blooms ix-xi (tropical America, Ord/83)

Honey flow
Honey yield - often mixed with that of G. polygama (DOR, Ord/83)

Pollen
P DOR

Honey no data

212 Gouania polygama (Jacq.) Urban;   Rhamnaceae

bejuco lenatero, rattan (Es/CUB);   bejuco de indio (Es/DOR);   jaboncillo
(Es/ELS, PAM);   pie de pava (Es/GUM);   limpia dientes (Es/HOD);   rema
(Es/VEN);   liana savon (HAI)
Shrub, climber, <7 m;   fls white, small, in tassels
Distribution tropical C America, S America, Caribbean.   **Habitat** low
altitudes <1000 m;   forest edges and coastal scrub where climber covers
trees and shrubs

Nectar rating;   blooms, nectar flow
N1 DOR(Ord/64)
N2 ANT(Ord/83);   CUB(Ord/44;   Ord/56;   Ord/83);   DOR(Ord/83)
N ELS(Ord/83);   GUM(Ord/83);   HAI(Ord/83);   HOD(Ord/83);   PAM(Ord/83);
VEN(Ord/83)
Blooms ix-xi (tropical America, Ord/83).   **Nectar flow** short (CUB, Ord/44)

Honey flow
Honey yield "one of the most important plants" in many parts of ANT and C
America (Ord/83);   often mixed with that of G. lupuloides (DOR, Ord/83)

Pollen
P DOR

Honey: physical and other properties
Colour almost transparent (Ord/83).   **Pfund** light amber
Flavour and aroma characteristic

213 Grevillea robusta A. Cunn. ex R.Br.; Proteaceae

silky oak; grevilea (Es, Pt); helecho (Es/DOR); peineta (Es/GUM)
Tree, medium size, fern-like lvs, briefly deciduous; fls orange-yellow
Distribution tropical Asia, Africa, C America; subtropical Oceania, N
America; native to Australia. **Habitat** planted for protection of coffee
plants in mountainous regions, GUM
Soil wide range; well drained deep soil preferred. **Temperature** frost
resistant (-10°) when mature; young plants frost tender. **Rainfall** 400-
2500 mm; 700-1500 mm optimum; medium to high rainfall areas (AUS, Aus/83)

Economic and other uses
Fuel. **Timber.** **Land use** windbreak, shade, amenity

Warning
Brittle branches; seedlings can be a nuisance (Usa/80)

Nectar rating + honeybee species; blooms, nectar flow; composition
N1 INI[ac](Ded/57; Wak/81)
N2 DOR(Ord/83); GUM(Elm/52; Ord/83); TAN,tm(Smt/57)
N ?HOD(Ord/63)
Blooms ix-xii (BRA/RG, Jul/72); spring (C America, Ord/83); iv-vi, x-xii
(INI, Ded/57). **Nectar flow** each fl secretes for 3 days (Ded/57). **Nectar
secretion** abundant (Ded/57; Lov/57c). **Sugar concentration** [high] 15-79%
(Ded/57); 78% (Wak/81); [medium] 22-43% (Jul/72). **Sugar analysis**
(Ded/57; Han/72; Wak/81)

Pollen
P DOR; GUM; USA/FL. **Pollen yield** heavy (Aus/83). **Pollen grain**
illustrated and described (Smt/56a). **Reference slide**

Recommended for planting to increase honey production
AUS(Aus/83); INI/MAH where it makes evergreen, semi-evergreen forests
(Sub/79). Propagate by seed/cuttings; grows rapidly (Usa/80). See
Warning

Honey: physical and other properties
Colour reddish black (Lov/57c). **Pfund** dark amber (Cra/75)
Granulation rapid (Lov/57c)
Flavour pronounced (Cra/75); rank (Lov/57c)

214 Guaiacum officinale L.; Zygophyllaceae

lignum vitae
Tree, <10 m; fls blue, rarely white
Distribution tropical S America, Caribbean

Economic and other uses
Timber - one of the hardest woods available. **Land use** amenity. **Other
uses** medicinal, resin

Nectar rating; blooms, nectar flow
N1 ?BEL(Mul/79); ?HAI(Mul/78); JAM(Chp/70)
Blooms iv (tropical America, Ord/83)

Honey no data

215 Guiera senegalensis J. Gmelin; Combretaceae

n'guer (SEN)
Shrub/tree
Distribution tropical Africa.   **Habitat** fallow fields
Soil sandy

Economic and other uses
Fodder – browse for camels.   **Fuel.**   **Other uses** medicinal

Nectar rating + honeybee species
N1 SEN,tm(Ndi/74)
N SEN,tm(Dou/70)

Honey no data

216 Guizotia abyssinica Cass.; Compositae

niger
Herb, 50–150 cm, annual
Distribution tropical Africa, Asia; native to tropical Africa.   **Habitat**
cultivated crop plant
Soil poor soil tolerated.   **Rainfall** moderate, <1000 mm

Economic and other uses
Food – oil from seed.   **Fodder** – pressed cake; bird-seed.   **Other uses**
lamp oil from seed

Nectar rating + honeybee species; blooms, nectar flow; composition
N1 ETH[tm](Cra/73)
N3 INI/MAH,ac(Chu/80)
N BUM(Zma/80)
Blooms x–xii (BUM, Zma/80); ix–x (INI/MAH).   **Sugar concentration** [medium]
29% (Zma/80)

Honey flow
Honey yield [moderate] up to 30 kg/colony/season (BUM, Zma/80)

Pollen
P3 INI/MAH.   P BUM.   **Pollen value** good for colony development (BUM,
Zma/80).   **Reference slide**

Honey: physical and other properties
Pfund light amber (Zma/80)
Granulation slow
Flavour peculiar, bitter

217 Gymnopodium antigonoides (Robinson) Blake; Polygonaceae

dzidzilché (Es/MEX)
Shrub/tree, 5-8 m, deciduous, branching near base; forms pure stands,
"aguanales" (MEX); fls pale yellow-green, fragrant
Distribution tropical C America. **Habitat** dry rocky areas, often on slopes
(MEX/Yucatan); grows again after felling and burning; lowland rainforest
(MEX)
Soil shallow, rocky. **Rainfall** drought resistant

Economic and other uses
Fuel

Warning
Difficult to eradicate from soil; nuisance in agave-growing areas (Ord/83)

Nectar rating; blooms, nectar flow
N1 MEX(Ord/63; Ord/66; Ord/72; Ord/83; Saf/73; Smt/60; Wis/53)
Blooms iii-iv (MEX/YUC). **Nectar secretion** heavy (MEX, Saf/73)

Honey flow
Honey yield [high] 136 kg/colony/season, mixed with that of Viguieria
helianthoides (MEX, Smt/60); "most valuable plant for commercial
apiculture" in Yucatan peninsula (MEX, Ord/83); several crops/yr, after
good rains (MEX, Saf/73)

Honey: chemical composition
Water [medium] 18.0% (Rob/56)
Amylase 24. **HMF** <1 ppm

Honey: physical and other properties
Colour light (Ord/83). **Pfund** 64 mm, light amber (Cra/75; Rob/56)
Relative density 1.4171 (Rob/56)
Flavour delicate (Ord/83); characteristic (Rob/56). **Aroma** characteristic
(Rob/56)

218 Haematoxylum campechianum L.; Leguminosae

logwood; campeche (Es/CUB, DOR, MEX); palo de tinta (Es/MEX); campêche
(Fr/MAT, MAY);
tree/shrub, (15 m, spiny, forms pure stands "tintales"; fls golden-yellow
Distribution tropical S America, C America, Caribbean, Africa, native to C
America, Caribbean. **Habitat** near coast of Gulf of Campeche; on plains

and marshes of C America subject to periodic flooding; also dry coastal areas (MAY)
Soil well irrigated, moist. **Rainfall** – in HAI annual mean is <1500 m

Economic and other uses
Timber. **Land use** amenity. **Other uses** dyes from wood

Alert to beekeepers
Pollen inadequate for brood rearing (Ord/83)

Nectar rating + honeybee species; blooms, nectar flow
N1 ANT(Ord/83); CUB(Smt/60); GUY(Cra/73; Smt/60); ?HAI(Mul/78); JAM(Chp/70; Met/66; Ord/83); MAT(Bal/76); MAY(Bro/82; Cra/73); MEX(Ord/72)
N2 CUB(Ord/44); DOR(Ord/64); MEX/YUC(Ord/83; Saf/73)
N BAA(Clr/44); TRI(Lau/76)
Blooms x, xii, ii depending on rains (C America, Ord/83); ii-iii (HAI); iii, x-xii (GUY, Cra/73). **Nectar flow** 2-3 wks (JAM, Met/66); affected greatly by soil moisture; drought causes fl buds to wither and drop; rain at optimum time results in a long heavy flow

Honey flow
Honey yield [high] 200 kg/colony/season (tropical America, Ord/83)

Pollen
P1 MAT. **P** DOR; HAI. **Alert to beekeepers** pollen inadequate for brood rearing (Ord/83). **Reference slide**

Recommended for planting to increase honey production
CUB (Ord/44). Propagate by seed; grows rapidly; small flowering tree in 2nd yr. Recommended for low-lying ground

Honey: physical and other properties
Colour light (Cra/75; Ord/83); pearly white (Pel/76)

219 Hedychium coronarium Koen.; Zingiberaceae

camia (PHI)
Herb, perennial; fls white, large, fragrant
Distribution tropical Asia, America

Nectar rating + honeybee species; nectar composition
N1 PHI[ac,ad](Row/76)
Sugar concentration [medium] 24% (Row/76). **Sugar analysis** (Row/76)

Honey no data

220 Hedysarum coronarium L.; Leguminosae

French honeysuckle, Spanish sainfoin, sweetvetch; sulla (It)
Herb, 1-2 m, deep rooted, erect/prostrate, biennial/perennial; fls deep
red/purple, fragrant
Distribution temperate (warm) Europe, Africa, Oceania
Soil deep, rich, calcareous; also poor compact soil if it contains lime;
not acid, saline or stagnant soil. **Temperature** not winter-hardy to N of
Alps (Maz/82). **Rainfall** winter rain or irrigation preferred; drought
resistant

Economic and other uses
Fodder - hay/green fodder. **Soil benefit** - green manure

Nectar rating; blooms, nectar flow; composition
N1 ITA(Ric/78); MAQ(Far/79)
Blooms ii-v (MAQ). **Sugar analysis** of nectar (Bat/72)

Honey flow
Honey yield - in Italy, highest in Calabria, Sicily and Sardinia (Ric/77)

Pollen
P1 ITA. **Colour** of load grey (Ric/77). **Pollen grain** in honey is regarded
as indicative of Italian source, although it is also found in N African
honey (Ric/77). **Reference slide**

Honey: chemical composition
Water [medium] 15.3-20.3% (19 samples, Fin/74)
Sugars, total 71.30-79.30%. **Sucrose** [medium] 1.42-5.20% (Fin/74).
Sugars (as % of total): **glucose** 44.57% (Bat/73); 47.0% (Maz/59);
fructose 46.44% (Bat/73); 49.2% (Maz/59); **maltose** 3.9% (Maz/82); also
contents of **isomaltose, trehalose** and **gentiobiose** (Bat/73)
Ash [low] 0.060% (10 white samples, Fin/74); 0.044% (17 samples,
Pes/80); [medium] 0.169% (9 light amber samples, Fin/74)
Total acid 24.38, 32.19 meq/kg. **Free acid** [medium] 14.41, 21.94 meq/kg.
Lactone 9.97, 10.25 meq/kg
Amylase 15.85, 26.92. **HMF** 1.68, 1.04 ppm
Nitrogen 0.032% dry wt (Bos/78). **Amino acids,** free 0.145%, protein 0.120%

Honey: physical and other properties
Colour v light (Cra/75); yellowish (Far/79). **Pfund** 11-34 mm, white
(Bab/61; Fin/74; Ric/78); also 46-82 mm, light amber (Fin/74); white to
water white (Pia/81)
Optical rotation -26.36 deg (Bat/73). **Electrical conductivity** 0.000173
per ohm cm (Pes/80)
Granulation fine grain (Pes/80); becomes white but not hard (Pia/81)
Flavour mild (Cra/75); delicate but characteristic, in drought areas taste
is slightly acerbic like raw green beans (Pia/81). **Aroma** v slight, almost
none (Pes/80; Pia/81; Ric/78)

221 Helianthus annuus L.; Compositae

sunflower; tournesol (Fr); girasol (Pt/BRA); Sonnenblume (De); bunga
matahari, kembang srengéngé (In)

Herb, <1-3 m, annual/perennial; fls yellow, large capitulum, 1000-2000
florets/head on single-headed plant; nectary in fl, also extrafloral
nectaries in bract edges beneath fl head and in basal edges of laminae of
top lvs of stem (Frj/70)
Distribution temperate Europe, S America, N America, Asia; subtropical
Asia, Africa, Oceania, N America, S America; tropical Africa, Asia, S
America; native to N America and Mexico. **Habitat** cultivated crop plant;
well adapted to all tropical/subtropical savannah regions
Soil wide range but deep moisture-retentive soils preferred; "strong"
nitrogen-rich; some salinity/alkalinity tolerated. **Temperature** warm
climates preferred, occasional low temperatures tolerated; low temps
tolerated better than by soya bean. **Rainfall** some drought tolerated
especially when plant well established; intermittent rainfall preferred

Economic and other uses
Food - oil from seed; seed. **Fodder** - oilcake; stems and lvs,
fresh/silage; bird-seed. **Land use** amenity. **Other uses** oil for
varnish/soap; dried green stems/lvs for smoker fuel (How/79)

Warning; alert to beekeepers
Warning plants in high rainfall areas may be damaged by disease (Liz/76).
Alert to beekeepers swarming has been recorded when hive space is
insufficient for brood rearing (Cri/57)

Nectar rating + honeybee species; blooms, nectar flow; composition
N1 BUL(Sim/65); BUM(Zma/80); CAF/MAN(Smi/72); ETH[tm](Cra/73);
FRA(Lou/81); INI/MAH,ac(Chu/80); ITA(Ric/76); MOZ[tm](Cra/73);
PAK,ac(Pak/77); PAR(Bra/54a); ROM(Cir/77; Int/65); UGA[tm](Nsu/77);
URS(Ave/78)
N2 INO(Bee/77); ISR(Chi/65); SOU,tm(Mou/72); SOU/TVL,tm(And/73;
Cri/57); URS(Fed/55); ?YUG(Kon/77)
N3 AUS/VIC(Gom/73); GFR(Gle/77); USA/UT(Nye/71)
N ALG(Ske/72); BRA/RG[tm](Cor/70); NEZ(Wal/78); SPA(EUR,Maz/82);
ZIM,tm(Pap/73)
Blooms xii-iii (BRA, Jul/72); xi-xii, iv (BUM); vii-ix (GDR, Bec/67);
vii (HUN, Pet/77); vii-ix (ROM). **Nectar flow** each fl 2-3 days (AA1231/-
78). **Nectar secretion** (mg/fl/day) 0.33 (Han/80); 0.212-0.500 (means, 10
cvs, AA1231/78). Secretion is highest: during first 10 days of bloom
(AA127/65); at 09.00 h (AA969/79); from 10.00-14.00 h (Maz/82); in
plants with long daily exposure to light (experimental, AA165/79); at 80-
90 m from shelter belt (AA127/65). Effects of various fertilizers also
reported (AA1231/78). **Sugar concentration** [medium] 42% (Jul/72); 35-38%,
up to 60% in hot areas (Maz/82); 53.5% (Mog/58); 42.2% (Pek/78); 45.4%
(Pet/77); 38% (Zma/80); 33.3-48.9% (means, various cvs/yrs, AA1231/78).
Sugar value (mg/fl/day) [medium] 0.27 (Han/80); 0.097-0.192 (means,
various cvs/yrs, AA1231/78); 0.1135-0.2522 (means, 4 yrs, AA805/60); 0.11-
0.25 (Frj/70). **Sugar analysis** (Bat/73a; Wyk/52; AA171/61; AA753/75).
Potassium content and **fluorescence** AA491/80

Honey flow
Honey yield (kg/colony/season) [moderate] 2.5-15.0 (poor year, usually
double, ROM, AA577/81); 12 (URS/Rostov, AA127/65). **Honey potential**
(kg/ha) [moderate] 39.7 (BUL, Pek/78); 18.6-49.0 (10 cvs, BUL,
AA1231/78); 30-60 (GDR, Bec/67); 56-69 (HUN, Pet/77); 34-102 (ROM,
Apc/68; Cir/77); 56.7 (ROM, Bac/60); 34-140 (ROM, Cir/80); 43-63 (4

yrs, ROM, AA305/60);   24-63 (4 cvs, ROM, AA815/63);   *in URS, varies from*
13 (Bashkiria) to 27.4 (Ukraine, Fed/55)

Pollen
P1 FRA; INO; SOU/TVL.   **P2** AUS/VIC; GFR; INI/MAH; ITA; SOU/TVL.   **P3**
NEZ; PAK; ROM; UNY/UT; ZIM.   **Yield** 3.9-11.5 mg/10 fls (means for 13
cvs, BUL, AA1231/78); stamen yields 26 mg/day (Maz/82); high (Cri/57).
Chemical composition (Cir/80; Sta/74); 18.5% crude protein (AA1244/78).
Colour light yellow to dark orange (varies with cv, Wal/78).   **Pollen grain**
illustrated and described (Saw/81); [under-represented in honey] 11 000
grains in 10 g (Pes/80).   **Reference slide**

Honeydew produced ROM (Apc/68)

Recommended for planting to increase honey production
?NEZ, Wal/78; URS, Ave/78.   Propagate by seed.   Cultural notes
(Liz/76).   See **Warning; alert to beekeepers**

Honey: chemical composition
Water [medium] 14.70-18.58% (15 samples, Bac/61); 15.60-20.96% (28
samples, Iva/78)
Sugars, total 74.24-79.30% (Iva/78).   **Glucose** [medium] 34.72-42.33%
(Bac/65); 31.09% (Mur/76).   **Fructose** [medium] 34.75-40.28% (Bac/65);
41.16% (Mur/76).   **Sucrose** [medium] 1.32-3.60% (Bac/65); 0.00-6.65%
(Iva/78).   **Reducing sugars** 69.40-77.76% (Iva/78); 72.86% (Mur/76).
Dextrin 1.00-5.30% (Bac/65)
Ash [low] 0.06-0.27% (Bac/65); 0.04-0.15% (Iva/78); 0.32% (Mur/76); 0.09-
0.11% (5 samples, Pes/80)
pH 3.6-3.9 (Uni/83).   **Total acid** 16.0-48.0 meq/kg (Iva/78)
Amylase 8.3-38.5 (Bac/65); 8.0-20.4 (Iva/68).   **HMF** 0.19-9.41 ppm (Iva/78)
Nitrogen 0.237% dry wt (Bos/78).   **Amino acids,** free 0.212%, protein
0.217%.   **Protein** 0.0067-0.0070% (13 samples, Gen/67)
Other constituents - antibacterial properties reported (Pop/79a)

Honey: physical and other properties
Colour yellow or golden (And/73; Mur/66; Ric/78); egg-yolk yellow or
dark (Cra/75); golden with greenish tinge (Fed/55).   **Pfund** amber
(Fed/55); Pfund-Lovibond grade (Aub/83)
Relative density 1.425-1.452 (Bac/65).   **Optical rotation** -0.65 to -3.10
deg (Iva/78).   **Electrical conductivity** (per ohm cm) 0.000306 (Iva/78);
0.000307-0.000347 (5 samples, Pes/80)
Granulation rapid (Mot/64); fine (Pes/80); soft (Ric/78)
Flavour mild (And/73); mild but characteristic (Cra/75); distinctive,
rather like butter (Uni/83).   **Aroma** strong (Cra/75); fairly strong (Pes/80)

222 Heliconia aurantiaca Ghiesb.;   Heliconiaceae
syn Heliconia brevispatha Hook.

platanillo (PHI)
Herb, large, perennial
Distribution tropical America, Asia

Nectar rating + honeybee species; nectar composition
N1 PHI[ac,ad](Row/76)
Sugar concentration [medium] 25% (Row/76).   **Sugar analysis** (Row/76)

Honey no data

223 Hevea brasiliensis Muell. Arg.; Euphorbiaceae

para rubber, plantation rubber
Tree, deciduous; fls yellow; male and female fls in same inflorescence;
extrafloral nectaries (only) on young leaf petioles and fleshy scales of
young shoots
Distribution tropical Asia, Africa, C America, S America; native to
Brazil.   **Habitat** cultivated crop plant in equatorial regions
Rainfall high rainfall areas only; high humidity

Economic and other uses
Other uses rubber from latex; oil from seeds

Alert to beekeepers
Pollen may be inadequate for brood rearing (Mae/73).   Pesticides often
kill bees on this crop (Deo/72)

Nectar rating + honeybee species; blooms, nectar flow
N1 INI/KAR[ac](Ini/73); INI/KER,ac(Deo/72; Dev/77; Hol/65; Kha/59);
SRI,ac(Kud/81)
Nectar flow completely extrafloral (Mae/73); 30-43 days (INI/KER,
Mae/75); colonies migrated to plantations during iii for extrafloral 2-wk
flow (INI/KAR, Ini/73).   **Nectar secretion** continuous while lvs are young
but ceases as soon as lvs are mature (Mae/73)

Honey flow
Honey yield (kg/colony/season) [high] 45 (INI/KER, Mae/73); [moderate]
4 (INI/KAN, Ini/73); 20-25 (INI/KER, Deo/72); 5-10 (INI/south, Yes/80);
half of honey from in INI is from rubber plantations (Bes/81). **Honey
potential** 3 kg/tree (INI/KER, Mae/73); recommended for further use by
beekeepers (Bes/81; Mae/73)

Pollen
Alert to beekeepers pollen may be inadequate for brood rearing (Mae/73).
Pollen grain sticky

Honey: chemical composition
Water [high] 25% (Fer/78)
Glucose [high] 40.7%.   **Fructose** [low] 27.2% [sic].   **Sucrose** [high]
4.0%.   **Reducing sugars** 67.9%
pH 4.6

Honey: physical and other properties
Colour translucent, straw (Fer/78); v clear (Mae/75).   **Pfund** water white
(Nig/83)

Granulation rapid (Nig/83);   [medium] starts after 2 mths (Fer/78)
Flavour v sweet (Mae/75)

224 **Hibiscus rosa-sinensis L.;   Malvaceae**

shoe-flower
Shrub; fls, wide range of colours, v showy
Distribution tropical America, Asia, Africa;   subtropical and temperate
Asia.   **Habitat** widely cultivated in tropics

Economic and other uses
Land use amenity.   **Other uses** dye

Nectar rating + honeybee species;   blooms, nectar flow;   composition
N1 INI/BIH,ac(Nai/76);   PAK,ac(Pak/77)
Blooms v-viii (BRA/SP, Caa/72).   **Sugar concentration** [low] 19-21%
(Caa/72).   **Sugar analysis** (Pec/61;   Row/76;   Vah/72;   AA538/67)

Pollen
P PAK.   **Pollen grain** illustrated and described (Sao/61).   **Reference slide**

Honey no data.   Hibiscus (species unknown) in KEN gives "water-white honey
which granulates almost at once" (Nih/83)

225 **Holigarna grahamii Hook. f.;   Anacardiaceae**

mothi ranbibata (INI)
Tree, <15 m; fls brown
Distribution tropical Asia.   **Habitat** evergreen forest

Economic and other uses
Timber

Nectar rating + honeybee species;   blooms, nectar flow
N1 INI/KAR,ac,ad,Trigona(Diw/64)
Blooms xii-i (INI/KAR)

Pollen
P INI/KAR

Honey no data

226 **Hymenaea courbaril L.;   Leguminosae**

locust (En/GUY);   West Indian locust (En/USA)
Tree, <30 m; buttress roots; fls whitish

Distribution tropical C America, Caribbean, S America; native to tropical America.   **Habitat** deciduous forests of tropical America; Pacific coasts

Economic and other uses
Timber ?termite resistant.   **Other uses** resin from trunk and twigs (copal)

Nectar rating; blooms, nectar flow
N1 GUY(Cra/73)
N2 tropical America (Ord/83)
Blooms ii-iv (tropical America, Ord/83)

Recommended for planting to increase honey production
Tropical America (Ord/83).   Blooms before tree is mature.   Propolis collected by bees

Honey no data

227 Hymenaea stilbocarpa Hayne; Leguminosae

paquio (Es/BOL)
Tree, v large; fls rosy
Distribution tropical S America

Economic and other uses
Food - fruit contains an edible mealy substance.   **Timber**

Nectar rating + honeybee species; blooms, nectar flow
N1 BOL[tm](Kem/80)
Blooms x-xii (BOL)

Honey no data

228 Hyptis suaveolens (L.) Poit.; Labiatae

oregano (Es/ANT, HOD); mastranto (Es/VEN); bamburral, guanxuma-de-cheiro, marolo (Pt/BRA)
Herb, 1-2 m; fls bluish
Distribution tropical C America, Caribbean, S America
Soil wide range

Nectar rating + honeybee species; blooms, nectar flow; composition
N1 VEN(Cra/73)
N ANT(Ord/83); BRA,tm(Ord/83); HOD(Ord/83)
Blooms iii (BRA/CS); iv (south BRA, Ord/83); viii-x (Antilles); ix-ii
(VEN).   **Sugar concentration** [medium] 30-37% (Caa/72)

Honey flow
Honey yield "more important than has been realized; major source during dearth periods" in Antilles (Ord/83)

Honey no data

229 Hyssopus officinalis L.; Labiatae

hyssop; hysope officinale (Fr/ALG)
Herb, 20-60 cm, perennial; fls blue/violet, rarely white
Distribution temperate S Europe, (Med) Africa. **Habitat** dry hills, rocky
ground, screes; elsewhere locally naturalized from gardens

Economic and other uses
Food - aromatic pot-herb; tea from lvs. **Other uses** oil from lvs for
perfume

Nectar rating
N1 POL(Dem/64a); URS(Fed/55)
N ALG(Ske/72); URS(Glu/55)

Honey flow
Honey potential [moderate] >400 kg/ha (Dem/64a)

Pollen
P1 URS. **Reference slide**

Honey: chemical composition
Glucose 36.6% of total sugars (Maz/64). **Fructose** 54.6%. **Sucrose** 1.2%.
Maltose 6.0%. **Fructomaltose** 1.6%

Honey: physical and other properties
Electrical conductivity 0.000144/ohm cm (Vor/64)
Flavour aromatic (Fed/55)

230 Ilex glabra (L.) A. Gray; Aquifoliaceae

inkberry; lowbush gallberry (En/USA)
Shrub, 1.5-2.0 m, evergreen; fls white, dioecious
Distribution subtropical N America. **Habitat** around ponds/lakes and in
swampy low-lying pinelands of Gulf and S Atlantic states, USA
Soil acid, not limestone; some moisture needed

Warning
Fruits toxic (Mot/64)

Nectar rating + honeybee species; blooms, nectar flow
N1 USA/NC(Lor/79)
N2 USA/AL,GA,NC,SC(Pel/76); USA/FL(Mor/56)
Blooms iv-v (USA/FL, Mor/58). **Nectar flow** best in dry weather (USA/FL,
Mor/56). **Nectar secretion** abundant on some soils, moderate on sandy
ridges (USA/FL, Mot/64)

Honey flow
Honey yield [high] usually 30-50, max 135 kg/colony/season (USA, Lov/56)

Pollen
Pollen grain illustrated and described (Ada/76).   **Reference slide**

Honey: chemical composition
Water [medium] 15.4-19.6% (4 samples, age 8-20 mths, Whi/62)
Glucose [low] 27.45-32.24%.   **Fructose** [medium] 39.63-40.89%.   **Sucrose**
[low] 0.35-1.20%.   **Maltose** 6.42-10.44%.   **Higher sugars** 0.89-1.66%.
Melezitose 0.43% (1 sample, Whi/62)
Ash [medium] 0.072-0.247% (4 samples, Whi/62).
pH 3.81-4.75.   **Total acid** 10.13-23.48 meq/kg.   **Free acid** [medium] 8.89-
23.66 meq/kg.   **Lactone** 1.24-6.17 meq/kg
Amylase 12.5-21.4
Nitrogen 0.018-0.044%

Honey: physical and other properties
Colour v light (Cra/75);   slightly yellowish (Lov/56).   **Pfund** light amber
(Lov/56;   Pel/76);   17-42 mm, white to extra light amber (Mot/64;   Rof/75;
Whi/62)
Viscosity "heavy body" (Cra/75;   Pel/76)
Granulation slow (Lov/56;   Mot/64)
Flavour v mild (Cra/75;   Pel/76;   Roo/74);   aromatic after-taste (Lov/56)

231 Ilex integra Thunb.;   Aquifoliaceae
syn Ilex integrifolia Hort.

yerboso (Es/BOL)
Tree;   fls whitish
Distribution tropical S America

Nectar rating + honeybee species;   blooms, nectar flow
N1 BOL,tm(Kem/80)
Blooms v-vi (BOL)

Honey no data

232 Ilex theezans Mart.;   Aquifoliaceae

congonha (Pt/BRA)
Tree
Distribution subtropical S America

Nectar rating + honeybee species;   blooms, nectar flow;   composition
N1 BRA/SC,tm(Wie/80)
Blooms x-xi (BRA).   **Sugar concentration** [medium] 38-40% (Wie/80)

Pollen
P BRA

Honey: physical properties
Colour light (Wie/80)

233 Inga laurina (Sw.) Willd.; Leguminosae

guamá (Es/CUB, MEX, PUE); guamá de Puerto Rico, caspirol (Es/MEX)
Tree
Distribution tropical C America, Caribbean. **Habitat** cultivated for shade
in plantations; altitudes <1400 m

Economic and other uses
Land use shade

Nectar rating; blooms, nectar flow
N1 PUE(Ord/44; Ord/83; Ph1/14)
N2 CUB(Ord/83); MEX(Ord/72)
Blooms - flowering dependent on rainfall (Ord/83)

Honey flow
Honey yield 5 kg/day; "main contributor to honey crop" (GUM, Ord/83)

Pollen
P1 PUE. **P2** CUB

Honey no data

234 Inga micheliana Harms; Leguminosae

chalúm, cushin (Es/GUM)
Tree
Distribution tropical C America. **Habitat** cultivated for shade in coffee
plantations (GUM, Ord/83); altitudes <1800 m

Economic and other uses
Land use shade

Nectar rating
N1 GUM(Ord/83)

Honey flow
Honey yield "source of most of early spring honey" (GUM, Ord/83)

Honey no data

235 Inga vera Willd.; Leguminosae

guamá (Es/DOR, PUE); guava (Es/PUE)
Tree, <20 m, evergreen
Distribution tropical C America, Caribbean. **Habitat** humid/sub-humid
zones; widely cultivated in coffee/cocoa plantations in Caribbean and C
America; river banks and sheltered ravines; low altitudes
Soil wide range including limestone. **Temperature** humid tropics.
Rainfall moist areas but in PUE, tree grows on the dry south coast where it
shows some drought tolerance

Economic and other uses
Food – edible pulp around seeds. **Fuel.** **Timber.** **Land use** shade for
commercial crops

Nectar rating; blooms, nectar flow
N1 ?HAI(Mul/78); PUE(Phl/14)
N3 DOR(Ord/83); JAM(Ord/83); PUE(Ord/83)
Blooms for several mths during spring and summer (Antilles, Ord/83)

Honey no data

236 Ipomoea acuminata (Vahl) Roem. & Schultes; Convolvulaceae
syn Ipomoea leari Paxt.

blue dawn flower (En/INI)
Herb, climber
Distribution tropical Asia

Nectar rating + honeybee species; blooms, nectar flow
N1 INI/MAH,ac(Chu/80)
Blooms i-xii (INI/MAH)

Pollen
P2 INI/MAH

Honey no data

237 Ipomoea batatas (L.) Lam.; Convolvulaceae

sweet potato; batata doce (Pt/MOZ)
Herb, perennial/annual with trailing stems; fls dark pink
Distribution temperate (warm) N America, Europe; subtropical N America;
tropical C America, S America, Caribbean, Asia; native to S America.
Habitat cultivated crop plant especially in wetter regions of tropics
Soil wide range; sandy loam preferred. **Temperature** frost tender;
prolonged exposure to <10° is damaging. **Rainfall** moderately drought
resistant; minimum 900 mm, evenly distributed; high humidity undesirable

Economic and other uses
Food – tubers.   **Fodder** – tubers.   **Other uses** – for commercial starch

Nectar rating + honeybee species; blooms, nectar flow
N1 MOZ(tm)(Cra/73)
Blooms x–ii (tropical America, Ord/83); rainy season (ZIM, Wid/72)

Honey: chemical composition
Water [high] 24.3% (Lin/77)
Glucose [medium] 33.4%.   **Fructose** [medium] 37.0%.   **Sucrose** [medium] 0.1%
pH 3.9.   **Free acid** [medium] 30.9 meq/kg (39.8 after 1 yr)

Honey: physical and other properties
Colour earth yellow
Relative density 1.37

238 Ipomoea nil (L.) Roth; Convolvulaceae

batatilla, campanita (Es/COL)
Shrub, climber; fls pink/blue
Distribution tropical S America

Nectar rating; blooms, nectar flow
N1 COL(Cor/76)
N ELS(Woy/81)
Blooms x–xi (ELS)

Pollen
Colour white (Woy/81)

Honey no data

239 Ipomoea sidifolia Choisy; Convolvulaceae

white campanilla (En/USA); aguinaldo blanco, aguinaldo de Pascua
(Es/CUB); aguinaldo, campanilla (Es/GUM)
Shrub, climber, perennial, vigorous; fls white
Distribution tropical C America, Caribbean.   **Habitat** wild plant covering
fences, thickets and trees, also grows on uncultivated land
Soil wide range; red and calcareous soils preferred (CUB)

Nectar rating; blooms, nectar flow
N1 CUB(Ord/44; Ord/56); GUM(Elm/52)
N3 (Lov/56; Pel/56)
Blooms xi–ii (CUB).   **Nectar secretion** v heavy; best early morning to noon

Honey flow
Honey yield "principal source" – one third of total in CUB (Ord/44)

Pollen
Pl GUM

Honey: physical properties
Colour pearly white (Ord/44).   **Pfund** white (Lov/56)

240 Ipomoea triloba L.; Convolvulaceae

aguinaldo rosado (Es/CUB); campanilla morada (Es/MEX)
Herb, climber; fls violet-pink bells
Distribution tropical S America, C America, Caribbean.   **Habitat** coffee
plantations and fields (CUB); Pacific coasts; lakesides

Nectar rating; blooms, nectar flow
N1 CUB(Ord/44; Ord/56)
N2 CUB(Ord/83); MEX(Ord/72)
N ELS(Woy/81)
Blooms late x to mid xii (tropical America, Ord/83)

Pollen
Colour white (Woy/81)

Honey: physical and other properties
Colour pearly white (Ord/44); v light (Ord/83)
Flavour delicate (Ord/83)

241 Isoberlinia angolensis (Welw. ex Benth.) Hoyle & Brenan; Leguminosae

kapane, msanganza, mutoba (ZAM)
Tree, <20 m, semi-deciduous; fls white
Distribution tropical Africa

Economic and other uses
Timber.   **Other uses** tannin from bark; string from bark

Nectar rating + honeybee species; blooms, nectar flow
N1 ZAM[tm](Sto/82)
N2 ZAM,tm(Zam/79)
Blooms ix-xii (ZAM)

Honey no data

242 Isoglossa deliculata C.B. Clarke; Acanthaceae

kiesieblaar, sewejaarbossie (Af); hlalwane (SOU)
Herb; fls white

Distribution subtropical Africa;   native to S Africa.   **Habitat** areas of thorn bushes, krantzes;   valley bottoms, major river valleys of Natal and Zululand (SOU)
Temperature hot days and cool nights preferred

Economic and other uses
Fodder - heavily grazed by cattle

Nectar rating + honeybee species;   blooms, nectar flow
N1 SOU/NATAL,tm(Mou/72)
N SOU,tm(And/73)
Nectar flow short but intense (And/73);   fast, in a good yr a shallow super may be filled in 2.5 days;   stops abruptly if no rain during flowering period (Tod/74)

Honey flow
Honey yield [medium] 35-55 kg/colony/season, once every 7-10 yrs, and less in other yrs (SOU, Tod/74)

Pollen
P SOU

Honey: physical and other properties
Pfund water white (Tod/74)
Granulation slow if monoforal
Flavour mild but characteristic

243 Ixerbe brexioides A. Cunn.;   Saxifragaceae

tawari (NEZ)
Tree, <17 m, evergreen;   fls white, large, waxy
Distribution temperate Oceania;   native to New Zealand.   **Habitat** bush (NEZ) where number of trees is decreasing

Economic and other uses
Land use amenity

Nectar rating;   blooms, nectar flow
N1 NEZ(God/52)
N NEZ(Wal/78)
Blooms x-xii (NEZ)

Honey flow
Honey yield [high] 80 kg/colony/season (NEZ, Sip/82)

Pollen
Reference slide

Recommended for planting to increase honey production
NEZ (Wal/78)

Honey: chemical composition
Water high (Wal/78)
Fermentation on storage likely (Mao/82; Wal/78)

Honey: physical and other properties
Colour dull white (Wal/78).   **Pfund** white (Mao/82)
Other physical properties – froths
Flavour mild, v sweet (Mao/82; Wal/78)

244 Jacquemontia nodiflora G. Don; Convolvulaceae

campanitas (Es/DOR)
Herb, climber
Distribution tropical C America, Caribbean.   **Habitat** dry areas where it
covers other xerophytic vegetation; roadsides
Soil poor dry soil; not fertile damp soil.   **Rainfall** low rainfall areas;
drought resistant

Nectar rating; blooms, nectar flow
N1 DOR(Ord/64;   Ord/66)
N3 DOR(Ord/83)
Blooms ix-xi or xii (tropical America); ix-xii (DOR).   **Nectar secretion**
"plentiful for most of the day" (Ord/83)

Honey flow
Honey yield "one of the most valuable sources" in DOR (Ord/64)

Pollen
P DOR

Honey no data

245 Julbernardia globiflora (Benth.) Troupin; Leguminosae

muba (TAN); munondo (ZAM, ZIM)
Tree, <15 m, deciduous/semi-evergreen, no root nodules; fls whitish,
inconspicuous, fragrant, calyx and stalks hairy, brown
Distribution tropical Africa.   **Habitat** mixed deciduous woodland; co-
dominant with Brachystegia spiciformis over large areas of ZIM, MOZ and
countries north of them

Economic and other uses
Timber.   **Other uses** rope from bark used for hives, sacks etc (Pag/77);
hives made from bark (Tak/76); tannin and dyes from bark

Nectar rating + honeybee species; blooms, nectar flow
N1 TAN[tm](Smt/57;   Smt/60;   Tak/76);   ZAM[tm](Smt/59;   Sto/82;   Zam/79);
ZIM,tm(Mou/72)
N ZIM[tm](Pap/70)

Blooms i-v (southern Africa, Pag/77).   **Nectar flow** iv-vi, most dependable in areas with 1000-1300 mm rainfall (TAN, Smt/60)

Honey flow
Honey yield "one of the most important sources" in ZAM (Sto/82)

Honey: physical and other properties
Pfund extra light amber (Cra/75);   medium amber (Mou/72)
Relative density 1.4129 (Sil/76a)
Granulation slow, coarse (Cra/75);   coarse brown crystals (Sil/76a)

246 Julbernardia paniculata (Benth.) Troupin; Leguminosae

mucondo, mumué, omanda (ANA);   munsa (TAN)
Tree, <15 m, semi-evergreen;   fls creamy-white, calyx and stalks hairy, brown
Distribution tropical Africa.   **Habitat** dry evergreen forest
Soil plateau and escarpment soils;   sandy.   **Rainfall** dry areas

Economic and other uses
Timber for log hives (Ros/60).   **Other uses** strong rope from bark for thatching;   hives made from bark (Ros/60)

Nectar rating + honeybee species; blooms, nectar flow
N1 ANA,tm(Ros/60);   TAN,tm(Smt/60; Tak/76);   ZAM,tm(Smt/59; Sto/82; Zam/79)
Blooms iv (southern Africa, Pag/77);   iii-vi, vii (ZAM)

Honey: physical and other properties
Colour "clearer than Brachystegia honey" (Ros/60).   **Pfund** extra light amber (Cra/75)
Granulation slow, coarse (Cra/75)

247 Julbernardia unijugata J. Leon.; Leguminosae

Tree
Distribution tropical Africa.   **Habitat** deciduous woodland of western province, TAN

Nectar rating + honeybee species
N1 TAN,tm(Smt/57)

Honey no data

248 Julocroton triqueter Baill., Euphorbiaceae

Distribution subtropical S America;   tropical C and S America

Nectar rating + honeybee species; blooms, nectar flow
N1 BRA,tm(Bar/68)
Blooms xi-iii, during rainy season (BRA)

Honey flow
Honey yield "one of the most important sources" in BRA (Bar/68)

Honey: physical properties
Colour v light (Bar/68)
Viscosity low

249 Khaya senegalensis (Desv.) A. Juss.; Meliaceae

acajou du Sénégal, caïlcédrat (Fr/SEN)
Tree, large; fls whitish, small
Distribution tropical Africa; native to Africa. **Habitat** widespread in drier areas of W Africa

Economic and other uses
Fodder.   Timber.

Nectar rating + honeybee species; blooms, nectar flow
N1 SEN,tm(Ndi/74)
N2 GUS,tm(Sve/80)
Blooms iv-v (SEN)

Honey: physical and other properties
Flavour v acid (Ndi/74)

250 Knightia excelsa R.Br.; Proteaceae

honeysuckle (En/NEZ); rewarewa (NEZ)
Tree, <30 m; fls reddish
Distribution temperate Oceania; native to NEZ. **Habitat** pioneer species of native bush; confined to lowland forests of N Island and Marlborough Sounds (NEZ)

Economic and other uses
Timber.   Land use amenity

Nectar rating; blooms, nectar flow
N1 NEZ(Coo/67)
N NEZ(Wal/78)
Blooms x-xii (NEZ). **Nectar flow** not completely reliable, heavy in wet yrs (Wal/78)

Honey flow
Honey yield (high) 70-90 kg/colony/season (in good yr, God/52)

Pollen
P NEZ.   **Colour** light yellow (Wal/78).   **Reference slide**

Recommended for planting to increase honey production
NEZ (Wal/78)

Honey: chemical composition
Water [medium] 16.0% (K. excelsa "probable source", Kir/60)
Reducing sugars 87.4% of dry wt.   **Dextrin** 2.18% of dry wt
Ash [medium] 0.69% of dry wt
pH 4.00
Nitrogen 0.10% of dry wt.   **Colloids** 0.62% of dry wt

Honey: physical and other properties
Colour dark (God/52);   reddish orange (Wai/75).   **Pfund** amber (Mao/82a);
dark amber (Rob/56);   medium amber (Wal/78)
Viscosity "heavy body" (Cra/75; Wal/78).   **Optical rotation** –6.96 deg
(Kir/60)
Granulation slow, coarse (Cra/75; Mao/82; Wal/78)
Flavour rich, distinctive (Cra/75; Rob/56);   slightly burnt, evident in
blended honey (Cra/75; Wal/78);   sweet like toffee (God/52);   strong,
sweet (Kir/60).   **Aroma** strong (Wai/75)

251 Lagerstroemia parviflora Roxb.;   Lythraceae
syn Lagerstroemia lanceolata Wall. ex Dalz. & Gibbs

benteak;   nandi (INI)
Tree;   fls pale violet
Distribution tropical Asia

Nectar rating + honeybee species;   blooms, nectar flow
N1 INI/KAR,KER[ac](Kha/59)
Blooms iv-v (INI/KAR, KER)

Pollen
P INI/KAR, KER

Honey no data

252 Lavandula angustifolia Miller;   Labiatae
syn Lavandula officinalis Chaix;   Lavandula spica L.;   Lavandula vera DC.

lavender;   espliego (Es);   lavande fine, lavande vrai (Fr);   *lavanda (It)*
Shrub, <1 m, exceptionally 2 m, v aromatic;   fls lavender blue
Distribution temperate (Med) Europe, Asia, N America.   **Habitat** dry,
sunny;   widely cultivated, sometimes naturalized;   altitudes <2000 m;
often in coastal areas
Soil often calcareous

Economic and other uses
Land use amenity.   **Other uses** aromatic oil used in perfumery

Nectar rating + honeybee species;   blooms, nectar flow;   composition
N1 FRA(Lou/81);   URS(Fed/55)
N2 ITA(Ric/78);   YUG(Kul/59)
N3 ?USA/WA(Lov/56)
N SPA(Sau/82a);   URS(Glu/55);   ZIM[tm](Pap/73)
Blooms vi-viii (SPA).   **Sugar concentration** [medium] 38.17, 20.41% (2 yrs,
AA373/60);   14.0-32.9% (6 cvs, AA422/74).   **Sugar analysis** (Pry/44;
Jyk/52; AA753/75)

Honey flow
Honey yield [moderate] 5-20 kg/colony/season (FRA, Uni/83);   often mixed
with honey from L. latifolia (Sau/82a).   **Honey potential** (kg/ha)
[moderate] 183, 117 (experimental plot, 2 yrs, first more humid, BUL,
AA373/60);   50-100 (ROM, Apc/68; Cir/80)

Pollen
P1 FRA.   **Yield** 0.001 mg/10 fls (BUL, Sim/75).   **Pollen grain** illustrated
and described (Bab/63);   under-represented in monofloral honey sample
(Ric/78).   **Reference slide**

Honey: chemical composition
Water may be high (Cra/75)
Glucose [medium] 32.4-33.8% (8 samples, Gon/79).   **Fructose** [medium] 38.7-
40.9%.   **Sucrose** may be high (Cra/75)
Other constituents - trace of methyl anthranilate (Des/62).   Acceptable
characteristics for lavender honey (from this sp), and for lavandin honey
(256), have been published in the Proceedings of the International
Beekeeping Congress, 1975 (Int/75)

Honey: physical and other properties
Colour yellow (Bab/61);   golden (Cra/75).   **Pfund** dark amber (Cra/75);
extra light amber (Lov/56)
Viscosity "good body" (Lov/56)
Granulation v smooth, then almost like butter (Cra/75)
Flavour delicate (Glu/55)

253 **Lavandula dentata L.;   Labiatae**

Toothed lavender;   lavande (Fr/MOR)
Shrub, <1 m;   fls dark purple
Distribution temperate (Med) Europe, (Med) Africa.   **Habitat** dry, sunny;
thickets, open woods
Soil rocky clay and siliceous

Nectar rating;   blooms, nectar flow
N1 MOR(Cra/73)
N MOR(Mat/49)
Blooms iii-vi (EUR, Med)

Honey no data

254 Lavandula latifolia Medicus;   Labiatae

spike lavender;   alhucema (Es);   aspic, grande lavande, lavande mâle, spic (Fr)
Shrub, 30-60 cm, hybrizes readily with L. angustifolia;   fls purple, blue-violet
Distribution temperate (Med) Europe, (Med) Africa;   native to Europe.
Habitat hillsides and slopes;   <600 m altitude;   full sun
Soil light, well drained;   calcareous soil tolerated.   **Temperature** – damaged by frost.   **Rainfall** dry areas

Economic and other uses
Aromatic oil for perfumery

Nectar rating;   blooms, nectar flow;   composition
Nl FRA(Lou/80);   SPA (Sau/82a)
N ALG(Ske/72)
Blooms summer, vii onwards (SPA)

Pollen
Pollen grain illustrated and described (Bab/63)

Honey: physical and other properties
Colour fairly dark (Ske/72)

255 Lavandula stoechas L.;   Labiatae

French lavender;   cantueso (Es);   lavande maritime, lavande stéchas (Fr)
Shrub, <1 m;   fls violet
Distribution temperate (Med) Europe, (Med) Africa;   native to Europe.
Habitat hills, open thickets
Soil dry, stony, siliceous

Economic and other uses
Land use amenity

Nectar rating;   blooms, nectar flow
Nl SPA(Bab/63)
N FRA(Bab/63)
Blooms iii-vi (EUR, Med);   ii-v (FRA);   iii-viii (SPA)

Pollen
Pollen grain illustrated and described (Bab/63)

Honey no data

256 Lavandula angustifolia Miller x latifolia Medicus; Labiatae
syn see 252

lavandin (Fr)
Shrub, sterile hybrid; fls bluish-pink
Distribution temperate (Med) Europe, (Med) Africa.   **Habitat** cultivated
crop plant only;   dry, sunny areas

Economic and other uses
Aromatic oil used in perfumery

Nectar rating;   composition
N1 FRA(Bor/59);   MOR(Cra/73)
Sugar concentration [high] 38.4-66.8% (3 cvs, AA422/74).   **Sugar analysis**
(Bat/73a)

Honey flow
Honey potential [moderate] >150 kg/ha (URS, AA814/63)

Pollen
Pollen value fl produces either aborted pollen grains or none (Bab/63;
Lou/80).   **Pollen grain** illustrated and described (Bab/63);   v under-
represented in honey (Bab/63)

Honey: chemical composition
Glucose [medium] 32.3-34.1% (5 samples, Gon/79).   **Fructose** [medium] 36.2-
39.7%
Acceptable characteristics for lavandin honey have been published in the
Proceedings of the International Beekeeping Congress, 1975 (Int/75)

Honey: physical and other properties
Pfund white (Ric/78;   Ske/72)
Granulation v fine (Ric/78);   fine like butter (Ske/72)
Aroma strong (Ric/78)

257 Leonurus cardiaca L.; Labiatae

motherwort
Herb, 60-120 cm, perennial;   fls pinkish or white
Distribution temperate Europe (except extreme N and Med), Asia, N America;
native to Europe and N Asia.   **Habitat** wasteland, hedges, railways, etc

Economic and other uses
Medicinal

Nectar rating;   blooms, nectar flow;   composition
N1 URS(Fed/55)
N3 ROM(Cir/77);   USA/MO(Pel/76)
N USA/MD(Die/71;   Pel/76)
Blooms v-x (USA);   vi-vii (ROM);   mid vii-ix (ROM, Jua/64).   **Nectar**
secretion 0.12 mg/fl/day (Sim/80);   unaffected by drought (Pel/76).   *Sugar*
concentration [medium] 43.93% (Pek/78);   38.5% (Sim/80).   **Sugar value**
(mg/fl/day) [medium] 0.198 (Jua/64);   0.616 (Sim/80).   **Sugar analysis**
(Jua/64)

Honey flow
Honey potential (kg/ha) [moderate] 182 (BUL, Jua/64);   113.3 (BUL,
Pek/78);   230 (GDR, Bec/67);   230-400 (ROM, Cir/77)

Pollen
P ROM.   **Colour** white (How/79).   **Reference slide**

Honey: chemical composition
Sugars (as % of total, Maz/64):   **glucose** 31.8%;   **fructose** 57.7%;   **sucrose**
1.7%;   **maltose** 6.2%;   **fructomaltose** 2.6%

Honey: physical and other properties
Colour straw-coloured (Fed/55);   light (Lov/59)
Flavour mild (Lov/59)

258 Leonurus cardiaca L. subsp. villosus (Desf. ex Sprengel) Hyl.;   Labiatae

motherwort
Herb, 60-120 cm, perennial;   fls pinkish or white
Distribution temperate Europe;   native to Europe and N Asia.   **Habitat**
wasteland

Economic and other uses
Medicinal

Nectar rating;   blooms, nectar flow;   composition
N1 POL(Dem/64a)
Blooms mid vi-vii (POL, Jab/78).   **Nectar secretion** 0.22-2.40 mg/fl/day
(Jab/68).   **Sugar concentration** [high] 41.6-75.6% (Jab/68).   **Sugar value**
[medium] 0.06-0.78 mg/fl/day (Jab/68)

Honey flow
Honey potential [high] 610 kg/ha (POL, Jab/68)

Honey: physical and other properties
Colour light (Lov/56);   light orange (Pel/76)
Electrical conductivity 0.000086/ohm cm (Vor/64)

259 Leptospermum scoparium J. & G. Forst.;   Myrtaceae

manuka (En/AUS, NEZ);   red tea tree (En/NEZ)
Shrub/tree, <6 m, evergreen, aromatic;   fls white/pink
Distribution temperate Oceania;   native to New Zealand and Tasmania.
Habitat pioneer sp on dry land and swamps but destroyed by scrub clearance
(NEZ)
Soil infertile soil preferred

Economic and other uses
Timber.   Land use hedges, windbreaks

Warning; alert to beekeepers
Warning plant can be seriously damaged by a coccid insect Eriococcus
scoparium (NEZ, Rob/56).   **Alert to beekeepers** honey is thixotropic,
difficult to extract from combs (Mao/82; Pry/50; Rob/56)

Nectar rating; blooms, nectar flow; composition
N1 NEZ(Coo/67)
N3 AUS/SA(Pur/68);   AUS/VIC(Gom/73)
N NEZ(Wal/78)
Blooms vi-vii; ix-x (NEZ).   **Nectar flow** ix, x (NEZ, Wal/78).   **Nectar
secretion** ceases during periods of cold winds; also influenced by soil
type (Wal/78).   **Sugar analysis** (Pec/61)

Pollen
P2 AUS/VIC.   **P** NEZ.   **Yield** fair (Wal/78).   **Colour** muddy white (Wal/78)

Honey: chemical composition
Water [medium] 20.6% (Woo/76; Woo/76a)
Sugars, total 72.9% (70.9% after 44 days at 50°, Woo/76a).   **Glucose**
[medium] 31.4% (31.8%).   **Fructose** [medium] 36.4% (35.2%).   **Sucrose** [low]
0.9% (0.6%).   **Maltose** 2.5% (1.8%).   **Melezitose** 0.9% (0.5%).   **Turanose**
0.5% (1.0%)
pH 4.23 (4.25, Woo/76).   **Total acid** 46.4 (47.0) meq/kg.   **Free acid**
[medium] 38.0 (39.0) meq/kg.   **Lactone** 8.4 (8.0) meq/kg
Amylase 27.6 (amylochrome method), 27.5 (Codex method, Edw/75)
Nitrogen 0.063% (Woo/76).   **Amino acids**, free 839.2 (621.1) µM/100 g
(Woo/76a); also contents of individual acids (proline 30% of total).
Protein high, 1.0-1.2% (Pry/50)
Other constituents 47 volatile compounds present, 9 named (Woo/78a)

Honey: physical and other properties
Colour dark orange (Kir/60); varies with soil and intensity of flow
(Wal/78).   **Pfund** light amber (Cra/75; Rob/56); medium amber (Mao/82;
Pur/68); 69.4 mm light amber, 140 mm dark amber after 44 days at 50°
(Woo/76)
Viscosity (alert to beekeepers) thixotropic, difficult to extract from
combs (Mao/82; Pry/50; Rob/56); easier on some soils in some areas of
NEZ (Wal/78); "heavy body" (Sip/82)
Granulation slow, coarse (Nee/56; Rob/56)
Flavour distinctive, best in comb honey (Cra/75); of burnt sugar,
objectionable (Kir/60); strong (Mao/82); intensifies on extraction
(Rob/56); fairly strong, v sweet and cloying (Wai/75).   **Aroma** decreases
on extraction (Rob/56)

260 Lespedeza bicolor Turcz.; Leguminosae

Shrub, 2-3 m, perennial, coarse woody stems; fls bright pink
Distribution temperate N America, Asia; native to Korea.   **Habitat**
tropical highlands; altitudes <2000 m, humid/sub-humid zones; thickets
where forests have been cut, URS/Siberia
Soil most types including infertile rocky soil.   **Temperature** – damaged but
not killed by frost.   **Rainfall** 500-2500 mm with 3-5 mths max dry period

Economic and other uses
Fodder - lvs if cut young. **Fuel.** **Land use** amenity. **Soil benefit** erosion control; N-fixation; green manure. **Other uses** cover for game

Alert to beekeepers
Plant sometimes cut before flowering

Nectar rating; blooms, nectar flow; composition
Nl URS(Dan/75); URS(Glu/55)
Blooms late summer (USA). **Nectar flow** viii (URS, Dan/75); lengthy (Pel/76). **Alert to beekeepers** plants sometimes cut before flowering.
Sugar concentration [medium] 49% (Mog/58)

Honey flow
Honey yield "good in sylvosteppe zone of Far East region" (URS, Dan/75)

Pollen
Reference slide

Recommended for planting to increase honey production
USA (Pel/49). Propagate by seed. Suitable for rough ground. Shrub is cut to ground by frost but resprouts in spring. See **Alert to beekeepers**

Honey: chemical composition
Sugars: maltose, isomaltose, kojibiose, nigerose present (Waa/60)

261 Lespedeza cyrtobotrya Miq.; Leguminosae

Shrub, 2-3 m, perennial, coarse and woody
Distribution temperate Asia
Soil will grow in some soils where other legumes fail. **Temperature** - damaged but not killed by frost

Economic and other uses
Fodder. **Soil benefit** erosion control. **Other uses** cover for game

Nectar rating
Nl KOS(Bek/65); KOS(Bep/65)

Honey flow
Honey yield [moderate] 26 kg/colony/season, Lespedeza spp (Bep/65)

Recommended for planting to increase honey production
USA (Pel/49). Propagate by seed. Suitable for rough ground. Shrub is cut to ground by frost but resprouts in spring.

Honey no data

262 Leucas aspera Link; Labiatae

wild ocimum; tumba (INI)
Shrub
Distribution tropical Asia.   **Habitat** common forest weed (BUM)

Nectar rating + honeybee species; blooms, nectar flow; composition
N1 INI/KER,ac(Hol/65)
N BUM(Zma/80)
Blooms x-xi, after monsoon rains (BUM).   **Sugar concentration** [medium] 34%,
38% (BUM, Zma/80)

Pollen
Yield very low (BUM, Zma/80)

Honey no data

263 Ligustrum walkeri Decne.; Oleaceae

pungalam (INI)
Shrub
Distribution tropical Asia.   **Habitat** small evergreen patches of shaded
woods in ravines, "sholas"; altitudes <1500 m, INI/TAM

Nectar rating + honeybee species; blooms, nectar flow
N1 INI/TAM,ac(Chn/74)
Blooms v-vi (INI/TAM)

Honey flow
Honey yield "one of the most important and promising sources" (INI/TAM,
Chn/74)

Pollen
P1 INI/TAM

Honey no data

264 Linum usitatissimum L.; Linaceae

flax; lino (Es/ARG, MEX);   linho (Pt/MOZ)
Herb, 40-80 cm, annual; fls blue/white
Distribution temperate Europe, S America, Asia, N America; tropical
Africa, C America; origin uncertain.   **Habitat** cultivated crop plant, not
known wild but recorded as a casual in EUR

Economic and other uses
Fodder - linseed meal; oil-cake.   **Other uses** linseed oil; fibres for linen

Nectar rating + honeybee species; blooms, nectar flow; composition
N1 MOZ[tm](Cra/73)
N2 ARG(Ord/83);   MEX(Ord/83)
Nectar secretion 0.79 mg/fl/day (Han/80).   **Sugar concentration** [medium]
33% (Han/80);   25.5% (Pek/80);   49% (Van/41).   **Sugar value** [medium] 0.26
mg/fl/day (Han/80).

Honey flow
Honey potential (kg/ha) [moderate] 11.9 (BUL, Pek/80);   10 (ROM, Apc/68;
Cir/80)

Pollen
P2 ARG;   MEX.   **Colour** of load blue (Van/41);   greyish-blue (Ric/78).
Pollen grain illustrated and described (Nak/65);   under-represented, and
hardly every found, in Italian honey (Ric/78).   **Reference slide**

Honey: physical and other properties
Colour light (Ord/83).   **Pfund** white (Cha/48);   water white (Van/41)

265 Lippia nodiflora (L.) Michx.;   Verbenaceae
syn Lippia repens (Bertel.) Spreng.

cape vine, creeping Charlie, fogfruit, mat grass, matchweed (En/USA);
yerba de sapo (Es/CUB)
Herb, creeping;   fls blue
Distribution subtropical N America;   tropical C America, Caribbean, S
America;   native to USA/CA.   **Habitat** common weed of lawns, pastures,
groves, roadsides, levees, riversides and flooded areas;   near beaches of
Atlantic coast in C America
Temperature - damaged by frost

Economic and other uses
Soil benefit erosion control (USA/MS)

Nectar rating;   blooms, nectar flow;   composition
N1 USA/CA(Lov/77)
N CUB(Ord/83)
Blooms v-frost (USA/CA, Pel/76);   all yr, peak spring-summer (CUB,
Ord/44).   **Nectar secretion** steady and abundant on muck (deep, rich,
organic soil), but none on sand (USA/FL, Mot/64).   **Sugar concentration**
[medium] 26%, fairly dilute nectar collected by bees in hot months, viii-ix
(ISR, Eis/82)

Honey flow
Honey yield "good in parts of FL, and in LA, but not in TX" (USA, Mot/64);
"major source" in Matanzas (CUB, Ord/83)

Pollen
P tropical America.   **Colour** grey (Ord/83).   **Reference slide**

Honey: chemical composition
Water [high] 22.3% (1 sample, age 6 mths, Whi/62)

Glucose [medium] 31.61%.   **Fructose** [medium] 36.05%.   **Sucrose** [low]
0.45%.   **Maltose** 5.18%.   **Higher sugars** 0.59%
Ash [medium] 0.119%
pH 3.93.   **Total acid** 28.89 meq/kg.   **Free acid** [medium] 22.28 meq/kg.
Lactone 6.61 meq/kg
Amylase 24.0
Nitrogen 0.17%

Honey: physical and other properties
Colour light (Pel/76).   **Pfund** white or light amber (Mot/64);   27-34 mm,
white (Whi/62)
Viscosity "heavy body" (Mot/64; Pel/76)
Granulation fine (Mot/64);   rapid (Pel/76)
Flavour mild, distinctive (Mot/64);   delicate, characteristic (Ord/83)

266 Lippia triphylla (L'Her.) Kuntze;   Verbenaceae
syn Lippia citriodora Kunth

Distribution tropical S America.   **Habitat** upland forest, latitudes 19-23
degrees S, BRA; pasture and forest areas, COL
Temperature mean 24°

Economic and other uses
Food - tisane from lvs

Nectar rating + honeybee species;   blooms, nectar flow;   composition
N1 BRA[tm](Smt/60)
Blooms ix-i (BRA/SP).   **Sugar concentration** [medium] 35-45% (Caa/72)

Pollen
Pollen grain illustrated and described (Sao/61).   **Reference slide**

Honey no data

267 Liriodendron tulipifera L.;   Magnoliaceae

tulip poplar, tulip tree, yellow poplar (En/USA);   liriodendro (It)
Tree, <38 m, deciduous;   fls yellow-green and orange, tulip-like
Distribution subtropical N America, Oceania: temperate N America, Europe,
Oceania;   native to N America.   **Habitat** forested areas;   river valleys;
not on exposed ridges
Soil deep;   constant moisture required;   standing water seldom tolerated

Economic and other uses
Timber.   **Land use** amenity

Alert to beekeepers
Blooms early in season so beekeepers must make colonies strong to harvest
honey (USA/MA, Cao/79)

Nectar rating; blooms, nectar flow; composition
N1 URS(Fed/55); USA/MD(Die/71); USA/NC(Stp/54)
N2 ITA(Ric/78); USA/AL,DC,KY,NC,SC(Pel/76); USA/TN(Lit/54; Pel/76)
N USA/MA(Shw/50a)
Blooms iv-v (USA/MA). **Alert to beekeepers** blooms early in season so
beekeepers must make colonies strong to harvest honey (USA/MA, Cao/79).
Nectar secretion heavy, reservoirs of nectar frequently evident in fls,
which "rain" nectar in a breeze (Cas/79). **Sugar concentration** [low] 16.7%
at start of flowering, [medium] 35.9% on 2nd day (Cao/79)

Honey flow
Honey yield (kg/colony/season) [high] 35-45 (USA/DE, MD, VA); [moderate]
12.5 (mean, 6 yrs, USA/MD, Cao/79). **Honey potential** 0.9 kg/tree/season
(Cao/79); 1 kg/tree (URS, Fed/55)

Pollen
P USA/MA. **Pollen grain** illustrated and described (Ada/74). **Reference
slide**

Honeydew produced USA/MA (Cao/79)

Recommended for planting to increase honey production
USA/MA (Cao/79). First fls at 15 yrs. See **Alert to beekeepers**

Honey: chemical composition
Water [medium] 16.9-18.2% (4 samples, age 5-18 mths, Whi/62)
Glucose [low] 23.08-27.35%. **Fructose** [low] 32.74-36.11%. **Sucrose** [low]
0.14-1.11%. **Maltose** 9.63-14.64%. **Melezitose** 0.29% (1 sample). **Higher
sugars** 2.19-4.23% (4 samples)
Ash [medium] 0.308-0.620%
pH 4.21-4.65. **Total acid** 28.45-50.43 meq/kg. **Free acid** [medium] 26.15-
45.24 meq/kg. **Lactone** 2.30-7.10 meq/kg
Amylase 13.2-33.3 (3 samples, age 14-18 mths)
Nitrogen 0.052-0.098% (4 samples). **Protein** content, before and after
dialysis (Whi/67)

Honey: physical and other properties
Colour reddish - becoming less red but darker on storage (Cao/79); reddish
amber (Glu/55; How/79); dark red (Lov/56); reddish brown (Maz/82);
bright amber when fresh, becoming reddish with age (Roo/74). **Pfund** dark
brown/amber (Cra/75; Pel/76); 50-114 mm, light amber to amber (Whi/62)
Viscosity "heavy body" (Cra/75); "v thick" (Roo/74)
Flavour mild, quince-like (Cra/75); strong (Die/71; How/79); distinctive
(Lov/56). **Aroma** strong (Maz/82)

268 Litchi chinensis Sonner.; Sapindaceae
syn Nephelium litchi Cambess.

litchi, lychee; letchi (Fr/MAY)
Tree, <10 m, evergreen, v long life; fls whitish, insignificant
Distribution tropical Asia; subtropical Asia, N America, S America;
native to S China. **Habitat** cultivated crop plant; sheltered positions,
higher altitudes in tropics; submountainous regions (INI)

Soil well drained, acid, humus-rich; clay; heavy loam. **Temperature** mature trees are hardy, seedlings not frost tolerant. **Rainfall** poor fruiting in v humid tropics; irrigation needed (INI/UTT)

Economic and other uses
Food - fleshy aril of fruit

Nectar rating + honeybee species; blooms, nectar flow; composition
N1 CHN,ac also am(Bia/79; Kel/68; Mad/81; Tse/54); INI/BIH,ac(Nai/76); INI/PUN,ac(Chd/77); INI/UTT,ac(Cht/69; Koh/58; Raw/80); MAY(Bro/82; Cra/73); THA(Bur/82); TAI(Fan/52)
N2 SOU/TVL,tm(Bey/68); USA/FL(Ord/83)
N BUM(Zma/80); INI/BIH,ac(Nar/81)
Blooms iii (BUM); ii-iii (USA/FL, Ord/83). **Nectar flow** brisk and reliable (Nar/81). **Nectar secretion** - in morning only (BUM, Zma/80). **Sugar concentration** [high] >62% (BUM, Zma/80). Juice from damaged fruit also collected by bees (Ord/83)

Honey flow
Honey yield (kg/colony/season) [moderate] >20 (one area of BUM, Zma/80); 7-10 (INI/BIH, Nai/74); 23 (INI/UTT, Koh/58); 4.5-7.0 (TAI/Fan/52); also 27 kg/colony/15 days (INI/BIH, Nar/81); "very important source" in Kuantung and Fujian (CHN, Bia/79)

Pollen
P3 INI/PUN. **P** INI/BIH. **Pollen value** probably not an attractive source to bees (Frj/70). **Reference slide**

Honey: chemical composition
Water [medium] 18.5% (1 sample, Lov/67a)
Amylase 28.8 (after dialysis, Mau/71). **Invertase** 1.63 g glucose/100 g/h. **Acid phosphatase** 75.1 μmoles/100 g/h
Protein 0.18% (Mau/71)
Fermentation on storage - never if kept airtight (Koh/58)

Honey: physical and other properties
Colour light golden (Koh/58); light, may have reddish tinge, may be dark in dry yrs (Lov/67a). **Pfund** light amber (Lov/67a); light to dark amber (Mot/64; Ord/83); white (Zma/80)
Other physical properties - in dry yr, air bubbles take 3-4 mths to rise to top (Lov/67a)
Granulation medium, partial after several mths (Koh/58); slow (Lov/67a; Ord/83); rapid (Zma/80)
Flavour slightly acid (Ord/83)

269 Litsea stocksii Hook. f.; Lauraceae

phanashi, pisa (INI)
Tree, <5 m; fls with persistent cup-shaped perianth
Distribution tropical Asia. **Habitat** evergreen forest, Western Ghats, *INI; in small ravines, "sholas"*; altitudes <1500 m, INI/TAM

Nectar rating + honeybee species;   blooms, nectar flow
Nl INI/MAH[ac](Ded/53)
Blooms ix-x (INI/MAH);   xii-i (INI/MAH)

Honey flow
Honey yield "contributes about 10% of annual total" (INI/MAH, Ded/53)

Pollen
Pl INI/MAH.   **Pollen value** very rich source;   stimulates brood rearing and
swarming (Ded/53).   **Pollen grain** illustrated and described (Ded/53).
Honey may also contain pollens of Litsea polyantha and other related spp
(Ded/53)

Honey: physical and other properties
Pfund dark amber (Ded/53)
Granulation rapid
Flavour like molasses

270 Lonchocarpus pictus Pittier;   Leguminosae

majomo (Es/VEN)
Distribution tropical S America

Nectar rating;   blooms, nectar flow
Nl VEN(Cra/73)
N VEN(Ste/71)
Blooms iv (VEN)

Honey no data

271 Lonicera caerulea L.;   Caprifoliaceae

blue honeysuckle
Shrub, 30-130 cm;   fls yellow
Distribution temperate Asia.   **Habitat** moist wooded areas of Peaktu
plateau, altitude 1000 m, mainly in N Korea
Soil pumice-stone stratum covering rock base.   **Temperature** range -42 to
22.4°, fls frost resistant (KOS)

Economic and other uses
Food - fruit

Nectar rating;   blooms, nectar flow
Nl KOS(Bep/65)
Blooms v-vi (KOS).   **Nectar flow** 18 days (Bep/65)

Honey flow
Honey yield [moderate] 23 kg/colony/season, together with Vaccinium
uliginosum (KOS).   **Honey potential** [moderate] 277 kg/ha (KOS, Bep/65)

Honey no data

272 Lotus corniculatus L.; Leguminosae

bacon and eggs, birdsfoot trefoil; gemeine Hornklee (De); lotier
corniculé (Fr/ALG); trifoglio giallo, ginestrina (It)
Herb, perennial, long tap-root; fls yellow tinged red
Distribution temperate Europe, Asia, N America, S America, Oceania;
subtropical Africa; native to temperate EUR and Asia. **Habitat** cultivated
forage crop; on hilly, marginal and poorly drained land not suitable for
alfalfa (CAN)
Soil poor shallow dry soil tolerated; also waterlogged and saline
conditions. **Rainfall** drought resistant

Economic and other uses
Fodder - pasture; hay. **Land use** amenity, eg roadside ground cover

Nectar rating; blooms, nectar flow; composition
N1 CHL(Kar/56; Kar/60); FRA(Lou/81); IRN(Cra/73); ITA(Ric/78);
USA/VT(Med/54)
N2 CAF/ONT(Ada/79); HUN(Pet/77); URS(Fed/55)
N ALG(Ske/72); BRA,tm(Caa/72); EUR(Maz/82); NEZ(Wal/78)
Blooms x (BRA); v-ix (EUR); vi-viii (USA). **Nectar secretion** (mg/fl/day)
0.19 (Han/80); 0.33-0.55 (3 yrs, AA712/67). **Sugar concentration** [medium]
41% (Caa/72); 40% (Han/80); 19-39% (during one day, max at 15.00 h,
Mor/58); 35% (mean, Pek/77); 30.73% (Pet/77); 27.5-66.7% (3 yrs,
AA712/67); 5-18% (3 yrs, AA1244/77); see also Mue/82; [low] 15%, 26%
(Frj/70); 13.8-17.0% (Mog/58). **Sugar value** (mg/fl/day) [medium] 0.1-
0.221 (AA712/67); [low] 0.08 (Han/80). **Sugar analysis** (Bat/73a;
Kay/78; Maz/59; Pec/61)

Honey flow
Honey potential (kg/ha) [moderate] 1st crop 13.3, 2nd crop 12.9 (BUL,
Pek/77); 15-30 (GDR, Bec/67; ROM, Apc/68; Cir/80); 16-37 (3 yrs, POL,
AA712/67); 15-25 (URS, Fed/55)

Pollen
P1 FRA. **P3** ITA. **P** NEZ. **Colour** of load light brown (Han/80); light
grey (Ric/78). **Pollen grain** illustrated and described (Nak/65).
Reference slide

Honey: chemical composition
Sugars (as % of total sugars): **Glucose** 42.9, 48.2% (Maz/59); 33.8%
(Maz/64). **Fructose** 52.3, 50.3% (Maz/59); 54.8% (Maz/64). **Sucrose** 4.8,
1.5% (Maz/59); 4.6% (Maz/64). **Maltose** 4.7% (Maz/64). **Fructomaltose** 2.1%

Honey: physical and other properties
Colour light (Cra/75); greenish (Ske/72). **Pfund** white (Lov/56).
Viscosity "heavy body" (Lov/56). **Electrical conductivity** 0.000115/ohm cm
(Vor/64)
Granulation rapid (Ske/72)
Flavour like clover (Lov/56)

273 Ludwigia nervosa (Poir.) Hara; Onagraceae
syn Jussiaea nervosa Poir.

cariaquito (Es/VEN)
Herb
Distribution tropical S America

Nectar rating; blooms, nectar flow
N1 VEN(Cra/73)
N VEN(Ste/71)
Blooms vi-xii (VEN)

Honey no data

274 Lythrum salicaria L.; Lythraceae

purple loosestrife; rebel weed, spiked loosestrife, spiked lythrum
(En/USA); Blutweiderich (De); salicaire (Fr); salcerella, salicaria (It)
Herb, 50-150 cm, perennial; fls red-purple, garden cvs available
Distribution temperate N America, Europe (except extreme N), Oceania.
Habitat river and canal banks; wet areas
Soil moist soil preferred

Economic and other uses
Land use amenity

Nectar rating; blooms, nectar flow; composition
N1 IRN(Cra/73);   URS(Fed/55)
N2 POL(Dem/64a);   USA/MA,MI,NY(Lov/56)
N3 ITA(Ric/78);   ROM(Cir/77);   USA/MA,MI,NY(Pel/76)
N EUR(Maz/82);   NEZ(Wal/78);   USA/MA(Shw/50a)
Blooms x-i (NEZ); vii-frost (USA). **Nectar secretion** 0.27-0.64 mg/fl/day
(Han/80).   **Sugar concentration** [high] 62% (Cir/80);   52-72% (Han/80);
[medium] 28.9% (Pek/80).   **Sugar value** [medium] 0.19-0.36 mg/fl/day
(Han/80).   **Sugar analysis** (Maz/82; Pec/61; AA538/67)

Honey flow
Honey yield [moderate] 23 kg/colony/season (USA, Lov/64a).   **Honey
potential** (kg/ha) [moderate] 13.4 (BUL, Pek/80);   255-265 (EUR, Maz/82);   50-
100 (ROM, Apc/68);   50-200 (ROM, Cir/77; Cir/80)

Pollen
P1 ITA.   P NEZ; ROM; URS.   **Colour** of load yellow to green (Han/80);
load violet (Ric/78); yellow or green (How/79; Maz/82); greenish yellow
(Wal/78).   **Pollen grain** size related to staminal length (How/79).
Reference slide

Honey: chemical composition
Water [medium] 18.6% (1 sample, age 8 mths, Whi/62)
Glucose [medium] 21.24%.   **Fructose** [medium] 38.51%   **Sucrose** [low]
0.31%.   **Maltose** 5.87%.   Also contents as % of total sugars (Maz/64)
Ash [low] 0.083%

pH 3.88.   **Total acid** 31.27 meq/kg.   **Free acid** [medium] 21.91 meq/kg.
Lactone 9.36 meq/kg
Nitrogen 0.049%

Honey: physical and other properties
Colour reported as light (Cra/75; Lov/56; Pel/76); also dark (Cha/48;
Cra/75; Dem/64; Wal/78); dark yellow (Fed/55); yellow-green or colour
of machine oil (Lov/57f; Lov/64a).   **Pfund** 34-42 mm, extra light amber
(Whi/62)
Electrical conductivity 0.000306/ohm cm (Vor/64)
Flavour strong (Cra/75; Wal/78); aromatic, sharp (Fed/55); like molasses
(Lov/57f); in parts of Canada, reported to be affected by presence of
pollen of plant (Hol/70)

275 Machaerium eriocarpum Benth.; Leguminosae

tusequi (Es/BOL)
Tree, medium height, spiny
Distribution tropical S America

Nectar rating + honeybee species; blooms, nectar flow
N1 BOL,tm(Kem/80)
Blooms iv-vi (BOL)

Honey no data

276 Machilus macrantha Nees; Lauraceae

amai maram, kardel (INI)
Tree, >16 m; fls yellow
Distribution tropical Asia.   **Habitat** evergreen woods in ravines;
altitudes <1500 m, INI/TAM

Economic and other uses
Timber

Nectar rating + honeybee species; blooms, nectar flow
N1 INI/TAM,ac(Chn/74)
Blooms ii (INI/TAM)

Pollen
P1 INI/TAM

Honey no data

277 Mackenziea integrifolia (Dalz.) Brem.;   Acanthaceae
syn Strobilanthes perfoliatus T. And.

wahiti (INI)
Shrub
Distribution tropical Asia

Nectar rating + honeybee species;   blooms, nectar flow
N1 INI/MAH[ac](Ded/53)
Blooms x-i, every 7 yrs (INI/MAH)

Honey flow
Honey yield "very heavy" (INI, Ded/53)

Pollen
Pollen grain illustrated and described (Ded/53)

Honey: physical and other properties
Colour colourless, transparent (Ded/83)
Granulation rapid, uniform

278 Madhuca longifolia (Koenig.) J.F. Macbr. var. latifolia (Roxb.) A. Chev.;   Sapotaceae
syn Madhuca latifolia (Roxb.) A. Chev.

kadu-hippe, mahua (INI)
Tree, deciduous
Distribution tropical Asia.   **Habitat** dry forests, INI

Economic and other uses
Food - fls.   **Other uses** oils from seeds;   crushed fls yield light-coloured syrup used as supplementary feed for bees (Ini/81)

Nectar rating + honeybee species;   blooms, nectar flow
N1 INI/BIH,ac(Nai/76)
Blooms ii-iii (INI)

Pollen
P INI

Honey no data

279 Magnolia grandiflora L.;   Magnoliaceae

bay, bull bay tree, magnolia (En/USA)
Tree, evergreen; fls white, large
Distribution temperate N America, Oceania;   subtropical Oceania;   native to south-east N America.   **Habitat** widely cultivated as ornamental
Soil moist

Economic and other uses
Land use shade, amenity

Nectar rating; blooms, nectar flow
N1 USA/NC(Lor/79)
N3 USA/MS(Pel/76)
Blooms spring to midsummer (USA)

Pollen
Pollen grain illustrated (Lie/72)

Honey: physical and other properties
Pfund dark (Lov/56); v dark (Pel/76)
Flavour strong (Lov/56); strong, may be unpalatable (Pel/76)

280 Mahonia trifoliata (Moric.) Fedde; Berberidaceae

agritos, palo amarillo (Es/MEX)
Shrub, <4 m, forms large thickets; fls yellow, fragrant
Distribution subtropical N America, C America. **Habitat** steppe area of N
and central states of MEX, W of USA/TX and S of USA/NM; highways and road
verges; fallow ground
Rainfall <400 mm; drought tolerant

Economic and other uses
Food - fruit for jellies, cakes, wine; seed for coffee substitute. **Other
uses** dye and ink from wood

Nectar rating; blooms, nectar flow
N1 MEX(Ord/83)
Blooms i-iv (tropical America, Ord/83). Juice of fruit also collected by
bees, then combs in hive show reddish spots (Ord/83)

Honey flow
Honey yield "high" (Ord/83)

Pollen
P1 MEX.   **Reference slide**

Honey: physical properties
Pfund light amber (Ord/83)

281 Malus baccata (L.) Borkh.; Rosaceae

Siberian crab-apple
Tree, 9-12 m, deciduous; fls white
Distribution subtropical and temperate Asia

Nectar rating + honeybee species; blooms, nectar flow
N1 PAK,ac(Pak/77)
Blooms ii-iv (PAK)

Pollen
P PAK

Honey no data

282 Malus domestica Borkh.; Rosaceae

syn Malus communis Poir. in part; Malus sylvestris (L.) Miller var.
domestica (Borkh.) Mansf.; Pyrus malus L. in part

apple; pommier (Fr); macieira (Pt/BRA); Apfelbaum (De); melo (It);
eple (No); saib (PAK)
Tree, 6-8 m, deciduous; fls white flushed pink; many hybrids and cvs;
taxonomy difficult
Distribution temperate Europe, Asia, N America, S America, Oceania;
subtropical Africa, Asia, Oceania; native to Europe. **Habitat** cultivated
crop plant

Economic and other uses
Food – fruit; cider from fermented juice

Alert to beekeepers
Blooms early in season; honey stored by bees only in good foraging
conditions (UK, How/79; USA, Lov/77; Pel/76)

Nectar rating + honeybee species; blooms, nectar flow; composition
N1 ITA(Ric/78); PAK,ac(Pak/77)
N2 FRA(Lou/81); GFR(Gle/77); JAP(Sak/82); URS(Fed/55); USA(Pel/76)
N3 AUS/VIC(Gom/73); UK(How/79); USA(Pel/77)
N ALG(Ske/72); BRA(Wie/80); CAF/BC(Con/81); EUR(Maz/82); INI/HIM[ac]
(Sig/62); NOW(Lun/71); USA/MA(Shw/50a)
Blooms ix-xi (BRA); iv-v (EUR); v-vi (JAP). **Nectar secretion**
(mg/fl/day) 3.26-7.09 (Frj/70); 1.73-3.99 (4 vars, 3 yrs, AA1232/78);
also 1.3 ml/fl/day, secretion enhanced by extra potash but not by extra
nitrogen or phosphate (Frj/70). **Sugar concentration** [high] 30-65%, and in
hot places, eg ISR, 75-87% (Maz/82); [medium] 32.5-49.0% (3 vars,
Cir/80); 38.70-52.50% (2 vars, 2 yrs, Pet/72a); 45.71% (Pet/77); 50%
(Van/49); 40-45% (Wie/80); 27-58% (AA1232/78). **Sugar value** (mg/fl/day)
[medium] 1-3 (Maz/82); 0.596-1.388 (4 vars, AA1232/78). **Sugar analysis**
(Maz/59; Maz/82; Wyk/52; AA103/67; AA177/77)

Honey flow
Honey yield [high] 36 kg/colony/season (CAF/BC, Lov/77); 1.3-3.6
kg/colony/day (USA/ME, Lov/77). **Alert to beekeepers** blooms early in
season; honey stored by bees only in favourable foraging conditions (UK,
How/79; USA, Lov/77; Pel/76). **Honey potential** (kg/ha) [moderate] 20-30
(ROM, Apo/68); 30-42 (ROM, Cir/80); 20 (URS, Glu/55)

Pollen
P1 FRA; GFR. **P2** ITA; JAP. **P3** AUS/VIC. **P** ALG; BRA; CAF/BC; EUR;
INI/HIM; PAK; UK; USA/MA. **Yield** 1.7 mg/fl (Frj/70); min 0.020 mg/10
fls (BUL, Sim/75); 5.7-20.5 mg/10 fls (19.8-247.3 kg/ha, BUL, Sim/77a).
Pollen value high (Sta/74). **Chemical analysis** (Pec/61; Shp/80;
Sta/74). **Colour** of load pale yellow (Cir/80; How/79). **Pollen grain**
illustrated (Saw/81). **Reference slide**

Honeydew
Honeydew produced, and collected by bees from **Aphis pomi** De Geer,
Aphididae: flow 3-4 wks, honeydew analysis; honeydew produced by **Psylla
mali** Schmidberger, Psyllidae: heavy flow, sometimes visited by bees; also
produced by **Macrosiphum rosae** (L.), Aphididae: honeydew analysis (mid EUR,
Klo/65)

Honey: physical and other properties
Pfund amber (Ric/78)
Electrical conductivity 0.000212/ohm cm (Vor/64)
Granulation irregular (Ric/78)
Aroma of apples

283 Mangifera indica L.; Anacardiaceae

mango, Indian mango; mango (Es/CUB, ELS); manguier (Fr); mangga (In);
mavu (INI)
Tree, 9-12 m, evergreen; fls yellowish/reddish, fragrant, either male or
bisexual on the same inflorescence; 200-6000 fls in each panicle
Distribution tropical S America, C America and Caribbean, Africa, Asia;
subtropical Africa, Asia, N America; native to Asia. **Habitat** cultivated
crop plant; sun or part shade with protection from wind; best below 2000
m altitude
Soil well drained sandy soil preferred; also scrub/pine land. **Rainfall**
continuously wet climate lowers rate of successful pollination

Economic and other uses
Food - fruit. **Land use** shade, firebreak

Warning
Often a favoured habitat for mosquitoes

Nectar rating + honeybee species; blooms, nectar flow
N1 ?HAI(Mul/78); INI/KAR,KER[ac](Kha/59); JAM(Met/66); VEN(Cra/73)
N2 CUB(Ord/44; Ord/56); ELS(Woy/81); INO[ac](Bee/77); JAM(Ord/83)
N3 INI/MAH[ac](Chu/80); USA/FL(Pel/76)
N TRI(Lau/76); ZIM[tm](Pap/73)
Blooms i-iv, ix-xii (VEN, Ste/71); xi-iv, early and late cvs (USA/FL,
Mot/64). **Nectar secretion** reduced by cold and drought (USA/FL, Mot/64);
erratic (INI, Sig/62). Juice from damaged fruit may be collected by bees,
and flavours the honey (TRI, Lau/76). Bees reported to forage on lvs
(either for honeydew or ?for extrafloral nectar) (NEP, Cra/84)

Honey flow
Honey yield [moderate] 23 kg/colony/season (USA/FL, Lov/61e)

Pollen
P1 CUB.   P2 INO; JAM.   P3 INI/MAH.   P INI.   **Pollen grain** sticky
(Sig/62).   **Reference slide**

Honeydew produced INI (insect not specified, Sig/62); ?NEP (Cra/84)

Honey: physical and other properties
Colour reddish amber (Lov/61e); may be v dark (JAM, Ord/83).   **Pfund** amber
(Cra/75; Ord/83; Pel/76); extra light amber (Lov/62b)
Flavour "interesting" (Lov/61e); distinctive (Mot/64); bees may collect
juice from damaged fruit, which affects honey flavour (TRI, Lau/76)

284 Manihot glaziovii Muell. Arg.;   Euphorbiaceae

Ceará rubber, tree cassava
Tree, large
Distribution tropical Asia, Africa, S America;   subtropical S America;
native to southern BRA

Economic and other uses
Land use hedges, amenity.   **Other uses** ceara rubber

Nectar rating + honeybee species; blooms, nectar flow
N1 INI/KAR,KER[ac](Kha/59);   NIR[tm](Cra/73)
N TAN[tm](Smt/57)
Blooms spring-summer (tropical America, Ord/83)

Pollen
P INI/KAR, KER.   **Pollen grain** illustrated and described (Smt/54a)

Honey: physical and other properties
Pfund amber (Woy/81)
Viscosity "thick"
Flavour bitter, unpalatable (Smt/56a;   Smt/57)

285 Marquesia macroura Gilg;   Dipterocarpaceae

muvúca (ANA)
Tree, <25 m, evergreen, buttresses;   fls white to pale cream, fragrant
Distribution tropical Africa.   **Habitat** dry evergreen forest, miombo,
Kalahari woodlands and chipyas;   often locally dominant in plateau miombo

Economic and other uses
Timber   **Other uses** bark for end-stops for bark hives (Sto/82)

Nectar rating + honeybee species; blooms, nectar flow
N1 ANA,tm(Ros/60); ZAM[tm](Smt/59)
N2 ZAM,tm(Sil/76a)
N ZAM[tm](Sto/82)
Blooms vi-ix (ANA); v-i, usually vi-x (ZAM, Zam/79)

Honey: physical and other properties
Viscosity "consistency of stiff treacle" (Zam/79)

286 Marrubium vulgare L.; Labiatae

white horehound; hoarhound (En/USA); marrobio (It)
Herb, <45 cm, perennial; fls whitish, small
Distribution temperate Europe, Asia, N America, Oceania; subtropical
Oceania; native to Europe, from England, S Sweden and C Russia southwards.
Habitat waste ground; sheep trails USA/UT

Economic and other uses
Medicinal

Nectar rating; blooms, nectar flow; composition
N1 POL(Dem/64a); URS(Fed/55)
N2 USA/CA,KS,TX(Pel/76)
N3 AUS/SA(Pur/68); ITA(Ric/78)
N NEZ(Wal/78)
Blooms xii-iii (NEZ); vii-ix (USA); summer (USA/UT). **Nectar secretion**
0.41 mg/fl/day (Jab/68). **Sugar concentration** [medium] 36.6% (Jab/68);
21.1% (Mog/58); 27.1% (Pek/80a). **Sugar value** [medium] 0.15 mg/fl/day
(Jab/68). **Sugar analysis** (Pec/61). **Potassium content** and **fluorescence**
(AA491/80)

Honey flow
Honey yield may contribute 10% in Appenines in summer dry season (ITA,
Ric/78). **Honey potential** (kg/ha) [moderate] 50 (GDR, Bec/67); 233 (POL,
Jab/68); 50-60 (ROM, Cir/80)

Pollen
P3 ITA. P URS. **Pollen value** nil (Nye/71). **Pollen grain** illustrated
and described (Heu/71)

Honey: chemical composition
Glucose 26.6% of total sugars (Maz/64). **Fructose** 52.8%. **Sucrose** 1.9%.
Maltose 14.0%. **Fructomaltose** 2.7%. **Oligosaccharides** 2.0%

Honey: physical and other properties
Colour greenish (Wal/78). **Pfund** dark amber (Cha/48; Pel/76; Pur/68);
amber (Lov/56); medium amber (Wal/78)
Electrical conductivity 0.000085/ohm cm (Vor/64)
Granulation, then appears dark and dirty (Pur/68)
Flavour strong, resembling plant, or like horehound candy (Lov/56;
Lov/56d); pronounced (Pel/76; Wal/78)

287 Medicago falcata L.; Leguminosae
syn Medicago sativa L. subsp falcata (L.) Naeg. & Thell.

lucerne, sickle medick
Herb, perennial, hybridizes with M. sativa; fls yellow
Distribution temperate Europe, Asia. **Habitat** grassy places, waysides, banks
Temperature very hardy cvs available

Nectar rating + honeybee species; blooms, nectar flow
N1 IRN(Cra/73)
N2 INI/PUN[ac](Chd/77)
N3 URS(Fed/55)
Blooms ii-iii (INI/PUN); v-viii (EUR)

Honey flow
Honey potential (kg/ha) [moderate] 24-30 (GDR, Bec/67); 30 (ROM, Apc/68;
Cir/80)

Pollen
P3 INI/PUN. P URS. **Colour** of load grey-blue (Han/80). **Reference slide**

Honey no data

288 Medicago laciniata (L.) Mill.; Leguminosae

Herb, <40 cm, annual
Distribution temperate (Med) Europe, (Med) Africa, Asia

Nectar rating + honeybee species; blooms, nectar flow
N1 PAK,ac(Pak/77)
Blooms iii-iv (PAK)

Pollen
P PAK

Honey no data

289 Medicago lupulina L.; Leguminosae

black medick, hop clover, yellow trefoil
Herb, 5-60 cm, annual/biennial; fls yellow
Distribution temperate Europe, (Med) Africa, Oceania, S America, N America,
Asia; native to Europe, not extreme N. **Habitat** waste places, pastures
and meadows
Soil dry light soil; lime-rich preferred. **Rainfall** not drought tolerant

Economic and other uses
Fodder - early forage for sheep; straw

Nectar rating + honeybee species; blooms, nectar flow; composition
N1 PAK,ac(Pak/77)
N2 EUR(Maz/82); FRA(Lou/81)
Blooms iv-x (EUR); iv-v (PAK).   **Sugar analysis** (Pec/61)

Honey flow
Honey potential (kg/ha) [moderate] 24-30 (GDR, Bec/67); 30-40 (ROM, Cir/80)

Pollen
P2 FRA.   **P** PAK.   **Reference slide**

Honey no data

290 Medicago sativa L.; Leguminosae

alfalfa, lucerne; luzerne (En/CAF); alfalfa (Es/ARG, CHL, MEX); luzerna
(Pt/MOZ); erba medica (It)
Herb, <80 cm, perennial; winter-hardy; wilt- and drought-resistant cvs;
fls purple, white, greenish-yellow depending on subspecies
Distribution temperate Europe except N, Oceania, N America, S America,
Asia; subtropical Oceania, Africa, N America, Asia; tropical Africa,
Asia; native to Europe and southern central Asia; native to NW Iran.
Habitat cultivated crop plant but often naturalized; sun needed; steppe
region and irrigated zones of C Asia and Transcaucasia
Soil deep well drained alkaline soil preferred; limestone (SOU); not
heavy soils; rhizobia (bacteria) must be added to some soils; plant not
salt tolerant.   **Temperature** warm dry climates best; high temperatures
tolerated; plant hardy only at low RH; hardiness dependent on cv.
Rainfall short or moderate droughts tolerated; drought tolerance dependent
on cv

Economic and other uses
Fodder - pasture, hay, silage, lucerne meal.   **Soil benefit** N-fixation

Warning; alert to beekeepers
Warning not grown in Egypt as it harbours cotton pests during the dry
season (Why/53).   **Alert to beekeepers** in some areas pollen inadequate for
brood rearing (USA, Mcg/76)

Nectar rating + honeybee species; blooms, nectar flow; composition
N1 ARG(Lut/63; Per/80); AUS/SA(Pur/68); BEG(Grn/65); CAF/ONT(Tow/76);
CAF/QUE(Cha/48); CAF/SASK(Mec/58); CHN(Tae/54); FRA(Bor/59; Lou/81);
ITA(Ric/78); MOZ[Lm](Cra/73); PAK,ac(Pak/77); URS(Ave/78; Fed/55);
USA/CA(Jay/54; Pel/76; Van/41); USA/CO(Pel/76; Wio/58); USA/?IA,ID,?KS
(Pel/76); USA/ND(Les/54); USA/?NE,NM,NV,NY,OK(Pel/76); USA/UT(Nye/71);
USA/VT(Med/54); USA/WI,WY(Pel/76)
N2 AUS/NSW(Goo/47); AUS/VIC(Gom/73); CAF/ALTA(Hen/77; Wes/49);
CAF/QUE(Cou/59); INI/MAH[ac](Chu/80); MEX(Ord/72); SOU,tm(And/73);
URU(Rod/59); USA/AL(Bas/67); USA/CO(Wio/65); USA/OH(Bai/55);
USA/SD(Pel/76)
N3 AUS/QD(Bla/72); MEX(Ord/83)
N CHL(Roj/39); CZE(Svo/58); EUR(Maz/82); NEZ(Wal/78); OMA(Dut/77)

Blooms xii-ii (AUS/QD); vi-ix (EUR); viii-ix (INI/MAH); v-vii
(USA/OR). **Nectar secretion** 0.24-0.83 mg/fl/day (Han/80); v dependent on
soil moisture and temperature (Maz/82); secretion lower after low night
temp (AA415/59). **Sugar concentration** [medium] 18-48% (Cir/80); 14.7%
(Haa/60); 17-60% (Han/80); 20-25% (Mog/58); 30-60%, depending on soil
moisture (Nye/71); 27-33% (Shw/53); 28.6-44.6% (AA849/64); 40-57%
(AA850/64); 27.3-63.6% (AA347/77); 21-41% (AA694/77). **Sugar value**
[medium] 0.07-0.25 mg/fl/day (Han/80). **Sugar analysis** (Bat/72; Maz/59;
Maz/82; Wyk/52; AA156/55; AA493/66; AA582/77). **Potassium content** and
fluorescence (AA491/80)

Honey flow
Honey yield (kg/colony/season) [high] 56-112 (Mcg/76); 45-136 (USA,
Lov/77). **Honey potential** (kg/ha) [high] 473-1060 (CZE, AA849/64);
[moderate] 25-270 (GDR, Bec/67); 25-30, and irrigated 200 (ROM, Apc/68;
Cir/80); irrigated 260 (URS, AA336/83). Heaviest honey yields when M.
sativa is grown for seed and fields are left uncut (Nye/71); low rainfall
areas best for honey production (SOU, And/73); not reliable (NEZ, Wal/78)

Pollen
P1 AUS/VIC; FRA; URS. **P3** AUS/SA; AUS/QD; INI/MAH; SOU; USA/CA;
USA/CO. **P** CAF/QUE; NEZ; PAK; USA/UT. **Yield** 5.3 mg/fl (Maz/82);
heavy (AUS/VIC, Gom/73). **Pollen value** greater in dry hot regions, also
varies according to area and other crops nearby (Frj/70); bees prefer
other sources (SOU, And/73). **Alert to beekeepers** in some areas pollen
inadequate for brood rearing (USA, Mcg/76). **Chemical analysis** (Shp/79).
Colour lemon yellow but load auburn and hazel (Nye/71). **Pollen grain**
illustrated and described (Nye/71). Under-represented in honey
(Maz/82). **Reference slide**

Honeydew
Honeydew produced, and collected by bees from **Therioaphis trifolii** form
maculata (Buckton), Callaphididae; honey analysis (USA/CA, Whi/62)

Recommended for planting to increase honey production
URS(Ave/78). Propagate by seed. Useful in crop rotation schemes,
especially prior to cotton. See **Warning; alert to beekeepers**

Honey: chemical composition
Water [low] 14.4-17.5% (6 samples, age 7-15 mths, Whi/62); [medium] 18.6%
(Woo/76)
Sugars, total 79.4% (77.7% after 44 days at 50°, Woo/76a). **Glucose**
[medium, also low] 22.30% (Moh/82); 32.62-35.01% (Whi/62); 35.1% (33.2%,
Woo/76a). **Fructose** [medium, also low] 36.20% (Moh/82); 38.37-40.87%
(Whi/62); 34.8% (36.0%, Woo/76a). **Sucrose** [medium] 5.21-6.80% (Moh/82);
2.05-4.80% (Whi/62); 2.5% (0.8%, Woo/76a). **Reducing sugars** 71.60%
(Moh/82). **Maltose** 9.00% (Moh/82); 4.72-6.87% (Whi/62); 4.0% (5.1%,
Woo/76a). **Isomaltose** 0.27%, trehalose 1.92%, gentiobiose 0.24%, **raffinose**
0.17% of total sugars (Bat/73). **Melezitose** 1.6%, turanose 1.4% (Woo/76a)
Ash [low] 0.10% (Moh/82); 0.035-0.078% (Whi/62)
pH 5.5 (Moh/82); 3.60-4.05 (Whi/62); 3.80 (3.45 after 44 days at 50°,
Woo/76). **Total acid** (meq/kg) 17.81-33.89 (Whi/62); 15.5 (16.5,
Woo/76). **Free acid** (meq/kg) [low] 16.70 (Moh/82); 9.22-22.23 (Whi/62);
11.1 (12.5, Woo/76). **Lactone** (meq/kg) 3.24-12.06(Whi/62); 4.4 (4.4,
Woo/76)

Amylase 7.6-7.7 (Edw/75); 12.8 (after dialysis, Mau/71); 18.2 (Sce/66);
12.9-21.9 (Whi/62). **Sucrase** trace (Mau/71). **Acid phosphatase** 8.8
µmoles/100 g/h (Mau/71)
Nitrogen 0.025% dry wt (Bos/78); 0.018-0.039% (Whi/62); 0.18% (0.17%,
Woo/76). **Amino acids,** free 0.110% dry wt (Bos/78); 741.7 µM/100 g
(261.2, Woo/176a); protein 0.099% dry wt (Bos/78). Contents of
individual acids (proline 80% of toal, Woo/76a). **Protein** 0.0052-0.0065%
(4 samples, Gen/67); 0.18% (Mau/71)
Volatile compounds - 46 present, 13 named (Woo/78a)

Honey: physical and other properties

Colour light, hardly affected by heating, 44-79° (Stn/81). **Pfund** 27 mm,
white (Bla/72; Ric/78; Roc/68); white or extra light amber (Lov/56);
extra light amber to light amber (Pia/81); water white (Wal/78); <4 to 27
mm, water white to white (Whi/62); 12.1 mm, extra white (112.2 mm, amber
after 44 days at 50°, Woo/76); Pfund-Lovibond grade (Aub/83)
Viscosity 472.00 poise (Moh/82); "good body" (Cra/75; Lov/56; Van/41).
Optical rotation -36.50 deg (Bat/73); -6.27 deg (Moh/82)
Granulation rapid, hard fine grain (Bla/72); rapid (Cra/75; Pia/81)
rapid, fairly hard white grain (Gom/73); slow (Moh/82); often irregular
(Ric/78); dull appearance (Wal/78)
Flavour mild (And/73; Cra/75; Stn/81); unusual, slightly acid (Bla/72);
insipid (Gom/73); strong, characteristically irritates the throat
(Pia/81); flat, delicate (Wal/78). **Aroma** rather strong (Pia/81);
delicate (Ric/78)

291 Melaleuca leucadendron (L.) L.; Myrtaceae

belbowrie, broad-leaved tea-tree (En/AUS)
Tree, <25 m; fls white/yellow, fragrant. M leucadendron, which is often
confused with M. quinquenerva, does not grow in USA; all data from there
are entered under M. quinquenerva
Distribution tropical C America, Caribbean, S America; native to
Australia. **Habitat** coastal gullies where tidal water-courses occur and on
lower ground; grows well in exposed situations
Soil wide range but marshy ground preferred; brackish moist soil.
Temperature tree damaged by severe frost

Economic and other uses

Timber. **Land use** amenity. **Soil benefit** erosion control. **Other uses**
bark for fruit packing and ornamental wall-covering; oil from lvs;
beekeeper's smoker fuel from bark (Goo/47)

Nectar rating + honeybee species; blooms, nectar flow; composition

N1 MOZ[tm](Cra/73)
N2 AUS/NSW(Goo/47)
Blooms x-xi (AUS/NSW). **Nectar flow** annual, fairly consistent (AUS/NSW,
Goo/47). **Sugar concentration** [low] 8% (Zma/80)

Pollen

P AUS/NSW

Honey: physical and other properties
Colour dark (Goo/47)
Granulation rapid
Flavour strong

292 Melaleuca preissiana Schau; Myrtaceae
syn Melaleuca parviflora Lindl.

flat-leaved paperbark, moonah (En/AUS)
Tree
Distribution subtropical Oceania; native to Australia. **Habitat**
widespread in forest areas and swamps of south-eastern AUS/WA

Nectar rating; blooms, nectar flow
N1 AUS/WA(Col/62)
N AUS/WA(Smt/63)
Blooms i (AUS/WA). **Nectar flow** 3 wks (Smt/63); noted for its abrupt
start and finish (Col/62)

Pollen
P2 AUS/WA

Honey no data; but see 291, 293

293 Melaleuca quinquenervia (Cav.) S.T. Blake; Myrtaceae

paper bark tea-tree (En/MAY); cajeput, melaleuca, punk-tree (En/USA)
Shrub/tree; fls white or cream, sometimes red. M quinquenerva is the
only sp in USA, but M. leucadendron is often confused with it; data for M.
leucadendron from USA are entered here
Distribution subtropical N America, Oceania. **Habitat** coastal districts
usually on low swampy ground, also slopes in forested areas (AUS/QD)

Warning
Contact with trees causes allergy in some people (USA, Lov/77)

Nectar rating; blooms, nectar flow; composition
N1 AUS/QD(Bla/72)
N2 USA/CA(Pel/76); ?USA/FL(Pel/76)
N3 USA/FL(Ord/83)
N MAY(Bro/82); USA/FL (Hol/75)
Blooms iii-vii (AUS/QD); autumn-winter (USA/FL, Lov/77). **Nectar flow** 6
wks (USA/FL, Pel/76). **Nectar secretion** copious (USA/FL, Mot/64). **Sugar
analysis** (Vah/72)

Honey flow
Honey yield 1 shallow 10-frame super/colony/wk (USA, Pel/76); "substantial" (USA/FL, Lov/77)

Pollen
Pl AUS/QD. **Chemical analysis** 35.0-36.8% crude protein (AA1244/78)

Honey: physical and other properties
Pfund amber (Cra/75; Lov/55e; Mot/64; Pel/76)
Granulation rapid (Mot/64)
Flavour can be mild, distinct, also reported unpalatable (Cra/75; Pel/76); strong, unpleasant (Lov/56); strong, rather bitter, but loses objectionable flavour on stirring, heating or storing (Mot/64); a blend with only 5% of this honey is "inedible" (Tay/56). **Aroma** considerable, even penetrating (Cra/75); " v disagreeable" (Tay/56)

294 Melicoccus bijuga L.; Sapindaceae

Spanish lime; mamoncillo (Es/CUB); limoncillo (Es/DOR); mamón (Es/VEN); guinep (JAM)
Tree, <18 m; fls greenish-white, fragrant, male and female fls in separate racemes
Distribution tropical C America, Caribbean, S America; native to tropical America. **Habitat** cultivated crop plant

Economic and other uses
Food fruit; seed for roasting. **Land use** shade

Nectar rating; blooms, nectar flow
N1 JAM(Ord/83)
N2 CUB(Ord/56)
N CUB(Ord/83); DOR(Ord/83); VEN(Ord/83)
Blooms iii-iv (tropical America, Ord/83). **Nectar flow** heavy from old trees, but for only a short time (Ord/83)

Honey flow
Honey yield "one of the most important sources" in JAM (Ord/83)

Honey: physical and other properties
Colour quite dark (Mot/64; Ord/83)
Flavour acid but "pleasant" (Ord/83)

295 Melicoccus lepidopetala Radlk.; Sapindaceae

motoyoé (Es/BOL)
Tree; fls in terminal panicles
Distribution tropical S America; native to tropical America

Economic and other uses
Food - fruit

Nectar rating + honeybee species; blooms, nectar flow
N1 BOL[tm](Kem/71)

Blooms iv-v (BOL)

Honey no data

296 Melilotus alba Desr.; Leguminosae

sweet clover, white melilot, white sweet clover; biennial Bokhara clover
(En/AUS); bee clover, Bokara clover, melilot (En/USA); sweet white clover
(En/ZIM); meliloto, trébol de olor (Es/ARG); melilot blanc, trèfle hubam
(Fr/ALG); melilot (Fr/MAY); meliloto-branco, trevo-branco (Pt/BRA)
Herb, 30-150 cm, biennial (cv Hubam is annual); fls white, fragrant
Distribution temperate Europe, S America, N America, Asia, Oceania, (Med)
Africa; subtropical S America; tropical Africa.   **Habitat** cultivated crop
plant; waste places, persists in fields turned to other crops
Soil neutral to alkaline soil; lime preferred; rhizobia (bacteria) must
be added to some soils.   **Rainfall** >500 mm per season or irrigation
needed; some drought tolerance

Economic and other uses
Fodder - hay, pasture, seed; hay/silage (but toxic to livestock if poorly
harvested or if fermented) (Why/53).   **Land use** amenity.   **Soil benefit**
erosion control; N-fixation

Warning
Persists in fields turned to other crops; seed often harvested with
alfalfa seed; stalks can cause problems when harvesting wheat (Van/49);
hay/silage toxic to livestock if poorly harvested or if fermented (Why/53)

Nectar rating + honeybee species; blooms, nectar flow; composition
N1 ARG(Lut/63); CAF/BC(Con/81); CAF/NWT(Hat/81); CAF/QUE(Cha/48);
CHN(Mad/81); IRN(Cra/73); POL(Dem/64a); URS(Ave/78); Fed/55);
USA/AL(Pel/76); USA/CO(Pel/76; Wio/58; Wio/65); USA/IA,IL,KS,MI,MN,MO-
(Pel/76); USA/ND(Les/54; Pel/76); USA/SD(Pel/76)
N2 AUS/NSW(Goo/47); CAF/BC(Dav/69); FRA(Lou/81); ROM(Int/65);
USA/CA(Jay/54); USA/MD(Die/71); USA/UT(Nye/71)
N3 USA/CA(Van/41); ?USA/MS(Tat/56)
N ALG(Ske/72); ARG(Per/80); BRA/SP[tm](Caa/72); EUR(Maz/82);
ZIM[tm](Pap/73)
Blooms iv-vi, ix-xi (BRA/SP); vii-ix (ROM, Cir/80); vi-frost, mainly
midsummer (USA/UT, Van/49).   Flowering period 7 wks, each fl blooms 3.0-
4.5 days (POL, Dem/63a); fl period 100 days (ROM, Jua/64).   **Nectar
secretion** (mg/fl/day) 0.1 (Han/80); 1.09 (Sim/80).   **Sugar concentration**
[medium] 23-33% (Caa/72); 15.4-44.2% (Dem/63a); 35% (Han/80); 14.2-30.8%
(various dates/localities, Mog/58); 33% on wet soil, 55% on dry soil
(Nye/71); 40.3% (Pek/77); 42.21% (Pet/77); 36-48% (Shw/53); 24.8%
(Sim/80); 35% (Van/49); 45.0-52.5% (2 yrs, AA560/65); 57, 35% (2 yrs,
AA696/73); 55.2% (AA665/80).   **Sugar value** (mg/fl/day) [low] 0.04
(Han/80); 0.016 (Jua/64); 0.054 (Sim/80).   **Sugar analysis** (Jua/64;
Maz/59; Maz/80; Wyk/52)

Honey flow
Honey potential (kg/ha) [high] 26-678 (POL, Dem/63a); [moderate] 211.8
(BUL, Pek/77); 218, 180 (1st, 2nd crops, BUL, AA560/65); 200-500 (ROM,
Cir/80); 174 (ROM, Jua/64)

Pollen
P2 FRA; USA/CA; USA/UT. **P3** USA/CO. **P** ALG; BRA; CAF/QUE. **Yield**
good (Cir/80); abundant (Van/49); load size small-medium (Nye/71).
Colour of load yellow (Cir/80); load green-brown (Han/80). **Pollen grain**
illustrated and described (Sao/61). **Reference slide**

Recommended for planting to increase honey production
USA (Pel/76). Propagate by seed. Recommended for roadsides and railways
to prevent soil erosion and for eradicating obnoxious weeds by crowding
them out. See **Warning**

Honey: chemical composition
Water [medium] 18.8% (1 sample, 5 mths, Whi/62)
Glucose [medium] 33.72%. **Fructose** [medium] 36.77%. **Sucrose** [medium]
1.00%. **Maltose** 5.51%. **Higher sugars** 0.79%. Also contents as % of
total sugars (Maz/64)
Ash [low] 0.041% (Whi/62)
pH 3.65. **Total acid** 19.37 meq/kg. **Free acid** [medium] 15.62 meq/kg.
Lactone 3.75 meq/kg
Amylase 20.4
Nitrogen 0.010%

Honey: physical and other properties
Colour slightly green (Con/81); light (Pel/76). **Pfund** water white or
white (Lov/56); 4-8 mm, water white (Whi/62)
Viscosity "heavy body" (Lov/56). **Electrical conductivity** 0.000174/ohm cm
(Vor/64)
Granulation rapid (How/79; Pel/76); within a week of removing from hive
(USA/AK, Liv/84)
Flavour of cinnamon (Con/81); mild (Lov/56); mild, peppery (Pel/76)

297 Melilotus officinalis (L.) Pall.; Leguminosae

sweet clover, yellow melilot, yellow sweet clover; melilot officinal
(Fr/ALG); melilot jaune (Fr/CAF); erba vetturina, meliloto (It)
Herb, 40-250 cm, biennial; fls yellow, fragrant
Distribution temperate Europe, N America, (Med) Africa, Asia. **Habitat**
cultivated crop plant; weed of cultivated ground; waste places
Soil wide range, often clay or saline soil; neutral or alkaline soil;
fair amount of available lime needed. **Rainfall** dry areas; >450 mm per
season or irrigation needed; drought resistant

Economic and other uses
Fodder - pasture and hay; hay/silage (but toxic to livestock if poorly
harvested or if fermented) (Why/53). **Land use** amenity. **Soil benefit**
soil improvement, erosion control on banks and roadsides. **Other uses**
medicinal (source of coumarin)

Warning
Persists in fields turned to other crops; seed often harvested with
alfalfa seed; stalks cause problems when harvesting wheat (Van/49);
hay/silage toxic to livestock if poorly harvested or if fermented (Why/53)

Nectar rating; blooms, nectar flow; composition
N1 CAF/NWT(Hat/81);   CAF/QUE(Cha/48);   CHN(Mad/81);   IRN(Cra/73);
URS(Ave/78; Fed/55);   USA/CO(Wio/58;   Wio/65);   USA/ID,KS(Pel/76);
USA/ND(Les/54);   USA/SD(Pel/76);   USA/UT(Nye/71)
N2 FRA(Lou/81);   USA/MD(Die/71)
N3 ITA(Ric/78);   ROM(Cir/77)
N ALG(Ske/72);   USA/MA(Shw/50)
Blooms vii-ix (ROM); vi-frost (USA/UT).   **Nectar secretion** 0.110 mg/fl/day
(Sim/80); reduced by insufficient soil moisture.   **Sugar concentration**
[medium] 41.5, 43.6% (2 yrs, Dem/63a);   27.3-48.5% (various dates and
localities, Mog/58);   37.9% (Pek/77);   38-57% (Shw/53);   27.1% (Sim/80);
52% (Van/49).   **Sugar value** [low] 0.060 mg/fl/day (Sim/80).   **Sugar
analysis** (Bat/73a)

Honey flow
Honey potential (kg/ha) [moderate] 200 (ALG, Ske/72);   172.2 (BUL,
Pek/77);   23.5, 10.0 (POL, Dem/63a);   130-300 (ROM, Cir/77; Cir/80)

Pollen
P1 USA/CO; USA/UT.   **P2** FRA;   ITA.   **P** ALG;   CAF/QUE;   ROM;   USA/MA.
Yield abundant (Van/49).   **Colour** of load dark yellow (Han/80)

Recommended for planting to increase honey production
USA (Pel/76).   Propagate by seed.   Recommended for roadsides and railways
to prevent erosion and for eradicating obnoxious weeds by crowding them
out.   See **Warning**

Honey: physical and other properties
Colour light (Pel/76).   **Pfund** white or amber (Ske/72)
Granulation rapid (Pel/76)
Flavour mild, slightly peppery (Pel/76).   **Aroma** delicate, like vanilla
(Ske/72)

298 Metrosideros excelsa Sol. ex Gaertn.;   Myrtaceae
syn Metrosideros tomentosa A. Rich.

pohutukawa (NEZ)
Tree, <50 m; fls brilliant scarlet, buds white
Distribution temperate and subtropical Oceania;   native to New Zealand.
Habitat coastal but also inland (NEZ/Auckland Province)
Soil salt tolerant

Economic and other uses
Land use hedges if kept cut; windbreak

Nectar rating; blooms, nectar flow
N1 NEZ(God/52;   Rob/56)

N NEZ(Wal/78)
Blooms xi.   **Nectar flow** heaviest in dry season; shortened if high winds
damage fls (Wal/78).   **Nectar secretion** profuse (Mao/82)

Pollen
P NEZ.   **Colour** greenish-yellow (Wal/78).   **Reference slide**

Recommended for planting to increase honey production
NEZ(Wal/78)

Honey: physical and other properties
Pfund water white (Cra/75); white (Mao/82)
Granulation rapid, coarse (Cra/75; Mao/82)
Flavour unique salty flavour (Cra/75)

299 Metrosideros robusta A. Cunn.; Myrtaceae

northern rata (En/NEZ)
Tree, 30 m, hybridizes readily with M. umbellata; fls bright scarlet
Distribution temperate Oceania; native to New Zealand.   **Habitat** forests

Economic and other uses
Timber.

Nectar rating; blooms, nectar flow
N1 NEZ(Rob/56)
N NEZ(Wal/78)
Blooms xi-ii (NEZ)

Pollen
P NEZ.   **Yield** not heavy (Wal/78).   **Colour** creamy/dull white (Wal/78)

Recommended for planting to increase honey production
NEZ (Wal/78)

Honey: physical and other properties
Pfund white (Mao/82; Rob/56); water white (Wal/78)
Viscosity "light body" (Rob/56)
Granulation rapid, fine silky grain (Rob/56; Wal/78)
Flavour delicate but distinctive (Mao/82); mild (Wal/78)

300 Metrosideros umbellata Cav.; Myrtaceae
syn Metrosideros lucida (Forst. fil.) A. Rich.

southern rata (En/NEZ)
Tree, <20 m, hybridizes readily with M. robusta; fls bright crimson
Distribution temperate Oceania; native to New Zealand.   **Habitat** mountain
and sub-alpine areas on W coast and southern areas South Island and Stewart
Island, NEZ

Economic and other uses
Timber

Nectar rating; blooms, nectar flow
N1 NEZ(Rob/56)
N NEZ(Wal/78)
Blooms xi onwards through summer according to altitude (NEZ, Wal/78)

Honey: chemical composition
Water [medium] 16.8% (M. umbellata is "suggested source of sample", Kir/60)
Reducing sugars 92% of dry wt (Kir/60). **Dextrin** 1.80% of dry wt
pH 4.27. **Free acid** [high] 40 meq/kg
Nitrogen 0.16% of dry wt. **Colloids** 0.44% of dry wt

Honey: physical and other properties
Colour creamy white (Kir/60). **Pfund** water white (Cra/75); white (Mao/82)
Viscosity "light body" (Rob/56). **Optical rotation** −10.02 deg (Kir/60)
Granulation rapid, fine silky grain (Cra/75; Rob/56)
Flavour delicate but distinctive (Cra/75; Mao/82); v sweet (Kir/60)

301 Mikania scandens Willd.; Compositae
syn Mikania parkeriana DC.

bitter tally (En/GUY); climbing boneset, climbing hempweed, duckblind,
snowvine, wild potato vine (En/USA)
Herb, climber; fls white sometimes pink
Distribution temperate and subtropical N America; tropical S America.
Habitat low lying damp thickets and swampy areas (USA)

Nectar rating; blooms, nectar flow
N1 GUY(Cra/73)
N2 USA/LA(Pel/76)
Blooms iii–iv, x–i (GUY); viii–x (USA)

Honey flow
Honey yield "one of the best honey sources" in swampy sections of USA/LA
(Pel/76)

Pollen
P tropical America (Ord/83). **Yield** large (Ord/83). **Colour** bronze (Ord/83)

Honey: physical and other properties
Colour fairly light (Lov/63)

302 Mimosa scabrella Benth.; Leguminosae

bracatinga (Pt/BRA)
Tree/shrub, 10–12 m, thornless; fls cream, fragrant

Distribution tropical and subtropical S America; native to Brazil.
Habitat cool plains of south-east BRA; also in warmer drier areas;
altitudes <2400 m, humid, sub-humid, tropical highlands (BRA)
Soil wide range but must be well drained as wet soil stunts growth.
Rainfall 1000-2000 mm

Economic and other uses
Fuel.   Timber.   Land use "living" fence, shade in coffee plantations,
amenity.   **Soil benefit** green manure; N-fixation.   **Other uses** pulp for
paper

Nectar rating + honeybee species; blooms, nectar flow; composition
N1 BRA/RG,SC[tm](Wie/80)
N BRA/RG,SC[tm](Jul/70;   Jul/72)
Blooms vii-viii (BRA, Jul/72).   **Sugar concentration** [medium] 35-40% (Wie/80)

Pollen
P BRA/RG,SC.   **Pollen grain** described (Bah/73).   **Reference slide**

Honey: physical and other properties
Colour light (Bah/70); yellowish (Bah/73)
Granulation rapid, creamy consistency (Apd/83); slow (Bah/73)
Flavour characteristic, acid (Apd/83); mild (Bah/70); slightly bitter
(Wie/80)

303 Mimusops elengi L.; Sapotaceae

moulsari (INI)
Tree, evergreen; fls whitish
Distribution tropical Caribbean; subtropical N America, Asia; native to E
Indies.   **Habitat** evergreen forest (INI/MAH)

Economic and other uses
Food - fruit

Nectar rating + honeybee species; blooms, nectar flow
N1 INI/BIH,ac(Nai/76)
Blooms v-vi (INI/BIH)

Pollen
P INI/BIH

Recommended for planting to increase honey production
INI/MAH (Sub/79)

Honey no data

304 Monechma australe P.G. Meyer; Acanthaceae

perdebos (Af)
Distribution tropical Africa.   **Habitat** river banks in Kalahari (NAM);
Wiesskalk Plateau (NAM)

Nectar rating + honeybee species
N1 NAM[tm](Joh/73)

Honey flow
Honey yield "one of the best sources" in Gemsbok Park (SOU, Joh/73)

Honey no data

305 Moringa oleifera Lam.; Moringaceae
syn Moringa pterigosperma Gaertn.

horseradish tree; shevga, sohjan (INI)
Tree, <5 m, deciduous; fls white, insignificant, fragrant
Distribution tropical Asia, Africa, Caribbean; native to India.   **Habitat**
cultivated crop plant; escaped and naturalized in southern Africa

Economic and other uses
Food - lvs, roots, fls, pods.   **Land use** hedges, amenity.   **Other uses**
medicinal; ben oil from seeds for perfumery and lubrication

Nectar rating + honeybee species; blooms, nectar flow
N1 HAI(Mul/78);   INI/BIH,ac(Nai/76)
N3 INI/MAH[ac](Chu/80);   INO[ac](Bee/77)
Blooms i-iii (INI/BIH); spring and summer for a long period (tropical
America, Ord/83)

Pollen
P1 INI/BIH.   **P3** INI/MAH.   **Pollen grain** illustrated and described
(Nak/74;   Smt/56a)

Honey no data

306 Musa spp; Musaceae

banana, plantain; guineo, plátano (Es/DOR, HOD, NIA);   banano (Es/HOD,
NIA);   bananier (Fr/CEN);   pisang (In)
Herb, 3-9 m, tree-like stem formed from extended leaf petioles;   fls
yellowish-white or pinkish, monoecious.   Taxonomy difficult;   results for
M. paradisiaca, M. sapientum and their hybrids entered here
Distribution tropical S and C America, Caribbean, Africa, Asia.   **Habitat**
cultivated crop plant

Economic and other uses
Food - fresh and dried fruit;   flour and beer from fruit

Nectar rating + honeybee species; blooms, nectar flow; composition
N1 tropical America (Ord/83)
N2 DOR(Ord/64); INO[ac](Bee/77); NIA(Ord/63b)
N HOD(Ord/63)
Blooms all yr but chiefly in autumn and winter (tropical America, Ord/83);
all yr (BUM; USA/FL). **Nectar secretion** increased by rain and damp soil
conditions (Ord/83); not all of nectar accessible to bees (Mot/64).
Sugar concentration [medium] 27.4% (Fah/49); 25% (Zma/80). **Nectar
analysis** (Peo/74a; AA129,130/64; AA361/68)

Pollen
P1 INO. P DOR. **Yield** v abundant (Mot/64). **Pollen grain** v under-
represented in honey (Ric/83)

Honey: physical and other properties
Colour dark in JAM, light in AUS (Mot/64); dark (Ord/83)
Flavour astringent, like tamarind (Ord/83)

307 Myrospermum frutescens Jacq.; Leguminosae

tarara (Es/BOL)
Tree; fls golden
Distribution tropical S America

Nectar rating + honeybee species; blooms, nectar flow
N1 BOL,tm(Kem/80)
Blooms vii-ix (BOL)

Honey no data

308 Nephelium lappaceum L.; Sapindaceae

rambutan
Tree, <18 m; fls greenish-brown, small
Distribution tropical Asia; native to Malaysia. **Habitat** limited to
tropical lowlands; commonly found in villages
Rainfall high

Economic and other uses
Food - fruit. **Other uses** fat from seeds for candles

Nectar rating + honeybee species; blooms, nectar flow
N1 SIN(Kia/54; Smt/60)
N3 INO[ac](Bee/77)
Blooms iv-v (SIN)

Pollen
P SIN

Honey no data

309 Nicotiana tabacum L.; Solanaceae

tobacco; tabaco (Es)
Herb, <2.5 m depending on cv, annual; fls pink/red
Distribution tropical S America, C America, Africa, Asia; subtropical
Africa, N America; native to ?NW Argentina. **Habitat** cultivated crop
plant grown on a large scale at medium altitudes; not in areas with high
winds
Soil well drained, moderately fertile neutral soil preferred; varies with
cv. **Temperature** optimum 20-30° but up to 35° tolerated; frost-free
period of 100-120 days required. **Rainfall** moderate, 500-1000 mm, well
distributed throughout growing season

Economic and other uses
Lvs for tobacco

Alert to beekeepers
Plants may be cut before flowering is at its peak (USA, Pel/76)

Nectar rating + honeybee species; blooms, nectar flow; composition
N1 CAE[tm](Cra/73); USA(Ord/83)
N3 INO[ac](Bee/77); SOU/CAPE,NATAL,TVL,tm(And/73); USA/CT(Pel/76)
N NEZ(Wal/78); URS(Glu/55); ZIM[tm](Pap/73)
Blooms i-ii (NEZ); xi-iv (SOU); viii-frost (USA/CT). **Alert to
beekeepers** plants may be cut before flowering is at its peak (USA,
Pel/76). **Sugar analysis** (Waa/61a). Included in list of plants whose
nectar or pollen is injurious to bees (Eck/60), but no other mention of
toxicity to bees has been found

Honey flow
Honey yield [high] 40 kg/colony/season (USA, Ord/83)

Pollen
P1 INO. **P3** SOU/CAPE, NATAL, TVL. **P** ZIM. **Reference slide**

Honey: chemical composition
Water [medium] 17.00-17.96% (4 samples, Iva/78)
Sugars, total 73.56-74.50%. **Sucrose** [low] 0.18-2.56%. **Reducing sugars**
71.00-74.10%
Ash [medium] 0.10-0.12%
Total acid 22.0-33.0 meq/kg
Amylase 13.2-18.4. **HMF** 4.8-19.2
Other constituents - ?nicotine (Cra/73)

Honey: physical and other properties
Colour dark (Ord/83); v dark (Pel/76). **Pfund** dark amber (And/73; Wal/78).
Viscosity "v heavy body" (Pel/76). **Optical rotation** 0.00 to -3.25 deg
(Iva/78). **Electrical conductivity** 0.000302-0.000667/ohm cm (Iva/78)
Granulation slow (And/73; Pel/76)
Flavour burns the palate (And/73); bitter (Cra/73; Glu/55); strong,
unpalatable (Pel/76; Wal/78). **Aroma** musty (And/73)

310 Nyssa aquatica L.; Nyssaceae
syn Nyssa uniflora Wangenh.

big tupelo, black tupelo, cotton gum, tupelo gum, water tupelo (En/USA)
Tree, <30 m, deciduous; fls greenish, unisexual
Distribution subtropical N America; native to USA.   **Habitat** swamps and
river banks, southern USA; swamps on N coast of Gulf of Mexico, USA
Soil waterlogged for most of yr

Economic and other uses
Timber

Nectar rating; blooms, nectar flow
Nl USA/FL(Mor/56); USA/LA(Pel/76); USA(Smt/60)
Blooms iv (USA/FL, LA).   **Nectar flow** 10-14 days (USA/FL, Mor/56)

Honey flow
Honey yield 8 kg/colony/day (USA/LA, Pel/76)

Honey: chemical composition
Fructose high (Cra/75; Lov/66)

Honey: physical and other properties
Colour chartreuse tint (Lov/66); darker than N. ogeche honey (Rof/75).
Pfund white (Cra/75); light amber (Lov/66)
Viscosity "heavy body" (Lov/66)
Granulation slow (Cra/75)
Flavour mild (Cra/75; Lov/66); stronger than N. ogeche honey (Rof/75).
Aroma mild (Lov/66)

311 Nyssa ogeche Bartram; Nyssaceae

ogeche gum, ogeche plum, white tupelo, wild lime (En/USA)
Tree, <9 m, deciduous; fls greenish, unisexual
Distribution subtropical N America; native to USA.   **Habitat** swamps and
river banks, southern USA
Soil moist

Economic and other uses
Food - fruit "ogeche limes" for preserves

Alert to beekeepers
Shortage of summer pollen in good tupelo locations, so when flow finishes
colonies must be moved to pollen source (Pel/76)

Nectar rating; blooms, nectar flow
Nl USA/FL(Lov/56; Pel/76); USA/GA(Lov/56); USA/NC(Stp/54); USA/SC(Pel/76)
Blooms iv (USA/FL).   **Nectar flow** 10-14 days (USA/FL, Mor/56)

Honey flow
Honey yield [high] 37-54 kg/colony/season (USA/FL, Pel/76).   "leading
source" in south GA and north-west FL (USA, Lov/77)

Pollen
Alert to beekeepers shortage of summer pollen in good tupelo locations, so
when flow finishes colonies must be moved to pollen source (Pel/76).
Reference slide

Honey: chemical composition
Water [medium] 17.4-18.5% (6 samples, age 10-19 mths, Whi/62)
Glucose [low] 23.83-29.37%. **Fructose** [high] 42.25-44.26%. **Sucrose**
[medium] 0.94-1.31%. **Maltose** 6.89-8.53%. **Higher sugars** 0.82-1.22%
Ash [medium] 0.108-0.149%
pH 3.80-4.09. **Total acid** 30.27-45.14 meq/kg. **Free acid** [medium] 20.41-
30.58 meq/kg. **Lactone** 8.03-14.56 meq/kg
Amylase 15.8-19.1
Nitrogen 0.029-0.060%. **Protein** content before dialysis (Whi/62) and after
(Whi/67)
Other constituents - methyl anthranilate 0.05 µg/g (Whi/66)

Honey: physical and other properties
Pfund light amber (Lov/56); light amber to amber (Rof/75); 34-70 mm,
light amber (Whi/62)
Viscosity "good body" (Pel/76); "heavy body" (Rof/75)
Granulation slow, never if monofloral (Lov/56; Pel/76; Rof/75)
Flavour mild (Lov/56; Pel/76); mild but distinctive (Rof/75)

312 Nyssa sylvatica Marshall; Nyssaceae

black gum, highland black gum, pepperidge, sour gum (En/USA)
Tree, <30 m, deciduous; fls greenish, unisexual
Distribution subtropical and temperate N America; native to eastern USA.
Habitat hillsides or swampy sites (USA); coastal areas where protected
from wind (USA)
Soil wet or acid soil but tree can survive in drier areas; lime not
tolerated

Economic and other uses
Timber. Land use amenity

Nectar rating; blooms, nectar flow
N1 USA/NC(Lor/79; Stp/54)
N3 USA(Pel/76)

Honey flow
Honey yield "less than from other Nyssa sp, but useful inland from Gulf
coast" (USA, Pel/76)

Pollen
Pollen grain illustrated and described (Ada/72)

Honey no data

313 Olea africana Mill.; Oleaceae
syn Olea chrysophylla Lam.; Olea europaea L. subsp. africana (Mill.) P.S.
Green

wild olive; swartolienhout (Af); motlhware sigwana (BOT)
Shrub/tree, 5-18 m; fls greenish-white or whitish-cream, fragrant
Distribution tropical Africa. **Habitat** usually near water but also in open
woodland, among rocks or in mountain ravines (southern Africa)
Temperature frost tolerant. **Rainfall** drought resistant

Economic and other uses
Food - fruit. **Fodder** - browsed by stock but said to be astringent.
Fuel. Timber. Other uses medicinal

Nectar rating + honeybee species; blooms, nectar flow
N1 BOT[tm](Cra/73); ETH[tm](Cra/73); RWA,tm(Bau/66)
Blooms xi (RWA)

Pollen
P RWA

Honey no data

314 Onobrychis viciifolia Scop.; Leguminosae
syn Onobrychis sativa Lam.

esparcette, sainfoin; esparceta (Es/ARG); esparsette, sainfoin (Fr);
crocetta, lupinella (It)
Herb, 10-80 cm, perennial; fls rose pink
Distribution subtropical Africa; temperate S America, Asia, Europe, N
America; native to S Europe, W Asia. **Habitat** cultivated crop plant; not
at altitudes >300 m (UK)
Soil chalk/limestone areas; well drained soil; dry soil; pH not too
acid, 6.0-7.5 optimum. **Temperature** not winter-hardy in northern UK.
Rainfall drought resistant

Economic and other uses
Fodder - hay, pasture

Warning
Seriously affected by stem rot in USA (Lov/77)

Nectar rating; blooms, nectar flow; composition
N1 FRA(Lou/81); IRN(Cra/73); ITA(Ric/78); URS(Ave/78; Fed/55)
N2 HUN(Pet/77); ROM(Cir/80; Int/65); UK(How/79); YEA(Fie/80)
N3 POL(Dem/64a)
N ARG(Per/80); CZE(Svo/58)
Blooms vi-viii (ROM); v (UK); v-vi (URS). **Nectar flow** 10-14 days
(How/79). **Nectar secretion** 0.1-0.9 mg/fl/day (Maz/82); secretion at
temps 14-30°, optimum 22-25° (Frj/70); highest with full mineral and
phosphate fertilizers (AA316/57); secretion of plants on well fertilized
soil double that on soil with no fertilizers (AA291/56). **Sugar**

concentration [medium] 40-60% (Cir/80); 31.2% (Haa/60); 26-45% (Maz/82);
33.8% (Pek/77); 41.62% (Pet/77); 7.3-50.4% (AA130/72); 30-45%
(AA131/72); 42-52% (AA343/73); after temperature rise of 1° at night,
sugar concentration increased by 25% (AA722/72). **Sugar value** [low to
medium] 0.01-0.28 mg/fl/day (Maz/82). **Sugar analysis** (Bat/72; Bat/73a;
Maz/59; Maz/82; Wyk/52)

Honey flow
Honey yield (kg/colony/season) [high] up to 54.2 in Kazakhstan, 20-30 in
Ukraine (URS, AA686/77); 43.6 (mean wt gain, 2 hives, USA/MT, Dul/68);
high in central Italian Apennines (Ric/78). "Miel du Gâtinais" (France)
was largely from this source. **Honey potential** (kg/ha) [high] 500-600
(URS/Transcaucasia, Fed/55); [moderate] 65.5 (BUL, Pek/77); 120 (GDR,
Bec/67); 120-300 (ROM, Cir/80); 100 (ROM, AA655/70); 90-400 (URS, Ave/78)

Pollen
Pl FRA; ITA. **P** URS. **Yield** moderate (Cir/80). **Chemical analysis**
(Cir/80); pollen v oily (How/79). **Colour** of load dark brown,
consistency of load sticky/rubbery (Ric/78). **Pollen grain** illustrated and
described (Saw/82). **Reference slide**

Honey: chemical composition
Water [medium] 17% (Dul/68); 16.39% (Sac/55)
Sugars (as % of total sugars): **glucose** 41.89% (9 samples, Bat/73); 40.8-
42.9% (3 samples, Maz/59); **fructose** 50.26% (Bat/73); 51.3-55.0%
(Maz/59); 51.6% (Maz/64); **sucrose** 0.43% (Bat/73); 2.2-8.4% (Maz/59);
2.0% (Maz/64); **maltose** 3.41% (Bat/73); 4.8% (Maz/64); **isomaltose** 0.23%
(Bat/73); **fructomaltose** 2.0% (Maz/64); **trehalose** 1.57% (Bat/73);
gentiobiose 0.14%; **melezitose** 0.81%; **raffinose** 0.15%
Nitrogen 0.038% dry wt (Bos/78). **Amino acids,** free 0.180%, protein 0.130%
Fermentation likely (Pia/81)

Honey: physical and other properties
Colour yellow (Bab/61); deep yellow, bright and sparkling (How/79); v
clear, pale yellow (Lov/56); light yellow (Ric/78). **Pfund** light amber
(Cra/75); 7.5 mm, water white (Dul/68); white to extra light amber (Pia/81)
Optical rotation -27.90 deg (Bat/73). **Electrical conductivity** 0.000140
per ohm cm (Vor/64)
Granulation rapid, fine, solid consistency (Spo/50); regular, fine-grained
(Ric/78)
Flavour sweet, quite pronounced (Cra/75); characteristic (How/79); sweet,
like fruit (Pia/81); less sweet than other honeys, sometimes character-
istic (Spo/50). **Aroma** faint (Pia/81; Ric/78); delicate (Spo/50)

315 Opuntia engelmanii Salm-Dyck; Cactaceae

Indian fig, prickly pear (En/USA); nopal, tuna (Es/MEX)
Herb, cactus; fls yellow tinged red, large
Distribution tropical C America; subtropical N America. **Habitat** desert
areas of south eastern USA and MEX; increasingly common where heavy
grazing has occurred (USA)
Rainfall arid areas; drought resistant

Economic and other uses
Food - fruits.   **Fodder** - fruits

Nectar rating;   blooms, nectar flow
N1 USA(Ord/83)
N2 MEX(Ord/83);   USA/TX(Pel/76)
N MEX(Ord/72)
Blooms vi-vii (northern and central America, Ord/83).   **Nectar flow** brief, seldom more than 4-5 days (USA/TX, Pel/76)

Honey flow
Honey yield [high] 30 kg/colony/season (south-west USA, Ord/83);   about every 4 yrs (USA, Pel/76).   **Honey potential** high, especially during partial drought (USA/TX, Pel/76)

Pollen
P1 MEX;   USA.   **P** USA/TX.   **Yield** abundant (Pel/76).   **Pollen value** important (Ord/83).   **Pollen grain** illustrated and described (Nye/71)

Honey: physical and other properties
Pfund light amber (Cra/75;   Pel/76);   amber (Ord/83)
Viscosity high (Cra/75;   Ord/83).   **Other physical properties** - exhibits stringiness (Pel/76;   Pry/50;   Pry/52);   also dilatancy (Pry/50;   Pry/52)
Granulation - large crystals in clear liquid (Cra/75)
Flavour strong (Cra/75);   v rank (Pel/76)

316 Oxydendron arboreum (L.) DC.;   Ericaceae

sourwood, tree sorrel (En/USA)
Tree/shrub, 25 m (wild), 7-8 m (cultivated), deciduous;   fls white, fragrant
Distribution temperate N America;   native to southern USA.   **Habitat** sheltered semi-shaded areas;   steep rocky slopes in S Alleghenies (USA)
Soil acid humus-rich soil preferred

Economic and other uses
Fuel.   **Land use** amenity.   **Other uses** medicinal

Alert to beekeepers
Pollen inadequate for brood rearing, which ceases abruptly on this flow (Tea/54)

Nectar rating;   blooms, nectar flow
N1 USA/NC(Stp/54)
N2 USA/KY(Pel/76);   USA/NC(Lov/56;   Pel/76);   USA/TN(Lit/54;   Lov/56;   Pel/76)
Blooms vi-vii (USA/NC).   **Nectar flow** up to 21 days (Tea/54).   **Nectar secretion** cool nights and dry weather prolong flow (Lov/59c);   highest at altitudes >300 m (Tea/54)

Honey flow
Honey yield (kg/colony/season) [high] mean 27, max 102, heaviest every 4-5 yrs (USA, Lov/59c);   mean 34 (USA, Pel/76)

Pollen
Alert to beekeepers pollen inadequate for brood rearing, which ceases
abruptly on this flow (Tea/54)

Recommended for planting to increase honey production
USA/southern Alleghenies (Lov/55d).   Propagate by seed, layering,
cuttings.   See **Alert to beekeepers**

Honey: chemical composition
Water [medium] 16.6, 17.8% (2 samples, age 7, 15 mths, Whi/62)
Glucose [low] 25.48, 25.23%.   **Fructose** [medium] 40.73, 39.20%.   **Sucrose**
[low] 0.97, 0.85%.   **Maltose** 10.47, 11.38%.   **Higher sugars** 2.35, 2.29%
Ash [medium] 0.217, 0.259%
pH 4.65, 4.47%.   **Total acid** 16.13, 20.06 meq/kg.   **Free acid** [low] 14.89,
14.92 meq/kg.   **Lactone** 1.23, 5.14 meq/kg
Amylase 21.7, 15.6
Nitrogen 0.026, 0.014%
Other constituents - small amount of oxalic acid (Mei/71)

Honey: physical and other properties
Colour yellow (Mei/71); pinkish cast (Lov/59c).   **Pfund** almost water white
(Lov/56); 12-42 mm, white to extra light amber (Whi/62)
Viscosity "thick" (Mei/71); "heavy body" (Pel/76)
Granulation slow (Pel/76; Rof/75)
Flavour delicate, v slightly sour (Lov/56); distinctive (Lov/59c);
slightly piquant (Mei/71); mild (Rof/75)

317 Paliurus spina-christi Mill.; Rhamnaceae

Christ's thorn, Jerusalem thorn; marruca (It)
Shrub/tree, <3 m, deciduous; fls greenish-yellow, small
Distribution temperate Europe, (Med) Africa; native from S Europe to E
Asia.   **Habitat** hedges, roadsides, thickets, maquis and garigue
Temperature hotter areas.   **Rainfall** drier areas

Economic and other uses
Land use hedges (v resistant to grazing)

Nectar rating; composition
N1 ITA(Ric/78)
Sugar analysis (Bat/73a)

Honey flow
Honey yield - in ITA honey usually mixed with that from Erica sp (in
Grosseto) or Trifolium pratense (in Abruzzo) (Ric/78)

Pollen
P3 ITA.   **Colour** of load greenish-yellow (Ric/78).   **Reference slide**

Honey no data

318 Parkia biglobosa (Jacq.) Benth.; Leguminosae

locust bean tree; néré, nété (SEN)
Tree, <15 m, deciduous, dense spreading crown; fls red, small, in spheres
hanging on 38 cm stalks; not all fls fertile, only those nearest stalk
produce nectar
Distribution tropical Africa. **Habitat** savanna (IVO)
Soil deep heavy sand; also poor rocky soil. **Rainfall** 500-700 mm; humid,
sub-humid zones

Economic and other uses
Food - seeds dried for flour/flavouring; fruit pulp. **Fodder** - fruits.
Timber, but easily attacked by termites. **Land use** shade. **Other uses**
tannin from bark; mordant for indigo dyeing

Nectar rating + honeybee species
N1 IVO,tm(Bor/76; Dou/80); SEN,tm(Dou/70)

Honey flow
Honey yield "one of the most important tree sources" in IVO (Bor/76)

Honey no data

319 Parkinsonia aculeata L.; Leguminosae

Jerusalem thorn; horsebean (En/USA); retama (USA)
Tree, <10 m, thorny; fls bright yellow, numerous
Distribution tropical Africa; subtropical N America, S America, Africa;
native from SW USA to Argentina. **Habitat** desert grasslands and canyons
(USA); escaped and naturalized in southern Africa; coastal sandy sites;
in full sun
Soil dry sites; poor gravelly or sandy alluvial; salt tolerant;
waterlogging not tolerated. **Temperature** up to 36°; light frost
tolerated. **Rainfall** 200-1000 mm; drought resistant

Economic and other uses
Food - seeds. **Fodder** - pods and young branches. **Fuel**. **Land use** hedges
and "living fences", windbreak, amenity. **Soil benefit** cover for soil
conservation; erosion control

Warning
Thorny; reproduces easily from seed - can become a nuisance (Usa/80)

Nectar rating + honeybee species; blooms, nectar flow; composition
N1 MOZ[tm](Cra/73)
N2 USA/TX(Lov/61d)
N3 USA/CA,TX(Lov/56)
Blooms all summer (USA/TX). **Sugar concentration** [medium] 30.5%, fairly
dilute nectar collected by bees in hot months, viii, ix (ISR, Eis/82)

Pollen
P USA/CA. **Pollen grain** illustrated and described (Mag/78; Smt/56a).
Reference slide

Honey: physical properties
Pfund amber (Lov/56; Mot/64)

320 Parthenocissus quinquefolia (L.) Planch.; Vitaceae
syn Vitis parthenocissus quinquefolia (L.) Planch.

Virginia creeper; parrita (Es/CUB); tripas de iguana (Es/GUM);
Jungferurebe, wilder Wein (De)
Shrub, climber, deciduous, woody; fls greenish, small
Distribution temperate N America, Europe; subtropical N America; tropical
C America, Caribbean; native to Mexico and USA. **Habitat** woods, thickets,
by rivers and on rocky ground

Economic and other uses
Land use amenity

Nectar rating; blooms, nectar flow
N1 URS(Fed/55)
N2 USA/LA(Lie/72)
N3 GFR(Gle/77)
N USA(Ord/83)
Blooms vi-vii (USA)

Honey flow
Honey yield [moderate] 7-11 kg/colony/season, annual (Lov/59a)

Pollen
P3 GFR.   **Pollen grain** illustrated and described (Lie/72; Saw/82)

Honey: physical and other properties
Colour slightly reddish cast (Lov/59a).   **Pfund** light amber
Flavour distinctive

321 Persea americana Mill.; Lauraceae

avocado; aguacate (Es/DOR, ELS, HOD, MEX, NIA); avocatier (Fr/MAT);
abacateiro (Pt/MOZ)
Tree, <20 m, evergreen; fls yellow-green/cream, small
Distribution tropical C America, Caribbean, Africa; subtropical N America,
Africa; native to Mexico and C America. **Habitat** cultivated crop plant;
wind causes fl shedding
Soil wide range.   **Rainfall** >1500 mm (MAT); high humidity preferred

Economic and other uses
Food - fruit

Nectar rating + honeybee species; blooms, nectar flow; composition
N1 HAI(Mul/78); ?MAT(Bal/76); MOZ[Lm](Gra/77)
N2 DOR(Ord/64); ELS(Woy/81); MEX(Ord/83); NIA(Ord/63a)

N3 SOU/CAPE,NATAL,TVL,tm(And/73);   USA/CA,FL(Lov/56)
N CUB(Ord/44);   HOD(Ord/63);   PAR(Bra/59);   TRI(Lau/76)
Blooms viii-ix (BRA/SC, Caa/72); ii-iv (CUB); xi-i (ELS); ii-iv
(USA/FL).   **Nectar secretion** abundant in favourable conditions (Mot/64);
greatly affected by climate and soil moisture (And/73; Ord/44).   **Sugar
concentration** [medium] 44-49% (Caa/72; Sao/54); 15% (Van/41).   Citrus
crops, if adjacent, often preferred by bees as nectar source (Frj/70)

Pollen
P3 SOU/CAPE, NATAL, TVL.   P BRA, DOR.   **Pollen grain** heavy and sticky
(Frj/70); illustrated and described (Mag/78; Sao/61)

Honey: physical and other properties
Colour dark (Cra/75; Mot/64; Ord/83; Woy/81); v dark (Zma/80).   **Pfund**
dark amber (And/73)
Viscosity "heavy body" (Cra/75); "thick" (Woy/81)
Granulation slow (And/73)
Flavour and aroma strong (And/73; Ord/83)

322 Persea caerulea (Ruiz & Pavon) Mez.;   Lauraceae

aguacatillo (Es/VEN)
Tree
Distribution tropical S America.   **Habitat** valleys and humid regions (VEN)

Economic and other uses
Medicinal

Nectar rating; blooms, nectar flow
N1 VEN(Cra/73)
N VEN(Ste/71)
Blooms iii (VEN)

Honey: physical and other properties
Flavour bitter (Cra/73; Ste/71)

323 Petalidium linifolium T. Anders; Acanthaceae

lusernbos (Af)
Distribution subtropical and tropical Africa.   **Habitat** Weisskalk Plateau;
common in the Schwarzrand (NAM)

Nectar rating + honeybee species; blooms, nectar flow
N1 NAM[tm](Joh/73)
Blooms in winter (NAM)

Honey flow
Honey potential large-scale migration of hives to this flow suggested
(Joh/73)

Honey no data

324 Phacelia tanacetifolia Benth.; Hydrophyllaceae

phacelia; fiddle neck, valley vervenia (En/USA); phacélie (Fr);
Büschelschön (De)
Herb, 40-70 cm, annual; fls bluish-pink
Distribution temperate Europe, N America, Oceania; native to USA/CA.
Habitat cultivated plant, often especially for bees; naturalized elsewhere
Soil wide range but humus-rich sandy soil preferred. **Temperature** 5° of
frost tolerated

Economic and other uses
Fodder - either green or as silage. **Soil benefit** green manure

Nectar rating; blooms, nectar flow; composition
N1 CZE(Svo/58); GFR(Gle/77); POL(Dem/64a); URS(Glu/55)
N3 FRA(Lou/81); USA/CA(Pel/76)
Blooms vi-viii (EUR, Maz/82); early summer to autumn (USA). **Nectar
secretion** (mg/fl/day) 0.80-0.85 (Han/80); 1.04-1.62 (AA848/64); nectar
secreted at temperatures 10-31° (AA105/67); best at 16-24°, RH 55-70%
(AA341/64); secretion increased by application of complete N-K-P
fertilizer (AA264/58); increased by late spring rain (USA/CA, Lov/77).
Sugar concentration [medium] 28% (Cir/80); 40-43% (Han/80); 53.87%
(Pet/77); 34.0% (AA132/64); 15.4-36.0% (AA848/64); wet summer 28.7%, dry
summer 51.8% (AA767/65). **Sugar value** (mg/fl/day) [medium] 0.269, 0.288
(Bac/60); 0.31-0.36 (Han/80); 0.510 (Sim/75). **Sugar analysis** (Maz/82;
Wan/64; Wyk/52). **Amino acid analysis** (Bak/77; Yak/73; AA678/66)

Honey flow
Honey yield [moderate] 5-9 kg/colony/season (USA/CA, Lov/77). **Honey
potential** (kg/ha) [high] 300-1000 (CAF, Sza/82); 134-1129 (POL,
AA848/64); 300-1000 (ROM, Apc/68); [moderate] 214-496 (EUR, Maz/82); 400-
500 (GDR, Bec/67); 331 (POL, AA274/62); 361 (ROM, Bac/60); 100-340 (ROM,
Cir/80); 250-500 (URS, Fed/55)

Pollen
P2 GFR. **P3** FRA. **Yield** 0.5 mg/fl (Maz/82); good (Cir/80); abundant
(Van/41). **Colour** of load dull brown (Cir/80); load dark blue (Han/80;
Van/41). **Reference slide**

Recommended for planting to increase honey production
URS (Ave/78); YUG (Kul/59). Propagate by seed; blooms in 6 wks (NEZ,
Bes/81a). Widely recommended for wasteland, orchards or as cover crop;
also for providing flow for same season, in crops of (biennial) sweetclover
(How/79)

Honey: chemical composition
Sugars (as % of total, Maz/64): **glucose** 35.0%; **fructose** 49.7%; **sucrose**
7.5%; **maltose** 4.9%; **fructomaltose** 2.9%. **Galactose** present (Wan/64)
Amino acids 0.000489% dry wt, also contents of 9 individual acids (Mos/65)

Honey: physical and other properties
Colour sometimes light green (Glu/55; Lov/60a; Pel/76). **Pfund** amber
(Cra/75; Pel/76); may be white (Glu/55)
Viscosity "flows freely" (Cra/75). **Electrical conductivity** 0.000094/ohm
cm (Vor/64)
Granulation rapid (Cra/75)
Flavour mild (Pel/76)

325 Phaseolus multiflorus Lam.; Leguminosae

white kidney bean (En/SOU)
Herb, tall twining perennial but grown as annual; fls red or white,
depending on cv
Distribution subtropical Africa; temperate Europe; tropical S America;
native to S America. **Habitat** cultivated crop plant; Transvaal Highveld
(SOU)
Soil dolomitic (SOU)

Economic and other uses
Food - beans

Nectar rating + honeybee species; blooms, nectar flow
N1 SOU/TVL,tm(And/73; Joh/75)
N URS(Fed/55; Glu/55)
Blooms i-iii (SOU/TVL). **Nectar secretion** v dependent on adequate moisture
(And/73; Joh/75)

Honey flow
Honey yield "second most important source" in SOU/TVL (Joh/75). sometimes
"good" on Highveld (SOU, And/73)

Pollen
P3 SOU/TVL. **Yield** little if any collected (Joh/75). **Reference slide**

Honey: physical and other properties
Colour light (Joh/75). **Pfund** white (And/73; Fed/55)
Granulation slow, smooth to medium grain (And/73)
Flavour mild (And/73); little (Fed/55); honey from red fls v sweet
(Glu/55). **Aroma** little (And/73)

326 Phellodendron amurense Rupr.; Rutaceae

Amur cork tree; Amur lombardy poplar; arbore de plută (Ro)
Tree, 15 m, deciduous; fls yellow-green, dioecious
Distribution temperate Europe, Asia. **Habitat** forests of URS/Far East

Economic and other uses
Timber. Land use amenity. **Other uses** bark medicinal; insecticide from
seeds

Nectar rating; blooms, nectar flow; composition
N1 URS/Far East (Glu/55; Pem/65)
N ROM(Cir/80)
Blooms v-vi (ROM). **Nectar secretion** max at temp 21 to 23° and RH 44 to 58% (Pem/65). **Sugar analysis** (Pem/65)

Honey flow
Honey yield - together with Tilia amurensis, T. mandschurica and T. taqueti, 75-80% of honey in URS/Far East (Pem/65); also important in Ussuri region and N parts of European Russia (URS, Glu/55). **Honey potential** [moderate] 100-150 kg/ha (ROM, Apc/68; Cir/80)

Pollen
Reference slide

Honey no data

327 Phlebophyllum kunthianum Nees; Acanthaceae

kurinji (INI)
Distribution tropical Asia; native to India. **Habitat** very common throughout Kodaikanal hill country, INI/TAM

Nectar rating + honeybee species; blooms, nectar flow
N1 INI/TAM[ac](Chn/74)
Blooms x-xi, every 12 yrs (INI/TAM)

Pollen
P1 INI/TAM

Honey no data

328 Piscidia piscipula (L.) Sarg.; Leguminosae

dogwood (En/JAM); guamá candelón (Es/CUB); barbasco (Es/MEX); bois enivré (Fr/GUD, MAT)
Tree, medium height; fls pink
Distribution tropical C America, Caribbean, S America; subtropical N America. **Habitat** hills on dry rocky coasts of tropical America; common on Keys, USA/FL

Economic and other uses
Bark and lvs yield narcotics used in fishing

Nectar rating; blooms, nectar flow
N1 BEL(Mul/77)
N2 CUB(Ord/83); GUD(Ord/83); JAM(Ord/83); MAT(Ord/83); MEX(Ord/70; Ord/83)

Blooms iii (BEL); iii-v (tropical America, Ord/83).   **Nectar flow** "one of best in W Indies but abundant in few areas" (Mot/64)

Honey: physical and other properties
Pfund extra white (Mul/77); amber (Ord/83)
Granulation slow or none (Mot/64)

329 Pithecellobium arboreum (L.) Urb.; Leguminosae

Tree, <20 m; fls greenish-white
Distribution tropical C America, Caribbean.   **Habitat** forests and river banks

Economic and other uses
Timber

Nectar rating; blooms, nectar flow
N1 ?HAI(Mul/78)
N2 CUB(Smt/60)
Blooms iii-iv (Antilles, Ord/83)

Pollen
Yield low (Ord/83)

Honey no data

330 Pithecellobium dulce (Roxb.) Benth.; Leguminosae

Madras thorn (En/USA); chiminango (Es/COL); jina extranjera (Es/DOR); guamúchil (Es/MEX)
Tree, <20 m, almost evergreen, thorny; fls white/yellowish
Distribution tropical S America, C America, Caribbean, Asia, Africa, Oceania; subtropical N America; native from southern California to Venezuela and Colombia.   **Habitat** widely planted and naturalized in tropics; some dry coastal areas of Africa; warmer drier areas of Philippines and India; arid/semi-arid areas; altitudes <1500 m
Soil wide range; oolitic limestone, clay and barren sand; waterlogging and salt tolerated.   **Temperature** shade and heat tolerated.   **Rainfall** 450-1650 mm; drought resistant; max dry period 4-5 mths

Economic and other uses
Food - pods; oil from seeds.   **Fodder** - lvs and twigs; pressed cake from seeds. **Fuel.   Timber.   Land use** hedges, windbreaks, shade, amenity. **Soil benefit** N-fixation.   **Other uses** oil from seeds; tannin from bark; gum

Warning
Irritant sap; seed germinates rapidly; thorny, infests pastures in HAW; branches and trunks break in high wind (Usa/80)

Nectar rating; blooms, nectar flow
N1 tropical America (Ord/83)
N2 DOR(Ord/83);   MEX(Wis/53);   USA/FL(Ord/83)
N COL(Ken/76)
Blooms xii-iv (Ord/83)

Pollen
P2 DOR;   tropical America;   USA/FL.   **Yield** abundant (USA/FL, Mot/64).
Pollen value high (Ord/83)

Recommended for planting to increase honey production
Tropical America (Ord/83).   Propagate by seed/cuttings.   Grows 1 m per yr;   fls in 2nd yr.   See **Warning**

Honey no data

331 Pithecellobium unguis-cati (L.) Mart.;   Leguminosae

cat's claw (En/JAM)
Shrub/tree, small, v thorny;   fls greenish yellow, fragrant
Distribution tropical S America, C America, Caribbean;   subtropical N America.   **Habitat** shrub-covered areas on low-lying coasts (Antilles, USA/FL, northern S America)

Economic and other uses
Food - fruit

Nectar rating;   blooms, nectar flow
N1 JAM(Ord/83)
Blooms iii-iv (Caribbean, Ord/83).   **Nectar secretion** abundant (Mot/64)

Honey: physical and other properties
Pfund light amber (Mot/64;   Ord/83)

332 Plectranthus gerardianus Wall. ex Benth.;   Labiatae

Shrub/under-shrub, 60-200 cm;   fls white spotted purple
Distribution subtropical Asia.   **Habitat** lower ranges of Himalayas (INI)

Nectar rating + honeybee species;   blooms, nectar flow
N1 PAK,ac(Pak/77)
Blooms viii-ix (PAK)

Honey flow
Honey yield main source in north-east PAK (Pak/77)

Pollen
P PAK

Honey no data

333 Plectranthus striatus Wall. ex Benth.; Labiatae

Shrub/under-shrub, 60-200 cm; fls white
Distribution subtropical Asia.   **Habitat** lower ranges of Himalayas (INI)

Nectar rating + honeybee species; blooms, nectar flow
N1 PAK,ac(Pak/77)
Blooms x-xii (PAK)

Honey flow
Honey yield main source in north-east PAK (Pak/77)

Pollen
P PAK

Honey no data

334 Polygonum persicaria L.; Polygonaceae

heartsease, lady's thumb, smartweed, spotted knotweed (En/USA)
Herb, 20-80 cm; fls pink
Distribution temperate N America, Europe; native to Europe.   **Habitat** weed
of cornfields and waste places, USA
Soil moist

Economic and other uses
Food - lvs for seasoning

Nectar rating; blooms, nectar flow
N1 URS(Fed/55);   USA/IA,IL,KS(Lov/56)
N2 USA/IA,IN,WI(Pel/76)
N URS(Glu/55);   USA/MA(Shw/50a)
Blooms midsummer to frost (northern USA, Pel/76).

Honey flow
Honey yield (kg/colony/season) [high] 45 plus winter stores (USA/IA,
Pel/76);   mean 23, max 136 (USA/KS, Lov/77);   mean 23, max 223 (USA/KS,
Roo/74)

Honey: physical and other properties
Colour varies with soil, weather, and is affected by heating (Lov/57d);
pinkish tinge (Roo/74).   **Pfund** light to dark amber (Fed/55);   white to
amber (Lov/58)
Flavour strong, but less so than buckwheat honey;   unusually mild in parts
of IL (USA, Lov/57d);   v strong (Roo/74).   **Aroma** v strong (Lov/56);
unpleasant particularly when fresh (Lov/57d)

335 Pongamia pinnata (L.) Pierre; Leguminosae
syn Pongamia glabra Vent.

sour fruit; hunge, karanji (INI)
Tree, medium height, deciduous; fls pale pink
Distribution tropical Asia, Oceania; subtropical N America; native to
India.   **Habitat** humid lowland tropics, also drier parts of INI; coastal
forests and tidal river banks (INI/south); shade tolerated well;
altitudes <1200 m
Soil wide range including sandy/rocky; highly salt tolerant, survives with
roots in salt water.   **Temperature** 0-50°, mature trees only.   **Rainfall** 500-
2500 mm; drought resistant

Economic and other uses
Fodder – lvs; pressed cake for poultry.   **Fuel.**   **Land use** shade,
afforestation, amenity.   **Soil benefit** erosion control; green manure.
Other uses roots/seeds as a fish poison; oil from seed for lamps etc;
bark fibres; medicinal; pesticides

Warning
Toxic seeds and roots.   Aggressive surface root system; suckers and
seedlings may run wild (Usa/80)

Nectar rating + honeybee species; blooms, nectar flow
N1 INI/BIH,ac(Nai/76);   INI/KAR,KER[ac](Kha/59)
N3 INI/MAH[ac](Chu/80)
Blooms iii (INI);   iv-vi (INI/MAH)

Pollen
P1 INI/BIH.   **P3** INI/MAH.   **P** INI/KAR, KER

Honey: physical and other properties
Colour dark (Mot/64)
Flavour sweet at first, with chalky after-taste (Mot/64)

336 Prosopis cineraria (L.) Druce; Leguminosae

mesquite
Tree, 5-9 m, evergreen, prickly; fls yellow
Distribution tropical and subtropical Asia.   **Habitat** low altitudes;
regions with hot dry winds
Soil alluvial; coarse sandy soil; alkaline, pH up to 9.8; black cotton
soil in open forest; dry stony land; moderately salt tolerant.
Temperature in shade 40-50° to -6°.   **Rainfall** 75-850 mm with long dry season

Economic and other uses
Fodder – browse.   **Fuel.**   **Timber.**   **Land use** shade, afforestation.   *Soil*
benefit erosion control; dune stabilization; increase of soil fertility
beneath canopy; organic manure

Warning
Prickly pestilential weed in sub-humid areas (Usa/80)

Nectar rating + honeybee species; blooms, nectar flow
N1 PAK,ac(Pak/77)
Blooms xii-iii (PAK)

Honey flow
Honey yield "important" in parts of PAK (Pak/77)

Pollen
P PAK

Honey no data

337 Prosopis farcta (Sol. ex Russell) J.F. Macbride; Leguminosae

mesquite
Shrub/tree, 0.3-3.0 m, prickly; fls creamy green
Distribution temperate (Med) Africa; subtropical Africa, Asia; native to
N Africa, E Med, Iraq, Iran, Afghanistan, Pakistan, URS/Transcaucasia and
Turkestan. **Habitat** open dry scrubland; mountainous areas
Soil deep alluvium with shallow ground- water preferred; also dry clayey
soil; untilled saline soil

Economic and other uses
Fodder. Fuel. Other uses tannin from roots

Warning
Noxious invasive weed in Transcaucasia (URS, Buk/76)

Nectar rating + honeybee species; blooms, nectar flow; composition
N1 PAK,ac(Pak/77)
Blooms iv-ix (PAK). **Sugar concentration** [high] 75% (Fah/49)

Honey flow
Honey yield "important" in PAK/NWFP (Pak/77)

Pollen
P PAK

Honey no data

338 Prosopis glandulosa Torrey; Leguminosae

mesquite; honey mesquite, honey-pod (En/USA); guajilla, uña de gato (USA)
Tree/shrub, 1.5-9.0 m, often multistemmed, straggly, deciduous, spiny; fls
pale yellow
Distribution subtropical N America, Africa, Asia, Oceania; tropical Asia,
Caribbean; native to north Mexico and south-west USA. **Habitat** dry
plains, mesas, canyons and hillsides (USA); altitudes 760-1520 m (USA)
Soil light, shallow; sandy. **Rainfall** v drought resistant

Economic and other uses
Fodder young shoots; pods, but not as exclusive diet for cattle
(Usa/79). **Other uses** gum

Warning
Do not feed cattle on pods only (Usa/79). Major pest of grassland in
southern USA; highly invasive especially in moist locations or good soils
(Usa/79)

Nectar rating + honeybee species; blooms, nectar flow
N1 PAK,ac(Pak/77); USA/AZ,CA,NM,?NV,?OK,?UT(Lov/56); USA/TX(Ord/83)
N2 USA/AZ,CA,NM,TX(Pel/76)
N3 INO(Bee/77)
N USA/AZ(Mof/81)
Blooms iii-ix (PAK); iv, vi-vii (USA/TX). Most fls produced when soil
moisture is low; fls shed during rain (Lov/56a). **Nectar secretion** higher
on sandy than on heavy soil (Pel/76)

Honey flow
Honey yield (kg/colony/season) [high] mean 27, max 90 (USA/TX, Lov/56a);
up to 90 (USA, Roo/74); main source in Punjab and Sind (PAK, Pak/77)

Pollen
P1 PAK. **Pollen value** high (Van/49)

Honey: physical and other properties·
Pfund 30-40 mm, white to extra light amber (Lov/56a); light amber
(Roo/74)
Granulation rapid (Roo/74)
Flavour mild, sweet (Lov/57a)

339 Prosopis juliflora (Sw.) DC.; Leguminosae

algaroba (En/AUS, USA); mesquite (En/AUS, SOU, USA); cashaw (En/JAM);
cupesí (Es/BOL); duitswesdoring (Af)
Tree/shrub, 3-12 m, deciduous, somewhat spiny; fls greenish white to light
yellow; often confused with other Prosopis spp; all USA records for this
sp now treated as P. glandulosa; records from southern Africa included
here, but see Buk/76
Distribution tropical C America, Caribbean, S America, Africa, Asia;
native to C America, Caribbean and northern S America. **Habitat** coastal;
planted in many arid areas; altitudes <1500 m
Soil wide range; sandy; rocky if root growth not impeded. **Temperature** v
warm climates preferred; some cvs not frost hardy. **Rainfall** 150-750 mm;
v drought resistant

Economic and other uses
Food - flour from pods. **Fodder** - pods. **Fuel. Timber. Land use**
shade. **Soil benefit** dune stabilization

Warning
Aggressive invader; should be grown only in v arid problem sites (Usa/80)

Nectar rating + honeybee species; blooms, nectar flow
N1 AUS/WA(Col/62);   BOL,tm(Kem/71);   HAI(Mul/78);   JAM(Met/66);
PAK,ac(Pak/77)
N2 NAM[tm](Joh/73);   SOU,tm(And/73)
N3 INO(Bee/77)
Blooms xi-xii (AUS/WA);   vii-viii (BOL);   i-iv (JAM);   iv-vi (PAK);   x-xii
(NAM);   viii-xii, peak x-xii (SOU)

Honey flow
Honey yield "important" in Punjab and Sind (PAK, Pak/77);   "heavy" (SOU,
And/73)

Pollen
P1 AUS/WA.   **P3** SOU.   **P** BOL;   PAK.   **Reference slide**

Honey: physical and other properties
Pfund v light amber (And/73)
Granulation medium (And/73)

---

340 Prosopis pallida (Humboldt & Bonpl. ex Willd.) Kunth;   Leguminosae

kiawe (HAW)
Tree, shrub when on sterile soils, 8-20 m, spines small/absent; fls
greenish yellow
Distribution tropical Oceania, Caribbean, S America;   subtropical Asia;
native to Peru, Ecuador, Colombia.   **Habitat** coastal;   naturalized in HAW
and PUE;   cultivated in INI and AUS;   altitudes <300 m
Soil wide range including old lava flows, coastal sand;   highly salt
tolerant.   **Rainfall** 250-1250 mm;   v drought resistant

Economic and other uses
Food - pods for syrup to use in drinks.   **Fodder** - lvs and pods.   **Fuel.**
Timber.   **Land use** windbreak, afforestation, amenity

Warning
May become invasive and form thickets.   Shallow-rooted, easily blown down
in storms (Usa/80)

Nectar rating
N1 HAW (Esb/80)

Honey flow
Honey yield [high] Puako region 227-363, Molokai Island 120-150 kg per
colony/season (HAW, Esb/80)

Honey: chemical composition
Water [medium] 17% (Eck/52)

Honey: physical properties
Pfund 1.9 mm, water white (2.1, 2.6 mm after 2, 22 h at 70°, Eck/52)

341 Prosopis pubescens Benth.; Leguminosae

mescrew, screwbean (En/USA); tornillo (Es/MEX)
Tree/shrub, 2-10 m, spiny; fls yellow
Distribution subtropical N America, Africa; tropical C America; native to
south-west N America, N Mexico.   **Habitat** lowland areas subject to frequent
flooding (SW, USA); steppe zone from Coahuila to Sonora (MEX)

Economic and other uses
Food - pods.   **Fodder** - pods for cattle.   **Fuel.**   **Timber**

Nectar rating
N1 USA/south-west(Pel/76)
N2 MEX(Ord/83); USA(Ord/83)

Honey flow
Honey yield more "reliable" than P. glandulosa (USA, Pel/76)

Honey: physical properties
Colour light (Lov/56).   **Pfund** white (Lov/59f); amber (Ord/83)

342 Prunus x yedoensis Matsum.; Rosaceae

flowering cherry, Japanese cherry; sakura (JAP)
Tree, deciduous; fls pink; nectary in fl, also 1-2 extrafloral nectaries
on leaf petiole (Inu/57)
Distribution temperate Europe, Asia; origin uncertain.   **Habitat**
cultivated as ornamental; not known wild; hills and fields (JAP)

Economic and other uses
Land use amenity

Nectar rating; blooms, nectar flow; composition
N1 JAP(Inu/57)
N3 UK(How/79)
Blooms early iv (JAP).   **Sugar concentration** [medium] 28.2% (Ech/72;
Ech/77).   **Sugar analysis** (Ech/72; Ech/77)

Pollen
P JAP; UK

Honey: chemical composition
Water [medium] 20% (1 sample, Waa/61)
Glucose [high] 40.39, 50.9% (2 samples, Waa/61).   **Fructose** [medium] 35.21,
44.01%.   **Sucrose** [low] 0.45, 0.56%.   **Reducing sugars** 78.07, 97.59%

Honey: physical and other properties
Colour golden (Inu/57); pale yellow (Waa/61)
Optical rotation -13.19 deg (Waa/61)
Granulation rapid

343 _Psidium guajava_ L.; **Myrtaceae**

guava; guayabo (Es/BOL); goyavier (Fr); goiabeira (Pt/BRA, MOZ); amrud
(PAK)
Tree/shrub, <8 m, evergreen, forms thickets, "guayabales", in tropical
America; fls white, fragrant
Distribution tropical America, Caribbean, Africa, Asia, Oceania;
subtropical N America, S America, Asia; native to C America. **Habitat**
widely cultivated in tropical Americas; warm coastal areas (BRA/SC);
altitudes <5000 m
Soil wide range; well drained or moisture retentive; temporary
waterlogging tolerated. **Temperature** high tolerated; not frost resistant

Economic and other uses
Food - fruit, fresh or for preserves; juice for drink. **Timber.** **Other
uses** medicinal, dyes, tannin

Warning
Troublesome weed in pastures; designated noxious weed in FIJ (Pus/68)

Nectar rating + honeybee species; blooms, nectar flow; composition
N1 BOL[tm](Kem/71); MOZ[tm](Cra/73); PAK,ac(Pak/77)
N2 INI/MAH[ac](Chu/80); SOU/TVL,tm(Bey/68)
N3 BRA/SC,tm(Wie/80)
N TRI(Lau/76)
Blooms viii-ix (BOL); ix-i (BRA/SC); iii-vi, ix-xii (INI/MAH); iv-v, ix-
x (PAK). **Nectar secretion** abundant all day (Mot/64). **Sugar conc-
entration** [medium] 28% (Wie/80). Juice from damaged fruit collected by
bees i-ii (BOL, Kem/80)

Honey flow
Honey yield "important" in NWFP and Punjab (PAK, Pak/77)

Pollen
P1 tropical America. **P2** INI/MAH. **P** BRA/SC; PAK. **Pollen value** "one of
the best pollen producing sp" (tropical America, Ord/83). **Colour** white
(Ord/83). **Pollen grain** illustrated and described (Sao/61; Smt/56a).
Reference slide

Honey no data

344 _Psoralea pinnata_ L.; **Leguminosae**

taylorina (En/AUS); blue pine weed (En/NEZ)
Shrub, 1-3 m; fls blue with white wings
Distribution subtropical Oceania, Africa; temperate Oceania; native to
South Africa. **Habitat** thrives in rough clay country and among secondary
growth (NEZ); naturalized (AUS/WA)

Warning
Declared weed in Albany district, and there are few large areas of _this sp_
(AUS/WA, Col/62)

Nectar rating; blooms, nectar flow
N1 AUS/WA(Col/62)
N NEZ(Wal/78)
Blooms ix-xi (AUS/WA); xi (NEZ). **Nectar flow** "heavy" (NEZ)

Pollen
P1 AUS/WA

Honey: physical and other properties
Pfund light to extra light amber (Cra/75)
Viscosity "good body"
Granulation slow, coarse
Flavour distinctive, reminiscent of coconut

345 Pterocarpus rotundifolius (Sond.) Druce; Leguminosae

roundleaf kiaat, round-leaved bloodwood, Transvaal kiaat (En/SOU);
blinkblaarbloom, dopperkiaat (Af)
Shrub/tree, <20 m, deciduous, gregarious; fls yellow, fragrant
Distribution tropical Africa; native to Africa. **Habitat** woodland and
wooded grassland; common at low-medium altitudes, <1500 m; along water
courses and riverine fringes (ZIM); lowveld, sour bushveld (SOU)
Soil poor sandy soil

Economic and other uses
Timber

Nectar rating + honeybee species; blooms, nectar flow
N1 MOZ[tm](Cra/73)
N3 SOU/TVL,tm(And/73)
Blooms ix-xii (southern Africa); early rainy season (ZIM); during hot dry
weather fl buds remain closed, opening only when it rains (Pag/77).
Nectar flow ?3 wks (Wid/72); intense and v short also irregular and
unreliable (Cri/57)

Honey flow
Honey yield occasionally 8 kg/colony/day (Cri/57)

Pollen
P3 SOU/TVL

Honey: physical and other properties
Colour dark red (Cri/57). **Pfund** medium amber (And/73)
Flavour strong (And/73; Cri/57)

346 Rabdosia coetsa (Buch. Ham. ex D. Don) Hara; Labiatae
syn Plectranthus coetsa Buch. Ham. ex D. Don

plectranthus; kalthunia, shain (INI)

Shrub/undershrub, 60-200 cm, gregarious; fls lavender-blue
Distribution temperate Asia; native to Asia. **Habitat** hillsides of
temperate Himalayas; lower ranges of Himalayas, 900-2500 m; common in
mixed forests and as under-shrub in oak forests (northern INI)
Soil stony slopes

Economic and other uses
Fodder - cut with grass for cattle

Nectar rating + honeybee species; blooms, nectar flow
N1 ?INI/HIM,ac(Sig/62); INI/MAH[ac](Chu/80); PAK,ac(Pak/77)
N2 INI/UTT,ac(Koh/58)
Blooms viii-xi (INI/northern); ix-ii (INI/MAH); x-xi (PAK). **Nectar flow**
unreliable but heavy. **Nectar secretion** heavy rains before flowering and
occasional rains during flow are essential (Sig/62); cold nights and clear
warm days ideal (Sig/62)

Honey flow
Honey yield [moderate] 7-9 kg/colony/season (INI, Sig/62); main source in
Himalayan PAK (Pak/77)

Pollen
P INI/HIM; PAK

Recommended for planting to increase honey production
INI (Sig/62). Especially for covering stony hillsides

Honey no data

347 Rabdosia rugosa (Wall. ex Benth.) Hara; Labiatae
syn Plectranthus rugosus Wall. ex Benth.

plectranthus; shain (INI)
Shrub, 60-200 cm, deciduous, aromatic; fls white spotted purple
Distribution temperate Asia; native to Asia. **Habitat** temperate
Himalayas, hill slopes, even with a sharp gradient, altitudes 1800-3000 m,
INI/HIM, KAS and PAK
Soil shallow; stony; waterlogging not tolerated. **Rainfall** heavy rain
not tolerated

Economic and other uses
Soil benefit erosion control

Nectar rating + honeybee species; blooms, nectar flow
N1 INI/HIM,ac(Rah/40; Sig/62); INI/KAS,ac(Sar/73; Sha/76; Sin/71);
PAK,ac(Pak/77; Shi/77)
N INI/HIM,ac,also am(Goy/74; Ham/73; Sig/48); INI/KAS,ac(Sha/79)
Blooms viii-x (INI/KAS); ix-x (PAK/Swat). **Nectar flow** mid-ix to mid-x
(INI/HIM). **Nectar secretion** highest with good rains in viii and max
rainfall ix-x, together with cold nights and clear warm days (Sig/48);
cold wind due to early snow at high altitudes shrivels fls and stops
secretion (Sin/71)

Honey flow
Honey yield (kg/colony/season) [high] 50 (INI/KAS, Sha/79); mean 9-18, max 41, 90% of surplus honey in autumn (INI/KAS, Sin/71); [moderate] 9 (INI/HIM, Sig/48)

Pollen
P INI/HIM; INI/KAS; PAK. **Colour** cream (Sig/48)

Honey: chemical composition
Water [medium, also low] 17.5-19.0% (Sha/79); 15.0% (Sig/48); 14.87% (8 samples, Sig/62)
Glucose [medium] 38.4% (Sig/48); 35.35% (Sig/62). **Fructose** [medium] 40.0% (Sig/48); 41.21% (Sig/62). **Sucrose** [medium] 3.30% (Sig/48); 2.15% (Sig/62)
Ash [medium] 0.310%
Free acid [medium] 31.20 meq/kg (Sig/62)

Honey: physical and other properties
Colour may be greenish (Sha/79). **Pfund** water white (Sha/79); water white to light amber (Shr/59); almost water white (Sig/62)
Granulation fine (Sha/79; Sig/48); uniform, then colour creamy white (Shr/59); "looks like buffalo butter" (Sig/62)
Flavour mild (Sha/79); mild, improves on granulation (Sig/48)

348 Rhamnidium glabrum Reiss.; Rhamnaceae

turere (Es/BOL)
Tree, small
Distribution tropical S America

Economic and other uses
Food - small sweet berries

Nectar rating + honeybee species; blooms, nectar flow
N1 BOL,tm(Kem/80)
Blooms viii-ix (BOL). **Nectar secretion** copious (Kem/80)

Honey no data

349 Rhigozum trichotomum Burch.; Bignoniaceae

driedoring (Af)
Shrub, spiny
Distribution tropical Africa; native to Africa. **Habitat** Kalahari, Wiesskalk Plateau and Schwarzrand (NAM, SOU)
Rainfall arid areas

Nectar rating + honeybee species; blooms, nectar flow
N1 ?SOU[tm](Joh/73)

N2 NAM[tm](Joh/75a)
Blooms after rain (SOU).   **Nectar flow** light, after rains in NW Cape (SOU, Joh/73)

Honey no data

350 Rhizophora mangle L.;   Rhizophoraceae

red mangrove;   mangle rojo (Es)
Tree, 24–30 m, ?evergreen, stilt-like aerial roots;   fls white
Distribution tropical Caribbean.   **Habitat** calm bays into which rivers flow gently;   shallow water
Soil mud flats;   deep black muds usual but sand and carbonate soils colonized;   regular flushing with sea or freshwater required for optimal growth;   highly salt tolerant.   **Rainfall** >1000 mm

Economic and other uses
Fuel.   Timber.   Soil benefit coastal protection, binds and builds sand and soil.   **Other uses** reserves for aquiculture of fish;   tannin;   resins; wood pulp

Warning
Mangrove swamps are often breeding sites for mosquitoes

Nectar rating
N1 JAM(Met/66)

Honey no data

351 Rhus glabra L.;   Anacardiaceae

sumac (En/CAF,USA);   red sumac, scarlet sumac, smooth sumac (En/USA)
Shrub, <10 m;   fls greenish-yellow;   dioecious
Distribution subtropical N America;   temperate N America;   native to N America.   **Habitat** abandoned land and burned-over areas (eastern USA); hillside pastures and along stone walls (USA/CT)
Soil glacial morraine (USA/CT)

Nectar rating;   blooms, nectar flow;   composition
N1 USA/MD(Die/71)
N2 CAF/BC(Dav/69)
N USA/MA(Shw/50)
Blooms vi–vii (CAF/BC;   eastern USA).   **Nectar secretion** v free on hot clear days, ceases on cloudy, foggy or cool days (Roo/74).   **Sugar concentration** [medium] 42.0–58.5% (Shw/53);   44% (AA906/76)

Honey flow
Honey yield [high] 20–45 kg/colony/season (USA/CT, Roo/74)

Pollen
Pollen grain illustrated and described (Ada/76)

Honey: physical and other properties
Colour golden when pure (Roo/74).   **Pfund** amber (Con/81)
Viscosity "v heavy" (Roo/74)
Granulation [slow] does not granulate but goes waxy (Roo/74)
Flavour bitter when fresh, becomes mild but rich after few mths.   **Aroma**
almost none

352 Rhus taitensis Guill.; Anacardiaceae

tavahi (NIU, TON)
Shrub, abundant on Niue (Pacific) where it is a secondary forest sp
Distribution tropical Asia, Oceania; native from Philippines and Malaysia
to the Society Islands.   **Habitat** fernland areas and forests (NIU)
Soil shallow, porous, overlying coral rock.   **Temperature** 25° is annual
mean for NIU.   **Rainfall** 2035 mm is annual mean for NIU

Nectar rating
N1 NIU(Wao/82);   TON(Wao/82)

Honey no data

353 Rhus typhina L.; Anacardiaceae

staghorn sumac (En/USA)
Tree, <8 m, gauntly branched, dioecious
Distribution temperate N America; native to N America.   **Habitat** abandoned
land and burned-over areas (eastern USA)

Economic and other uses
Land use amenity.   **Other uses** tannin

Nectar rating; blooms, nectar flow; composition
N1 USA/NH(Hol/75)
Blooms vi-vii (eastern USA).   **Nectar secretion** 0.22 mg/fl/day (Sim/80).
Sugar concentration [medium] 38.5-53.5% (Shw/53);   44.3% (Sim/80);   33%
(AA906/76).   **Sugar value** [medium] 0.176 mg/fl/day (Sim/80)

Honey flow
Honey potential [moderate] 30-60 kg/ha (ROM, Cir/80)

Pollen
Pollen grain illustrated and described (Ada/76)

Honey: physical properties
Pfund amber (Cha/48)

354 Robinia pseudacacia L.: Leguminosae

false acacia, white acacia; black locust, honey locust, white locust,
yellow locust (En/USA); acacia blanca (Es/ARG); faux acacia, robinier
(Fr); falsche Akazie, Robinie (De); kikar (INI)
Tree, 13-35 m, deciduous, thorny stipules; many cvs listed (Kee/83); fls
white, fragrant
Distribution temperate Asia, N America, S America, Oceania; subtropical
Asia, Africa, N America; native to eastern N America. **Habitat** temperate
deciduous forests; widely naturalized in EUR; steppes, plains; banks
and steep hillsides; valleys and urban areas (INI/KAS); subtropical
highlands; gravel ridges, morraines (USA); wasteland where other species
have failed
Soil wide range tolerated; light sand if not too acid; pure quartz sand
and gravel; waterlogging for long periods not tolerated. **Temperature**
frost damages young growth but degree varies with cv. **Rainfall** 1000-1500
mm with 500-700 mm in growing season; humid regions of eastern USA;
drought tolerant

Economic and other uses
Food - fls in fritters. **Fodder** - lvs, especially for goats (Alb/78);
toxic to livestock (Why/53). **Fuel. Timber** - many uses including vine
posts and props. **Land use** hedges, windbreaks, shade, afforestation,
amenity. **Soil benefit** erosion control; improves poor soil; N-fixation

Warning; alert to beekeepers
Warning toxic to livestock (Why/53). Reseeds freely and also produces
root suckers, sometimes becomes a pest (Why/53). **Alert to beekeepers**
blooms early in season so beekeepers must make colonies strong to harvest
honey (INI/KAS, Sha/72; USA, Lov/77; YUG, Kon/77)

Nectar rating + honeybee species; blooms, nectar flow; composition
N1 AFG(Cra/73); BEG(Grn/65); BUL(Jur/65); CHN(Mad/81); CZE(Svo/58);
ETH(Cra/73); FRA(Lou/81; Mar/81); HUN(Kee/77; Kee/77a); INI/KAS,ac
(Sar/73; Sha/72; Sha/76); ITA(Ric/78); JAP(Sak/82); PAK,ac(Cra/73);
ROM(Cir/77; Int/65); URS(Glu/55); YUG(Kon/65; Kon/77; Kul/59)
N2 CAF(Cou/59); GFR(Gle/77); USA/AL,IN(Pel/76); USA/MD(Die/71);
USA/NC(Stp/54); USA/OH(Bai/55)
N3 SOU,tm(And/73); USA/CA,TX(Pel/76)
N ARG,?tm(Per/80); CAF/BC(Con/81; Dav/69); INI/KAS,ac(Sha/79);
LUX(Poo/65); NEZ(Wal/78); URS(Fed/55); USA/MA(Shw/50; Shw/50a; Pel/76)
Blooms v-vi (EUR, FRA, ROM, USA, YUG); v, lasting for 8-15 days, varies
with cv (HUN, Kee/83). **Nectar flow** 10-12 days, cvs can give a succession
of flows (HUN, Kee/77a); 3 days (INI/KAS, Sha/72); 8-14 days (ROM,
Int/65). **Alert to beekeepers** blooms early in season so beekeepers must
make colonies strong to harvest honey (INI/KAS, Sha/72; USA, Lov/77; YUG,
Kon/77). **Nectar secretion** (mg/fl/day) 2, but only 20% of available nectar
was collected by bees (Kee/77a); other published results range from 1.59
to 3.7 (Maz/82; AA129/72; AA213/78); secretion best at high temperatures
(Maz/82); optimum 27° (Sha/72). **Sugar concentration** [high] 33.0-62.3%
(48 cvs, Hal/77); means 34-59%, max 67% (Maz/82); other published results
range from 20 to 63% (Ded/57; Pet/72; AA271/56; AA129/72; AA213/78).
Sugar value (mg/fl/day) [high] 0.76-4.0 (Cir/77); other published results
range from 0.95 to 2.3 (Kee/83; Han/80; Maz/82; Sim/75; AA213/78).
Sugar analysis (Bat/72; Maz/82; Sad/60; Wan/64; Wyk/52; AA678/66;
AA213/78)

Honey flow

Honey yield (kg/colony/season) [high] 40-80 (CZE, Svo/58); 18, every 3 or
4 yrs (USA/MD, Lov/77); 80 (YUG, Kon/77; Kul/59); 8-10 kg/colony/day
(ROM, Sad/60); >10 kg/colony/day (YUG, Kul/58); 50-60% of all honey
produced in HUN (Kee/83a); 50% in YUG (Kul/58). **Honey potential** (kg/ha)
[high] 1000 (GDR, Bec/67); 200-1600 (ROM, AA815/63); 371 from trees age 6
yrs, increasing to 418 at 15 yrs, then decreasing (ROM, Kee/77); other
published results for ROM range from 48 to 1550 (Bac/60; Cir/77; Sad/60;
AA766/65; AA129/72); also 0.44 kg/tree (EUR, Maz/82)

Pollen

P1 FRA; JAP; URS; YUG. **P3** GFR; SOU. **P** EUR; INI/KAS; NEZ; ROM.
Pollen yield 0.01 mg/10 fls (BUL, Sim/75); small loads collected by bees
(EUR, Maz/82). **Pollen value** high (EUR, Sta/74). **Chemical analysis** low
protein content, 14.1% of dry matter (Maz/82). **Colour** of load light to
dark grey (Han/80; Maz/82); pale yellow (Wal/78). **Pollen grain**
illustrated and described (Ada/76; Ayt/71; Saw/81); 14 000 grains/10 g
honey, under-represented (ITA, Mal/77; Pes/80); generally under-
represented in honey in FRA but much higher in honey from HUN (Alb/78).
Reference slide

Honeydew

Honeydew produced in some yrs, eg during vi-vii in 1959 and 1960, when
extra 10-12 kg honey/colony was attributed to secretion mainly from **Aphis
medicaginis** Koch, Aphididae, also from **Parthenolecanium corni** (Bouché),
previously Eulecanium corni robiniarium (Douglas), Coccidae

Recommended for planting to increase honey production

HUN (Kee/83a); NEZ(Wal/78); URS(Ave/78). Propagate by root/softwood
cuttings or by grafting or by seed. Good for growing on slag heaps, spoil
banks, roadsides and railway banks; also for snow-fencing (Kee/83a). Few
pests or diseases (Kee/83a). Not sensitive to air pollution therefore
good for towns and industrial areas (EUR, Maz/82). See **Warning; alert to
beekeepers**

Honey: chemical composition

Water [medium] 15.2-20.4% (34 samples, Iva/78); 15.8% (1 sample, age 13
mths, Whi/62); other published results range from 14.5 to 20.4% (Bac/65;
Cer/64; Dus/67; Mal/77; Pae/77; Sha/79)
Glucose [low] 29.02% (Ech/77); 24.49% (Tou/80); 24.34% (Whi/62); other
results 23.7 to 39.9% (Bac/65; Bat/73; Cer/64; Gon/79; Pae/77).
Fructose [high, also medium] 41.42% (Ech/77); 43.02, 42.84% (Tou/80);
43.29% (Whi/62); other results 30.1 to 47.9% (references as for
glucose). **Sucrose** [medium, also low] 1.01% (Ech/77); 2.20, 2.07%
(Tou/80); 0.63% (Whi/62); other results 0.15 to 13.41% (Bac/65; Bat/73;
Bon/66; Cer/64; Iva/78; Pae/77). **Maltose** 6.51% (Bat/73); 5.44%
(Ech/77); 10.14% (Whi/62). **Isomaltose** 0.40% (Bat/73). **Trehalose**
2.98%. **Gentiobiose** 0.27%. **Raffinose** 0.27%. **Melezitose** 1.35-3.89%
(Pae/77). **Erlose** present (Bel/79). **Dextrin** 1.45-5.93% (Bac/65)
Ash [low] 0.04-0.21% (Iva/78); 0.043% (Whi/62); other results 0.017 to
0.80% (Bac/65; Cer/64; Pae/77; Pes/80). Contents of elements (Cer/64;
Var/70)
pH 3.68 (Ech/77); 4.30 (Whi/62); other results 3.56 to 4.3 (Dus/72;
Pae/77; Sha/79). **Total acid** (meq/kg) 12.99-28.03 (Pae/77); 9.88
(Whi/62). **Free acid** (meq/kg) [low] 10.53-16.71 (Pae/77); 7.64
(Whi/62). **Lactone** (meq/kg) 0.5-6.0 (Pae/77); 2.15 (Whi/62); other
results for acid contents (Cer/64; Mal/77)

Amylase 5.2-14.8 (Iva/78); 7.5 (Whi/62); other results 2.5 to 17.9
(Bac/65; Bon/66; Mal/77; Pae/77). **Invertase** 3.9-5.8 (Gontarski 1957
method, Dus/67); also Bon/66. **Glucose oxidase** 214 units/100 ml honey
(Ech/75). **Peroxide number** 17.5-32.2 µg/g/h (Dus/67); also Dus/72. **HMF**
0.19-10.98 ppm (Iva/78); also Mal/77; Pae/77
Nitrogen 0.009, 0.011% dry wt (Bos/78); 0.19% (Whi/62). **Amino acids**,
free 0.037, 0.060%, protein 0.035, 0.036% dry wt (Bos/78). **Protein** 0.20-
1.90% (Cer/64); 0.24% (Ech/75). Lipid composition (Pop/79a)
Fermentation on storage unlikely, yeast count low (Maa/73)
Vitamins 260 ppm (180 ppm after 30 min at 50°, Ech/77)
Compounds probably contributing to flavour (Wab/80)

Honey: physical and other properties
Colour pale yellow (Kee/77a); v clear (Pes/80); water clear, yellow tinge
if not monofloral (Ric/78). **Pfund** 4-8 mm, water white (Whi/62; also
Lov/56; Pia/81; Sha/79)
Relative density 1.414-1.435 (Bac/65); 1.4080-1.4440 (Cer/64). **Viscosity**
"heavy body" (Cra/75). **Optical rotation** -36.90 deg (Bat/73); -3.3 to
-0.7 deg (Cer/64). **Electrical conductivity** 0.000095-0.000208 ohm/cm
(Iva/78); other results are within this range (Dus/67; Pae/77; Pou/70)
Granulation slow, may take yrs (Dem/64; Kee/77a; Pes/80); small grain
(Fed/55); large, slightly transparent crystals (Pia/81)
Flavour sweet (And/73; Cra/75); mild (Kee/77a); delicate, sweet like
mature fruit (Pia/81); strong (Sha/79); analysis of flavour components
(Wab/80). **Aroma** slight (And/73; Cra/75; Pes/80); reminiscent of
flower, not persistent (Pia/81); strong (Sha/79)

355 Rosmarinus officinalis L.; Labiatae

rosemary; romero (Es); romarin (Fr); rosmarino (It)
Shrub, 0.5-2.0 m, evergreen, aromatic; fls violet-blue/whitish
Distribution temperate (Med) Europe, (Med) Africa, Asia; native to (Med)
Europe and Asia Minor. **Habitat** grows wild in coastal areas of Med
Europe; hillsides <1500 m (SPA); <500-600 m (FRA)
Soil calcareous preferred

Economic and other uses
Food - lvs for flavouring. **Fuel.** **Land use** amenity. **Other uses**
essential oil for medicinal and cosmetic uses

Nectar rating; blooms, nectar flow; composition
N1 FRA(Lou/81); MOR(Cra/73; Lav/76); SPA(Lav/76; Sau/82c); TUN
(Foo/79; Lav/76)
N2 YUG(Kul/59; Lav/76)
N3 ITA(Ric/78)
N ALG(Lav/76; Ske/72); MOR(Mat/49); ZIM(Pap/73)
Blooms xi-iii (ALG); iv-v, then sporadically (FRA); iii-iv most important
(SPA); autumn and for about 40 days in spring (YUG). **Nectar flow** iv-v
(SPA). **Nectar secretion** (mg/fl/day) 0.8-1.1 (various dates, Eis/80);
0.770 (Sim/80). **Sugar concentration** [medium] 38-40% (Cir/80); 24.7-38.4%
(Eis/80); 62.5% (Fah/49); 39.1% (Sim/80); also 21-45% in nectar from
honey sacs of bees (Hae/53). **Sugar value** (mg/fl/day) [medium] 0.27-0.40
(Eis/80); 0.731 (Sim/75); 0.942 (Sim/80). **Sugar analysis** (Bat/73a;
Pec/61)

Honey flow
Honey yield [high] 60 kg/colony/season (YUG, Lav/76);   5-7 kg/colony/day
(FRA, Hae/53).   **Honey potential** [moderate] 100-130 kg/ha (ROM, Cir/80)

Pollen
P1 FRA.   P3 ITA.   **Reference slide**

Honey: chemical composition
Water [low] usually <17.5% (Int/75 - see note at end of section);
refractive index 1.5007 (Moh/82).
Glucose [medium] 36.9-38.5% (7 samples, Gon/79);   24.95% (Moh/82).
Fructose [medium] 39.0-41.3% (Gon/79);   32.90% (Moh/82).   **Sucrose** [medium]
1.49-3.04% (Moh/82).   **Reducing sugars** 76.58%.   **Maltose** 10.10%.
Raffinose 11.90% [sic] (Moh/82);   but usually 0% (Int/75).   **Erlose** present
(Pou/70);   usually 0.8-3.0% (Int/75).   **Melezitose** 0% (Int/75)
Ash [low] 0.11% (Moh/82);   usually <0.10% (Int/75)
pH 6.1 (Moh/82);   6.40 (Pan/59).   **Total acid** range 8.7-19.1 meq/kg
(Int/75).   **Free acid** (meq/kg) [low] 13.10 (Moh/82);   range 4.0-11.0
(Int/75).   **Lactone** range 1.0-10.4 meq/kg (Int/75)
Amylase not less than 10, usually 10-20 (Int/75)
Nitrogen 0.012% dry wt (Bos/78).   **Amino acids**, free 0.040, protein 0.060%
dry wt
Acceptable characteristics for this honey have been published in the
Proceedings of the 25th International Beekeeping Congress 1975 (Int/75);
reprinted in French (Lav/76; Uni/83); and in Spanish (Sau/82c).   Some of
the data are quoted above and below

Honey: physical and other properties
Colour clear (Ske/72).   **Pfund** white (Bab/61);   water white (Lov/56)
Viscosity 166.00 poise (Moh/82).   **Optical rotation** -10.33 deg.
Electrical conductivity not above 0.00025/ohm cm (Int/75)
Granulation medium (Moh/82);   usually rapid, fine (Int/75)
Flavour and aroma distinctive (Int/75)

356 Roystonea regia (Kunth) O.F. Cook;   Palmae

royal palm;   palma real (Es/ANT, CUB, DOR, HOD, NIA, PUE)
Tree, 12-21 m, evergreen;   fls cream, small, monoecious
Distribution tropical Asia, S America, C America, Caribbean;   subtropical N
America;   native to Antilles and USA/FL.   **Habitat** planted throughout the
tropics, often forming large palm groves;   low altitudes (CUB);   damp
areas, river banks
Soil moist

Economic and other uses
Fodder - fruit for pig-feed.   **Land use** shade, amenity

Nectar rating;   blooms, nectar flow
N1 CUB(Ord/44;   Ord/56);   DOR(Ord/64);   PUE(Ord/44)
N2 ANT(Ord/83);   CUB(Ord/66);   NIA(Ord/63a);   PUE(Ord/83)
N HOD(Ord/63)
Blooms all yr but profusely x-xi (CUB).   **Nectar flow** autumn (Antilles,
Ord/83).   **Nectar secretion** copious (Ord/44)

Pollen
P1 ANT; CUB; PUE. P DOR; NIA. **Yield** copious (Ord/44). **Pollen value** high, good for brood rearing (Ord/83). **Colour** cream-white (Ord/44)

Honey: physical and other properties
Colour golden yellow (Ord/44). **Pfund** light amber (Cra/75; Mot/64) **Flavour** characteristic (Cra/75; Ord/83); strong (Mot/64). **Aroma** characteristic (Ord/83)

357 Rubus spp [R. fruticosus L.]; Rosaceae

blackberry, bramble
Shrub, deciduous; fls pale pink, white lavender. Data entered here are stated to be for R. fruticosus L.; this is an extremely variable and complex aggregate of microspecies
Distribution temperate Europe, Oceania; native to Europe, Asia, (Med) Africa. **Habitat** cultivated crop plant; wild in hedges, fields and roadsides; dense clumps on common land, wasteland and heaths; forms undergrowth in woods; low altitudes
Soil chalk, poor acid sand, rich alluvium, clay; swamp areas in Auckland (NEZ)

Economic and other uses
Food - fruit for dessert, preserves and wine-making. **Other uses** stems for binding straw in skep-making, and in thatching (How/79)

Warning
Troublesome weed in NEZ (How/79)

Nectar rating; blooms, nectar flow; composition
N1 IRR(Hil/68)
N2 NEZ(Wal/78)
N3 BEG(Cra/84)
N EUR (Maz/82); UK(How/79)
Blooms xi-i (NEZ); vi-viii and on until frost (UK). **Nectar secretion** (mg/fl/day) 4-6 (Maz/82); 0.84-2.9 (Sim/76); 3.8-6.1, higher on day after cool night (AA760/75); 4.5 (AA902/80). **Sugar concentration** [medium] 35-45% (Cir/80); 12-49% (Maz/82); 38.1-46.6% (Sim/76); 22.6-33.9% (AA760/75); 24.8% (AA902/80). **Sugar value** (mg/fl/day) [medium] 1.9-3.4 (Maz/82); 0.32-0.66 (Sim/76); 1.0-2.0 (AA760/75); 1.04 (AA902/80). **Sugar analysis** (Maz/82)

Honey flow
Honey potential (kg/ha) [moderate] 6-29 (BUL, Sim/76); 5-26 (EUR, Maz/82); 6-26 (POL, AA760/75); 30-50 (ROM, Cir/80)

Pollen
P2 NEZ. P UK. **Colour** dull greenish-white (Wal/78). **Pollen grain** illustrated and described (Saw/82)

Honeydew produced NEZ (Wal/78)

Honey: physical and other properties
Colour light (Cra/75). **Pfund** white (Mao/82); water white (Rob/56)
Granulation slow (Cra/75); appears dull, waxy (Rob/56)
Flavour delicate, and like clover (Rob/56)

358 Rubus idaeus L.; Rosaceae

raspberry; framboisier (Fr); Himbeere (De); lampone (It); bringebaere
(No)
Shrub, deciduous; fls white
Distribution temperate Europe, Asia, Oceania, N America, (Med) Africa;
native to E and S Asia. **Habitat** cultivated crop plant; in N hemisphere
grows wild in high hilly or heath country; often grows in burned-over
areas; Carpathian mountains (ROM); taiga (URS/Siberia)
Soil acid soil preferred by wild raspberry; light neutral soil for cvs;
cultivation difficult on heavy alkaline soil

Economic and other uses
Food - fruit

Nectar rating; blooms, nectar flow; composition
N1 BEG(Grn/65); GFR(Gle/77); ROM(Cir/77; Int/65); URS(Ave/78; Fed/55;
Gri/77; Kov/65)
N2 FRA(Lou/81); HUN(Pet/77); ITA(Ric/78); POL(Dem/64a); USA/NH(Hol/75)
N3 BEG(Cra/84); UK(How/79)
N ALG(Ske/72); EUR(Maz/82); NEZ(Wal/78); NOW(Lun/71; Thy/65)
Blooms v-vii (ROM); vi-vii for 15-25 days (URS/W Siberian Depression).
Nectar flow annual and reliable (Gri/77). **Nectar secretion** (mg/fl/day) 17-
22 (Han/80; Maz/82); 4.00-5.95 (AA317/57); 3.8-14.1 (AA607/79); 7.03
(AA902/80); secretion starts before fl opens, and bees forage between
petals (Gri/77); secretion affected by cv, soil, climate (Frj/70); best
on day after cool night (Sim/76); reduced by boron deficiency (Hol/75).
Sugar concentration [medium] 34-42% (Han/80; Maz/82); 49.33% (Pet/77);
30.15, 35.5% (AA450/61); 24.2% (AA132/64); 47, 49% (cultivated,
AA495/66); 37.8-59.2% (Sim/76); 33.1% (AA902/80). **Sugar value**
(mg/fl/day) [high] 1.9-6.7 (Sim/76); [medium] 0.7-1.0 (Cir/77); 1.410,
0.723 (AA450/61); 2.407 (AA902/80). **Sugar analysis** (Bat/73a; Maz/59;
Maz/82; Pec/61; Wyk/52). Juice from fruit also collected (UK, How/79)

Honey flow
Honey yield (kg/colony/season) [high] 60 min, 130 max (URS/Krasnoyarsk,
Gri/77); "v valuable" in Siberia and Priuralie zones, also N and C
European URS (Ave/78). **Honey potential** (kg/ha) [moderate] 59, 116
(cultivated, BUL, AA495/66); 47-270 (BUL, Sim/76); 117-122 (EUR,
Maz/82); 20 (GDR, Bec/67); 40-100 (POL, AA760/75); 50-200 (ROM,
Cir/77); 40, 50 (ROM, AA450/61); 70 (ROM, AA815/63); 200 (URS, Ave/78);
50-70 (URS, Fed/55); 99-274 (URS, Gri/77); 33-100 (4 yrs, URS, AA661L/75)

Pollen
P1 ITA. **P2** AUS/VIC; FRA. **P3** GFR. **P** ALG; NEZ; NOW; ROM; UK;
URS. **Yield** (mg/10 fls) max 0.006 (Sim/75); 5.8-11.4 mg (6 cvs,
AA1243/78). **Chemical analysis** (Gob/81); nitrogen content 4.6% (Maz/82).

Colour of load greyish white (Ave/78); load yellow-orange (Cir/80); load light to dark grey (Han/80; Maz/82); white (How/79; Wal/78). **Pollen grain** illustrated and described (Ada/74). **Reference slide**

Honey: chemical composition
Water [medium] 15.35-20.82% (10 samples, Bac/65); 18.7, 19.5% (Dus/67)
Glucose [medium] 32.90-38.80% (Bac/65). **Fructose** [medium] 34.96-41.34%.
Sucrose [medium] 1.43-5.20%. Also contents as % of total sugars (Maz/64). **Dextrin** 1.80-5.10% (Bac/65)
Ash [medium] 0.08-0.41%
Amylase 10.9-38.5 (Bac/65). **Invertase** 11.5, 17.2 (Gontarski 1957 method, Dus/67). **Peroxide number** 158, 277.5 µg/g/h
Protein 0.0019-0.0047% (6 samples, Gen/67)

Honey: physical and other properties
Pfund white (Wal/78)
Relative density 1.412-1.447 (Bac/65). **Electrical conductivity** (per ohm cm) 0.00023, 0.00027 (Dus/67); 0.000209 (Vor/64)
Flavour and aroma mild (Cra/75; Wal/78)

359 Rubus ulmifolius Schott.; Rosaceae

bramble; ronce (Fr/ALG)
Shrub, 1.5-2.0 m; cvs without prickles available; fls pink
Distribution temperate Asia, (Med) Europe, (Med) Africa. **Habitat** hillsides, hedges and wadis (dry riverbeds) ALG

Economic and other uses
Food - fruit

Alert to beekeepers
Honey may granulate v rapidly, even in the hive (Ske/72)

Nectar rating + honeybee species; blooms, nectar flow
N1 PAK,ac(Pak/77)
N ALG(Ske/72)
Blooms v-ix (ALG); ix-x (PAK)

Honey flow
Honey yield 21.5 kg/colony/12 days (ALG, Ske/72); "important" in north-east PAK (Pak/77)

Pollen
P PAK

Honeydew produced ALG (Ske/72)

Honey: physical and other properties
Pfund white (Ske/72)
Granulation (alert to beekeepers) may be v rapid, even in the hive (Ske/72)

360 Sabal florida Becc.; Palmae

palma cana (Es/CUB)
Tree, grows in large groups, "canales" (CUB)
Distribution tropical Caribbean.   **Habitat** central savanna (CUB)
Soil clay

Nectar rating
N1 CUB(Ord/83)

Honey: chemical composition
Fermentation likely (Ord/83)

Honey: physical properties
Pfund white

361 Sabal palmetto (Walt.) Lodd. ex Schultes;   Palmae

cabbage palm, palmetto, swamp cabbage, thatch palm (En/USA)
Tree, <26 m; fls whitish-yellow, small
Distribution subtropical N America; tropical Caribbean.   **Habitat**
prairies, marshes, pinelands and hammocks (USA/FL); USA/GA, NC, SC,
especially common on coast and coastal islands; dominant palm; widely
planted
Soil sandy; salt tolerant

Economic and other uses
Food – central bud but its removal kills the tree. **Fuel.**   **Timber.**   **Land
use** amenity.   **Other uses** lvs for roofing

Alert to beekeepers
Honey likely to ferment even in capped cells of comb (Lov/65a; Mot/64)

Nectar rating; blooms, nectar flow
N1 USA/FL(Lov/65a)
Blooms summer, chiefly vii (southern USA); iv-vi (subtropical/tropical
America).   **Nectar secretion** yields well every 3 yrs (Pel/76); nectar
abundant on damp soils, absent on dry ones (Ord/83)

Honey flow
Honey yield [high] mean 13, max 45 kg/colony/season (USA, Lov/65a)

Pollen
Reference slide

Honey: chemical composition
Water [high] can be v high (Lov/55e); 19.7% (1 sample, age 13 mths, Whi/62)
Glucose [medium] 31.20% (Whi/62).   **Fructose** [medium] 37.96%.   **Sucrose**
[low] 0.63%.   **Maltose** 6.25%.   **Higher sugars** 0.99%
Ash [low] 0.084%
pH 3.61.   **Total acid** 44,94 meq/kg.   **Free acid** [medium] 37.62 meq/kg.
Lactone 6.97 meq/kg

Amylase 20.1
Nitrogen 0.099%
Fermentation (alert to beekeepers) – likely, even in capped cells of comb
(Lov/65a; Mot/64)

Honey: physical and other properties
Colour light yellowish (Pel/76).   **Pfund** light amber (Mot/64); 27–34 mm,
white (Whi/62)
Viscosity "thin body" (Pel/76)
Flavour and aroma mild (Mot/64)

362 Saccharum officinarum L.; Gramineae

sugar cane; caña de azúcar (Es/BOL/DOR); cana-de-açucár (Pt/BRA)
Herb, 3 m, perennial; fls insignificant but in large panicle
Distribution tropical S America, C America, Caribbean, Asia, Africa,
Oceania; subtropical Africa, N America, Oceania; temperate (Med)
Europe.   **Habitat** cultivated crop plant
Soil must be fertile.   **Rainfall** high, or irrigation required

Economic and other uses
Food – sucrose from stem for sugar and molasses; cane sucked as sweetmeat

Nectar rating + honeybee species
N1 (sap not nectar) tropical America (Ord/83)
N2 BEL(Mul/79)
N BOL(Kem/80); BRA/SP,tm(Ama/55; Fle/63; Ker/57); COL(Ken/76);
DOR(Ord/64); GHA,tm(Gor/64); GUS,tm(Sve/80); MAY(Cra/82); REU(Cra/82);
SOU,tm(And/73)
Nectar secretion none (Kem/80; Ord/52); but bees forage on sap exuding
from cut stems and burnt canes (Cra/82; Lau/76; Ord/83); sap collected x-
iv when nectar is unavailable (AUS, And/73; USA, Mot/64); also juice
collected from sugar mills (COL, Ken/76; GUS, Sve/60; MAY, Cra/82).
Sugar concentration of sap 18–24% (BOL, Kem/80); 10% (MAY; Cra/82)

Honey flow
Honey potential [moderate] calculated as possibly 1.25 kg/ha (MAY, Cra/82)

Pollen
P3 SOU.   **Yield** none (Kem/80; Ord/52); no evidence of collection by bees
(MAY, Cra/82); "sometimes eagerly collected" (SOU, And/73)

Honeydew
Honeydew produced, and collected by bees: from **Melanaphis sacchari**
(Zehntner), previously Aphis sacchari, Aphididae (HAW, Ken/76); also ?from
Perkinsiella saccharicida Kirkaldy, Delphacidae (used to be in HAW, ?now in
S America, Ken/76).   Honeydew also produced (insect not specified) in ANA
(Por/74); ?COL (Ken/76)

Honey: chemical composition
Sucrose high (Cra/82)

Honey: physical and other properties
Colour dark (Mot/64)
Granulation rapid (Cra/84)
Flavour strong, distinctive

363 Salix alba L.; Salicaceae

white willow; sauce blanco (Es/BOL)
Tree, <25 m, deciduous; dioecious
Distribution temperate Europe, Asia, Oceania, N America, (Med) Africa;
tropical S America. **Habitat** lowland regions; cultivated in NE of N
America; banks of rivers, lakes and marshes (BOL)

Economic and other uses
Timber. **Land use** amenity. **Other uses** osiers for baskets; source of
salicin (glucoside)

Alert to beekeepers
Blooms early in season so beekeepers must make colonies strong to harvest
honey. Salix honeydew honey is not suitable as winter food for bees (ROM,
Fra/65)

Nectar rating + honeybee species; blooms, nectar flow
N1 URS(Glu/55)
N BOL,tm(Kem/71); ROM(Cir/80)
Blooms spring (BOL); iv-v (EUR). **Alert to beekeepers** blooms early in
season so beekeepers must make colonies strong to harvest honey

Honey flow
Honey yield "important" source in all regions of URS except E Siberia and
Far East (Glu/55). **Honey potential** (kg/ha) [moderate] 100-120 (ROM,
Apc/68); 100-150 (ROM, Cir/80)

Pollen
P ROM. **Yield** fair (Cir/80). **Chemical analysis** (Gob/81; Shp/80).
Colour of load lemon yellow (Cir/80)

Honeydew
Honeydew produced, and collected by bees: from **Tuberolachnus salignus**
(Gmelin), Lachnidae - flow may be v heavy (mid EUR, Klo/65); honey yield
from Salix spp up to 20 kg/colony (ROM, Cir/80); also from **Pterocomma
salicis** (L.), Aphididae - secretion "high", visited by bees (Salix spp, mid
EUR, Klo/65). **Alert to beekeepers** Salix honeydew honey is not suitable as
winter food for bees (ROM, Fra/65)

Honey no data

364 Salix caprea L.; Salicaceae

goat willow, great sallow, pussy willow, withy; Salweide (De); salicone
(It)

Shrub/tree, <9 m, deciduous; dioecious
Distribution temperate Europe, Oceania, Asia, S America.   **Habitat** widely
planted; woodlands, hedgerows; underwood in coppices; along water-
courses and by dams (AUS)
Soil wide range

Economic and other uses
Fuel.   Timber.   Land use amenity.   **Soil benefit** conservation and river
control.   **Other uses** tannin from timber; source of salicin (glucoside)

Warning; alert to beekeepers
Warning listed as a noxious weed in Waikato (NEZ, Wal/78).   **Alert to
beekeepers** blooms early in season so beekeepers must make colonies strong
to harvest honey.   Salix honeydew honey is not suitable as winter food for
bees (ROM, Fra/65)

Nectar rating; blooms, nectar flow; composition
N1 URS(Ave/78;   Fed/55)
N2 AUS/VIC(Gom/73);   FRA(Lou/81);   GFR(Gle/77);   ITA(Ric/78);   ROM(Cir/77)
N NEZ(Wal/78)
Blooms viii-ix (AUS/VIC);   iii-iv (ROM);   iv-v (URS).   **Nectar secretion**
(mg/fl/day) female 0.0245-0.0286, male 0.029 (Maz/82).   **Sugar concentra-
tion** [high] female 67-79%, male 66-69% (Maz/82);   female 79.0%, male 64.1%
(AA396L/71);   [medium] 22.83, 40.26% (2 yrs, Pet/72);   [low] 15-20%
($\overline{\text{Han}}$/80).   **Sugar value** [low] 0.05-0.1 mg/fl/day (Cir/77);   female 0.0165-
0.226, male 0.019-0.021 (Maz/82).   **Sugar analysis** (Maz/59;   Maz/82;   Pec/61)

Honey flow
Honey yield 2-4 kg/colony/day (URS, Fed/55).   **Honey potential** (kg/ha)
[moderate] 100-200 (ROM, Apc/68);   150-200 (ROM, Cir/77);   150 (URS, Ave/78)

Pollen
P1 AUS/VIC;   GFR;   ITA;   URS.   **P2** FRA.   **P** NEZ;   ROM.   **Yield** heavy
(AUS/VIC);   0.001 mg/10 fls (BUL, Sim/75).   **Chemical analysis** (Cir/80;
Shp/80;   Sta/74);   rich in protein (Maz/82).   **Colour** of load yellow
(Han/80;   Maz/82).   **Pollen grain** illustrated and described (Saw/82).
Reference slide

Honeydew
Honeydew produced and collected by bees: from **Tuberolachnus salignus**
(Gmelin), Lachnidae - honey yield from Salix spp up to 20 kg/colony (ROM,
Cir/80);   also from **Pterocomma salicis** (L.), Aphididae - secretion "high",
visited by bees (Salix spp, mid EUR, Klo/65).   Honeydew produced URS
(insect not specified, Glu/55).   **Alert to beekeepers** Salix honeydew honey
not suitable as winter food for bees (ROM, Fra/65)

Recommended for planting to increase honey production
NEZ (Wal/78);   URS (Ave/78).   See **Warning;   alert to beekeepers**

Honey: physical and other properties
Colour golden yellow (Fed/55);   v light (Gom/73).   **Pfund** light amber
(Wal/78)
Viscosity "thin" (Gom/73)
Granulation fine (Fed/55)
Flavour mild (Gom/73);   distinctive, "swampy but not unpleasant" (Wal/78)

365 Salix nigra Marshall; Salicaceae

black willow (En/USA);   sauce negro (Es/MEX)
Tree, <30 m, deciduous;   dioecious
Distribution temperate N America, Europe;   native to N America.   **Habitat** commonly planted in Europe

Economic and other uses
Timber.   **Land use** amenity

Alert to beekeepers
Blooms early in season so beekeepers must make colonies strong to harvest honey

Nectar rating; blooms, nectar flow
N1 USA/LA(Lov/56)
N2 MEX(Ord/83)
Blooms i-iii (C and N America, Ord/83).   **Alert to beekeepers** blooms early in season so beekepeers must make colonies strong to harvest honey

Honey flow
Honey yield [high] 45 kg/colony/season (USA/LA, Lov/56)

Pollen
P MEX.   **Pollen grain** illustrated and described (Ada/72;   Lie/72)

Honey: physical and other properties
Pfund extra light to light amber (Lov/56)
Flavour "weedy" (Lov/56)

366 Salvia apiana Jepson; Labiatae

white sage (En/USA)
Shrub, 90-250 cm;   fls bluish-white, white
Distribution subtropical N America.   **Habitat** altitudes <800 m;   canyons, mountain slopes, dry plains;   important component of the chaparral and coastal sage scrub, USA/CA

Nectar rating; blooms, nectar flow
N1 USA/CA(Van/41)
N2 USA/CA(Lov/56)
N3 USA/CA(Pel/76)
Blooms iv-vi (USA/CA)

Honey flow
Honey yield [high] 40-54 kg/colony/season (USA/CA, Lov/77)

Pollen
P2 USA/CA

Honey: physical and other properties
Pfund clear white (Lov/56);   water white (Van/41)

Granulation [slow], does not granulate (Lov/56); does granulate, unlike
other Salvias (Roo/74)
Flavour mild (Lov/56)

367 Salvia leucophylla Greene; Labiatae

purple sage, silver sage, white-leaved sage (En/USA)
Shrub, 30-120 cm, perennial; fls rose-lavender
Distribution subtropical N America. **Habitat** dry mountain foothills, <600
m, not extending inland beyond coastal ranges; coastal sage scrub (USA/CA)

Nectar rating; blooms, nectar flow
N1 USA/CA(Van/41)
N3 USA/CA(Pel/76)
Blooms v-vi (USA/CA). **Nectar secretion** highest after 250 mm rain in
winter followed by clear warm spring; ceases during drought (Roo/74)

Pollen
P2 USA/CA

Honey: physical and other properties
Pfund clear white (Lov/77)
Granulation [slow] does not granulate
Flavour mild

368 Salvia mellifera Greene; Labiatae

ball sage, black sage, button sage (En/USA)
Shrub, 60-180 cm, perennial, often forms extensive pure stands (USA/CA);
fls bluish-white
Distribution subtropical N America. **Habitat** coastal USA/CA from San
Francisco southwards; slopes and benches <600 m; important component of
the chaparral and coastal sage scrub (USA/CA)
Soil sandy/rocky

Nectar rating; blooms, nectar flow; composition
N1 USA/CA(Van/41)
N2 USA/CA(Pel/76)
Blooms iv-vii (USA/CA). **Nectar secretion** highest after copious winter
rain following a period of drought (Van/41). **Sugar concentration** [medium]
45% (Van/49)

Pollen
P2 USA/CA. **Colour** bluish (Van/41)

Honey: physical and other properties
Pfund water white (Lov/77; Van/41)
Granulation [slow], does not granulate (Lov/77)
Flavour mild

369 Salvia nemorosa L.;   Labiatae

Herb, 30-60 cm;   fls usually violet but can be pink/white
Distribution temperate Europe, Asia.   **Habitat** waysides, grassy places and meadows

Nectar rating;   blooms, nectar flow;   composition
N1 POL(Dem/64a)
N EUR(Maz/82)
Blooms v-vii (EUR);   mid vi-vii (POL, Jab/68).   **Nectar secretion** 0.07-0.039 mg/fl/day (Jab/68).   **Sugar concentration** [medium] 57.9% (Jab/68); 36.1% (Pek/80).   **Sugar value** [medium] 0.05-0.25 mg/fl/day (Jab/68). **Sugar analysis** (Maz/82)

Honey flow
Honey potential (kg/ha) [moderate] 300 (ROM, Apc/68;   Cir/80);   243 (POL, Jab/68);   190-600, mixed with S. officinalis (EUR, Maz/82)

Pollen
P3 EUR.   **Yield** low (Maz/82)

Honey: chemical composition
Sugars (as % of total, Maz/64):   **glucose** 27.7%;   **fructose** 56.4%;   **sucrose** 1.5%;   **maltose** 14.0%

Honey: physical properties
Electrical conductivity 0.000320/ohm cm (Vor/64)

370 Salvia officinalis L.;   Labiatae

sage;   salvia (It)
Herb/undershrub, 20-70 cm, aromatic;   fls pink/bluish-lilac
Distribution temperate Europe.   **Habitat** widely cultivated as a culinary herb;   warm dry areas preferred;   banks, stony places;   coastal region (YUG)

Economic and other uses
Food - lvs for flavouring, wine-making and as a tea

Nectar rating;   blooms, nectar flow;   composition
N1 POL(Dem/64a);   URS(Fed/55)
N2 ITA(Ric/78);   YUG(Kul/59)
Blooms v-vii (EUR);   vi (POL, Jab/68).   **Nectar secretion** (mg/fl/day) 3.20 (Jab/68);   2.55 (Sim/80).   **Sugar concentration** [medium] 47-60% (Cir/80); 30.7% (Jab/68);   37.7% (Sim/80);   36.5% (AA194/76).   **Sugar value** (mg/fl/day) [medium] 0.98 (Jab/68);   2.260 (Sim/75);   2.427 (Sim/80). **Sugar analysis** (Bat/73a;   Maz/82;   AA753/75)

Honey flow
Honey yield "important" in Ukraine and south (URS, Fed/55);   "important" on coast of YUG (Maz/82).   **Honey potential** (kg/ha) [moderate] 65-340 (BUL, AA194/76);   >400 (POL, Dem/64a);   302 (POL, Jab/68);   200-400 (ROM, Cir/80);   190-600, mixed stand with S. nemorosa (EUR, Maz/82)

Pollen
P3 ITA.   Reference slide

Honey: chemical composition
Glucose [medium] 34.41% (Mur/76).   **Fructose** high (Maz/82); 40.56%
(Mur/76).   Also contents of these sugars, and **sucrose, maltose,
fructomaltose,** as % of total sugars (Maz/59; Maz/64)
Ash [medium] 0.38% (Mur/76)

Honey: physical and other properties
Colour light (How/79); pale yellow (Mur/76).   **Pfund** amber (Fed/55);
white (Pel/76)
Relative density 1.4320 (Mur/76).   **Electrical conductivity** 0.000298/ohm cm
(Vor/64)
Granulation slow (How/79; Maz/82)
Flavour "aromatic" (Fed/55)

371 Sapindus emarginatus Vahl; Sapindaceae

soapnut
Tree, 10 m, evergreen; fls white
Distribution tropical Asia

Economic and other uses
Land use shade, amenity.   **Other uses** fruits used as soap substitute

Nectar rating + honeybee species; blooms, nectar flow
N1 INI/AND,KAR,ORI,TAM,ac(Kri/70)
Blooms x-xii (INI)

Honey flow
Honey yield 1.4-2.3 kg/tree/season, 25-50% of total honey yield in some
areas of Andhra Pradesh (INI, Kri/70)

Pollen
P3 INI/AND, KAR, ORI, TAM

Recommended for planting to increase honey production
INI (Kri/70)

Honey: physical and other properties
Colour light golden (Kri/70)

372 Sapindus laurifolius Vahl; Sapindaceae

soapnut; antuvala, antuwal, noraikayi (INI)
Tree; fls white
Distribution tropical Asia

Economic and other uses
Medicinal; fruits used for soap substitute

Nectar rating + honeybee species; blooms, nectar flow
Nl ?INI/KAR[ac](Kri/80);   INI/KAR,KER[ac](Kha/59)
Blooms x-i, peak mid-xi (INI/KAR, KER)

Pollen
P INI/KAR, KER

Recommended for planting to increase honey production
INI/KAR (Kri/80)

Honey no data

373 Sapindus mukorossi Gaertn.;   Sapindaceae
syn Sapindus detergens Roxb.

soapberry, soapnut; retha (INI)
Tree, deciduous; fls greenish-white
Distribution subtropical Asia, Oceania; native to China.   **Habitat**
cultivated in sub-Himalayan tracts and lower hills <1200 m; common tree in
Kangra Valley (INI/HIM)
Soil deep soil preferred.   **Temperature** from below 0 to summer max of 39°
(INI/HIM).   **Rainfall** 1800 mm or more preferred

Economic and other uses
Land use amenity.   **Other uses** fruits used for soap substitute

Nectar rating + honeybee species; blooms, nectar flow; composition
Nl INI/HIM, ac also am (Goy/74; Rah/41; Sig/48)
Blooms v (INI); spring (INI/HIM).   **Nectar flow** 2 wks (INI, Sig/62);
often reduced by bad weather which causes mass shedding of the loosely held
fls (Sig/48).   **Sugar concentration** [medium] 23.0-31.0% (Shr/58)

Honey flow
Honey yield [moderate] 6.8-9.1 kg/colony/season (INI, Sig/62)

Recommended for planting to increase honey production
INI (Sig/62).   Propagate by seed

Honey: chemical composition
Water [medium] 15.43, 17.0% (sample perhaps not monofloral, Sig/62)
Glucose [medium] 35.40%.   **Fructose** [medium] 41.80%.   **Sucrose** [high] 5.39,
4.94%
Ash [medium] 0.30, 0.14%
Free acid [medium] 37.44 meq/kg

Honey: physical and other properties
Colour may be light golden (Cra/75).   **Pfund** water white (Cra/75);   water
white to white (Sig/62)
Granulation fine, then colour milky white (Sig/40)
Flavour mild (Cra/75; Sig/62).   **Aroma** develops on storage (Sig/48)

374 <u>Sapindus saponaria L.; Sapindaceae</u>

soapberry (En/USA); isotoubo (Es/BOL)
Tree, medium height; fls white/yellowish panicles
Distribution tropical C America, Caribbean, S America; subtropical N
America. **Habitat** hammocks and keys (USA/FL)

Economic and other uses
Fruits used for soap substitute; crushed seeds as a fish poison

Nectar rating + honeybee species; blooms, nectar flow
N1 BOL,tm(Kem/74)
N2 BEL(Mul/79)
Blooms vi (BEL); iii (BOL); x-ii (tropical America, Ord/83)

Pollen
Pollen grain illustrated and described (Heu/71; Mag/78)

Honey: physical and other properties
Flavour bitter, objectionable (Mul/79)

375 <u>Sapindus trifoliatus L.; Sapindaceae</u>

antuwal (INI)
Tree; fls white
Distribution tropical Asia

Economic and other uses
Medicinal

Nectar rating + honeybee species; blooms, nectar flow
N1 INI/KAR,KER,ac(Kha/59)
Blooms x-i (INI/KAR, KER)

Honey no data

376 <u>Satureia montana L.; Labiatae</u>

mountain savory, winter savory
Herb/sub-shrub, 10-40 cm; fls white or pale purple
Distribution temperate (Med) Europe; native to S Europe and N Africa.
Habitat coastal areas presenting typical "Krass" (?limestone) formation
(YUG); mountainous regions (YUG/Dalmatia); dry banks and rocky areas
Soil dry limey soil preferred, v poor soil tolerated

Economic and other uses
Food – lvs for flavouring

Nectar rating; blooms, nectar flow
N1 GRC(Adm/54; Nic/55)
N2 FRA(Lou/81); YUG(Adm/54; Kul/59)
Blooms vii-ix (EUR); vii-x or xi (YUG, Vod/53)

Honey flow
Honey yield "important" on N and S mainland GRC, Crete, Aegean and Ionian Islands (Nic/55)

Pollen
P2 FRA.   **P3** ITA

Honey: physical and other properties
Colour yellow (Bab/61); greenish (Vod/53)

377 Scaevola frutescens (Mill.) Krause; Goodeniaceae

veloutier (Fr/CHG)
Distribution tropical Asia.   **Habitat** characteristic plant of tropical beach jungle; on some islands in Indian Ocean eg Diego Garcia; Chagos archipelago, where it lines beaches
Soil derived from coral rock

Nectar rating
N1 CHG(Sil/69)

Honey flow
Honey yield on Diego Garcia provides rest of honey not derived from Cocos nucifera (CHG, Sil/69)

Honey no data

378 Schefflera wallichiana Harms.; Araliaceae

doddabettu, pongabettu (INI)
Distribution tropical Asia.   **Habitat** valleys of Western Coorg; damp cardamom estates (INI/KAR)

Nectar rating + honeybee species; blooms, nectar flow
N1 INI/KAR, ac(Diw/72)
Blooms v (INI/KAR).   **Nectar flow** v reliable (Diw/72)

Honey flow
Honey yield about 60% of annual crop in Western Coorg (INI/KAR, Diw/72)

Pollen
P3 INI/KAR.   **Yield** low

Honey: physical and other properties
Granulation fine (Schefflera sp, probably S. wallichiana, Kri/80)

379 Schinus terebinthifolius Raddi; Anacardiaceae

Brazilian pepper, Mexican pepper; poivrier sauvage (Fr/MAY, REU)
Tree, 12 m, evergreen, vigorous; fls ivory-white, sightly fragrant, some
trees bear only male fls
Distribution subtropical N America; tropical Africa, Caribbean; native to
Brazil.   **Habitat** coastal, thrives in salt spray; widespread escape in
USA/FL; low-medium altitudes (MAY); covers large areas (REU; USA/FL)

Economic and other uses
Food – seeds as condiment.   **Fodder** – for goats, but toxic to cattle,
horses and birds (Mot/78).   **Timber.   Land use** hedges, shade, amenity.
Other uses resins and tannins from bark; toothpicks; medicinal

Warning
Fodder toxic to cattle, horses and birds (Mot/78).   Rapid aggressive
growth; designated noxious weed in HAW (Mot/78).   In USA/FL "fruit may
cause enteritis in children and pets; also skin and respiratory irritation
when plant is in bloom" (Mot/64)

Nectar rating; blooms, nectar flow; composition
N1 BER(Har/75); MAY(Bro/82; Cra/73);   REU(Cra/82); USA/FL(Ord/83)
Blooms iii-iv (MAY); vii-x (USA/FL).   **Nectar flow** autumn (BER).   **Sugar
analysis** (Vah/72)

Honey: physical and other properties
Pfund medium amber (Mot/64); amber (Ord/83)
Flavour distinctive, peppery (Mot/64); spicy (Mot/78).   **Aroma** slightly
pungent (Ord/83)

380 Sclerocarya caffra Sond.; Anacardiaceae

marula; canho (MOZ)
Tree, 10-15 m, deciduous; fls yellow and red, inconspicuous, monoecious or
dioecious
Distribution tropical Africa; native to southern Africa.   **Habitat** open
woodland and bush at low to medium altitudes, southern Africa
Temperature damaged by frost

Economic and other uses
Food – fruit for jellies, conserves and alcoholic drinks; seeds eaten as
raw nuts or cooked.   **Fodder** – fruit.   **Other uses** oil from seeds; medicinal

Nectar rating + honeybee species
N1 MOZ,tm(Cra/73)
N SOU,tm(Cri/57)

Pollen
P SOU.   **Yield** low

Honey: physical and other properties
Colour light (And/73).   **Pfund** (almost) water white to light amber (Cri/57)
Flavour "rich but mild" (And/73; Cri/57)

381 Scrophularia nodosa L.; Scrophulariaceae

knotted figwort; scrofulaire noueuse (Fr/ALG)
Herb, 30–150 cm, perennial; fls green and purplish-brown
Distribution temperate Europe, (Med) Africa.   **Habitat** usually damp and
shady places

Nectar rating; blooms, nectar flow; composition
Nl POL(Dem/64a);   URS(Glu/55)
N ALG(Ske/72)
Blooms vi–viii (ALG).   **Nectar secretion** (mg/fl/day) mean 17.84, max 24.44
(Dem/64a);   21.37 (Jab/68).   **Sugar concentration** [medium] 25.7%
(Jab/68).   **Sugar value** [high] 5.48 mg/fl/day (Jab/68).   **Sugar analysis**
(Pec/61)

Honey flow
Honey yield "important" in European URS (Glu/55).   **Honey potential** (kg/ha)
[high] 899 (POL, Dem/64a);   1068 (POL, Jab/68);   [moderate] 400–500 (ALG,
Ske/72)

Pollen
P ALG.   **Colour** of load yellow (Ske/72).   **Reference slide**

Honey: chemical composition
Sugars (as % of total, Maz/64):   **glucose** 32.0%;   **fructose** 51.9%;   **sucrose**
2.8%;   **maltose** 8.2%;   **fructomaltose** 5.1%

382 Serenoa repens (Bartr.) Small; Palmae

saw palmetto (En/USA)
Shrub, dwarf scrub sp; fls yellowish white, small, fragrant
Distribution subtropical N America;   native to USA/FL.   **Habitat** USA:
uncultivated pastures, pineland in NC, SC and coasts of Gulf of Mexico to
eastern TX;   hammocks, scrub and sand dunes
Soil salt tolerant

Nectar rating; blooms, nectar flow
Nl USA/FL(Lov/56;   Mor/58;   Smt/60)
Blooms iv–vi (USA/FL).   **Nectar secretion** highest in dry seasons (USA,
Lov/65a).   Bees sometimes collect berry juice (Mot/64)

Honey flow
Honey yield [moderate] mean 5, max 14 kg/colony/season (USA, Lov/65a);   "in
commercial quantities" (USA/FL, Mot/64)

Pollen
P USA/FL.   **Pollen value** low (Lov/65a).   **Colour** bright yellow (Lov/65a).
Pollen waxy (Lov/65a).   **Reference slide**

Honey: chemical composition
Water [low, also medium] 15.1, 18.0% (age 7, 8 mths, Whi/62)
Glucose [low] 30.88, 30.96%.   **Fructose** [medium] 37.40, 39.07%.   **Sucrose**
[low] 0.62, 1.04%.   **Maltose** 5.60, 7.36%.   **Higher sugars** 1.67, 1.70%
Ash [medium] 0.458, 0.245%
pH 3.89, 4.10.   **Total acid** 46.78, 35.71 meq/kg.   **Free acid** [medium]
31.48, 21.59 meq/kg.   **Lactone** 15.29, 14.12 meq/kg
Amylase 21.1, 7.7
Nitrogen 0.019, 0.024%

Honey: physical and other properties
Colour rich yellow (Lov/55e); sometimes dark if bees collect berry juice
(Mot/64).   **Pfund** 34-50 mm, light amber (Whi/62)
Viscosity "thick and waxy" (Cra/75); "heavy body" (Lov/65a)
Granulation rapid (Rof/75); slow, soft grain (Lov/65a)
Flavour pronounced (Cra/75); mild (Lov/56; Lov/65a); distinctive
(Lov/55e); sometimes strong, medicinal if bees collect berry juice
(Mot/64).   **Aroma** fragrant (Lov/65a)

383 Serjania triqueta Radlk.; Sapindaceae

bejuco cuadrado, bejuco de tres costillas, carretilla (Es/MEX)
Shrub, climber
Distribution tropical C America.   **Habitat** thickets (MEX); widely
distributed in C America
Temperature warmer regions of MEX

Economic and other uses
Medicinal

Nectar rating; blooms, nectar flow
N1 MEX(Ord/66)
N3 MEX(Ord/72)
Blooms viii (tropical America, Ord/83).   **Nectar flow** usually during the
last wk of viii and lasting for only a few days (Ord/83).   **Nectar**
secretion copious (Ord/83)

Honey no data

384 Sesamum indicum L.; Pedaliaceae
syn Sesamum orientale L.

sesame; ajonjolí (Es/CUB, DOR, HOD, MEX); wijen (In); gingelly, til (INI)
Herb, 40-200 cm depending on cv; annual; fls white, pink, mauve depending
on cv; nectary in fl, also extrafloral nectaries on either side of base of
fl stalk (Frj/70)

Distribution tropical C America, Caribbean, S America, Asia, Africa; subtropical N America, Asia, Africa; native to Africa. **Habitat** cultivated crop plant, especially in savanna zones; altitudes <500 m **Soil** fertile loam preferred; rocky limestone (USA/TX); waterlogging not tolerated. **Rainfall** 500-1100 mm; in low rainfall areas, irrigation required; short periods of drought tolerated by established plants

Economic and other uses

Food - seed; oil from seed; drink from seed. **Fodder** - oilseed cake. **Fuel** - stems. **Land use** shade plant in tobacco plantations. **Other uses** oil for lighting, soap, paints, lubrication; medicinal

Nectar rating + honeybee species; blooms, nectar flow; composition

N1 BUM(Zma/80); INI/BIH,ac(Nai/76); URS(Fed/55)
N2 CUB(Ord/44); DOR(Ord/64); GRC(Nic/55); HOD(Ord/63); INO[ac](Bee/77); MEX(Ord/72)
N USA/NE,TX(Lov/77)
Blooms vi-vii (BUM); vii-viii (CUB); vii-x (tropical America). **Nectar secretion** higher on day after rain (Ord/44). **Sugar concentration** [medium] 38% (Zma/80)

Honey flow

Honey potential [moderate] 45 kg/ha (URS, Fed/55)

Pollen

P1 BUM; INI/BIH. **P3** INO. **P** DOR. **Pollen grain** illustrated (Cht/77)

Recommended for planting to increase honey production

CUB (Ord/44); tropical America (Ord/83). Propagate by seed; increase flowering period by 3-4 sowings at 15-day intervals (Ord/83). One of the few annual plants worth cultivating for bees in the tropics (Ord/83)

Honey: physical and other properties

Colour clear, light (Lov/65a); clear (Ord/83); dark (Zma/80). **Pfund** v light amber (Cra/75); light amber (Lov/77)
Granulation slow (Zma/80); fine (Lov/65a)

385 Sicyos deppei G. Don; Cucurbitaceae

bur cucumber, spiny cucumber (En/USA); chayotillo (Es/MEX)
Herb, climber, annual; fls pale yellow, monoecious
Distribution temperate N America; subtropical N America, C America.
Habitat river banks and low land subject to flooding; cornfields (USA/IN)
Soil rich

Nectar rating; blooms, nectar flow

N1 MEX(Ord/72; Wis/55)
N USA(Lov/77)
Blooms vi- late ix (Lov/77)

Honey: physical and other properties

Colour yellowish tint (Lov/56c). **Pfund** extra light amber to light amber (Lov/56c); white (Pel/76)
Flavour mild (Lov/56c)

386 Sicyos laciniatus L.;   Cucurbitaceae

chayotillo (Es/MEX)
Herb, climber;   fls greenish-white
Distribution subtropical C America

Nectar rating
N1 MEX(Ord/72;   Wis/55)

Honey no data

387 Sinapis alba L.;   Cruciferae
syn Brassica alba L.

white mustard;   weisse Senf (De);   moutarde blanche (Fr)
Herb, <80 cm, annual;   fls yellow
Distribution temperate Europe, (Med) Africa;   native to Med region, and
Crimea (URS)
Soil regions without chernozem (URS)

Economic and other uses
Food - seed for table mustard;   seedlings for salad.   **Fodder** - forage crop
for sheep.   **Soil benefit** green manure

Alert to beekeepers
Honey granulates rapidly;   do not leave in hive for winter (Ave/78)

Nectar rating;   blooms, nectar flow;   composition
N1 URS(Ave/78;   Fed/55)
N2 CZE(Svo/58)
N3 POL(Dem/64a)
N ROM (Cir/80)
Blooms v-x (ROM, Cir/80).   **Nectar secretion** 0.14-1.1 mg/fl/day (Han/80;
Maz/82);   0.20, 0.56 mg/fl/season (AA332/61);   diurnal changes discussed
(AA162, 163/79).   **Sugar concentration** [high] 19-68% (Han/80);   [medium]
28% (Haa/60);   up to 60% in dry season (AA332/61);   approx 50% (AA162,
163/79).   **Sugar value** (mg/fl/day) [medium] 0.133, 0.189 (Bac/60);   0.09-
0.42 (Han/80);   [low] 0.054-0.099 (AA656/70).   **Sugar analysis** (Pry/50;
Wyk/53;   Yak/73)

Honey flow
Honey potential (kg/ha) [moderate] 22-23 (EUR, Maz/82);   <50 (POL,
Dem/64a);   37 (ROM, Bac/60);   40-100 (ROM, Cir/80);   21-34 (ROM, AA656/70),
<100 (URS, Ave/78);   40 (URS, Fed/55);   81-110 (URS, AA332/61)

Pollen
P URS (AA530/80).   **Yield** fair (Cir/80).   **Colour** of load lemon yellow
(Cir/80).   **Reference slide**

Recommended for planting to increase honey production
URS (Ave/78)

Honey: physical and other properties
Colour and some other properties similar to those of Brassica napus honey
(Cra/75)
Granulation rapid (Wyk/53). **Alert to beekeepers** do not leave honey in
hive for winter (Ave/78)

388 Sinapis arvensis L.; Cruciferae

charlock, corn mustard, wild mustard; moutarde des champs, séneve (Fr);
Ackersenf (De)
Herb, <1 m, annual; fls yellow
Distribution temperate Europe, N America, (Med) Africa, Oceania; native to
?Med. **Habitat** widely distributed weed of cultivated land; wasteland
Soil good soil preferred

Warning
Persistent weed of agricultural land; harbours diseases of turnips and
other Brassica crops (How/79)

Nectar rating; blooms, nectar flow; composition
N1 URS(Glu/55)
N2 FRA(Lou/81); GFR(Gle/77)
N3 POL(Dem/64a); ROM(Cir/77)
N ALG(Ske/72); NEZ(Wal/78)
Blooms vi-vii (EUR); v-ix (ROM). **Nectar secretion** (mg/fl/day) 0.15-0.36
(Maz/82); 0.157 (Sim/80). **Sugar concentration** [high] 57.8-67.8%
(Dem/63a); 17-73% (Han/80); [medium] 30.1% (Pek/80); 27.2% (Sim/80).
Sugar value (mg/fl/day) [medium] 0.05-0.2 (Maz/82); 0.102 (Sim/80); [low]
0.04 (Cir/77); 0.097 (Jua/64). **Sugar analysis** (Jua/64; Pec/61)

Honey flow
Honey potential [moderate] 40 (GDR, Bec/67; ROM, Cir/77); 17-38 (POL,
Dem/63a); 42 (ROM, Jua/64); 40-100 (ROM, Cir/80)

Pollen
P1 GFR; URS. **P2** FRA. **P** ROM. **Yield** 0.5-0.66 mg/fl (Maz/82); 35-102
kg/ha (Maz/82). **Colour** of load greenish yellow (Han/80; Maz/82).
Reference slide

Honey: physical and other properties
Colour yellow (Glu/55). **Pfund** white to pale amber (How/79); white
(Ske/72); light amber (Wal/78)
Granulation rapid, in 3 days when exposed to light (How/79); normally
coarse (Lov/59a)
Flavour mild with slight sharpness (Lov/59a); mild, but when fresh may be
slightly hot like mustard (How/79); "none" (Glu/55); delicate but rather
hot (Wal/78)

389 Spondias mombin L.; Anacardiaceae

hog plum; jobo (Es/DOR, HOD, MEX, VEN); jocote (Es/MEX)
Tree, large; fls whitish
Distribution tropical Africa, S America.   **Habitat** river banks; deciduous
forests; dry and humid woodland; altitudes <800 m
Rainfall moderate

Economic and other uses
Food - fruit.   **Land use** hedges

Nectar rating; blooms, nectar flow
N1 HOD(Ord/63);   MEX(Ord/72);   VEN(Cra/73)
N2 DOR(Ord/64)
Blooms iv-v (tropical America, Ord/83); iv, ix (VEN)

Pollen
P DOR

Honey: physical properties
Pfund amber (Ord/83)

390 Stachys annua (L.) L.; Labiatae

annual yellow woundwort
Herb, <30 cm, annual; fls yellowish-white
Distribution temperate Europe; native to Med.   **Habitat** fields, waste places
Soil some lime required

Nectar rating; blooms, nectar flow; composition
N1 CZE(Svo/58);   URS(Fed/55;   Glu/55)
N3 ITA(Ric/78)
N YUG(Kul/59)
Blooms vi-x (EUR).   **Nectar flow** v important in southern Moravia and
southern Slovakia (CZE, Svo/58).   **Sugar concentration** [medium] 43.65%
(AA194/76).   **Sugar value** [medium] 0.119 mg/fl/day (Jua/64).   **Sugar
analysis** (But/73a; Jua/64; Maz/82).   Bees cannot reach all the nectar
(Glu/55)

Honey flow
Honey yield [high] 40-80 kg/colony/season (CZE, Svo/58).   **Honey potential**
(kg/ha) [high] 120-550 (GDR, Bec/67); [moderate] 88-335 (BUL, AA194/76);
120-150 (ROM, Apc/68; Cir/80); 88 (ROM, Jua/64)

Pollen
Reference slide

Honey: physical and other properties
Colour light yellowish (Maz/82).   **Pfund** v light amber (Cra/75); white to
extra light amber (Pia/81); white (Ric/78)
Viscosity "rather viscous" (Pia/81)
Granulation soft, fine (Cra/75); soft, fine-grained, then colour white
(Maz/82); fine to medium grain (Pia/81); regular (Ric/78)

Flavour none, unless slightly acid due to fermentation (Pia/81).   **Aroma** v slight (Pia/81);   strong (Ric/78)

391 Symphoricarpos albus (L.) S.F. Blake;   Caprifoliaceae
syn Symphoricarpos racemosus Michx.;   Symphoricarpos rivularis Suks.

snowberry, waxberry;   badger brush, wolfberry (En/CAF);   Schneebeere (De)
Shrub;   fls white, small, inconspicuous
Distribution temperate N America, Europe;   subtropical N America.   **Habitat** mountain states (only), subtropical USA

Nectar rating;   blooms, nectar flow;   composition
N1 CAF/BC(Dav/69);   GFR(Gle/77)
N2 USA/ID,MA,UT(Pel/76)
N3 BEG(Cra/84)
N CAF/ALTA,MAN(Pel/76);   POL(Dem/64a)
Blooms vi-vii (CAF/BC; POL; USA/WA);   v-vi (USA/UT).   **Nectar secretion** (mg/fl/day) 6.11 (Sim/80);   4.00 (AA902/80).   **Sugar concentration** [medium] 30.3, 36.70% (2 yrs, Pet/72);   34.1% (Sim/80);   34.4% (AA194/76);   32.8% (AA902/80).   **Sugar value** (mg/fl/day) [high] 5.81 (Sim/75);   4.69 (Sim/80);   [medium] 1.03 (AA902/80).   **Sugar analysis** (Pek/61;   Wyk/52; AA884/82)

Honey flow
Honey potential (kg/ha) [moderate] 270-458 (BUL, AA194/76);   200-400 (POL, Dem/64a);   100 (ROM, Apc/68);   200 (ROM, Cir/80)

Pollen
P3 GFR.   Reference slide

Honey: physical and other properties
Pfund white (Cha/48);   amber (Lov/62);   extra light amber (Pel/76)
Granulation slow (Pel/76)
Flavour mild (Lov/62)

392 Symplocos spicata Roxb.;   Symplocaceae

porinelli (INI)
Distribution tropical Asia.   **Habitat** evergreen woodland in small ravines, <1500 m altitude

Nectar rating + honeybee species;   blooms, nectar flow
N1 INI/TAM,ac(Chn/74)
Blooms x-xi (INI/TAM)

Honey no data

393 Syzygium aromaticum (L.) Merrill & Perry; Myrtaceae
syn Eugenia caryophyllus (Spreng.) Bullock & S.G. Harrison

clove tree; giroflier (Fr)
Tree, <14 m, evergreen; fls white or reddish
Distribution tropical Asia, Africa; native to Moluccas. **Habitat** maritime climate; cultivated crop plant
Soil deep, red, sandy, acid loam; good drainage essential; no water-logging. **Temperature** range is 25-34° in Moluccas. **Rainfall** 2200-3700 mm annually in Moluccas; not continually humid

Economic and other uses
Food - dried fl buds for flavouring and for oil of cloves. **Other uses** medicinal; perfumery

Nectar rating + honeybee species; blooms, nectar flow
N1 INI/KAR,KER[ac](Kha/59)
N MAE(Lat/54)
Blooms iv-v (INI/KAR, KER)

Pollen
P INI/KAR, KER

Honey no data

394 Syzygium cordatum Hochst. ex Krauss; Myrtaceae

water berry; curre, mecurre, umdoni (MOZ)
Tree/shrub, 8-15 m; fls creamy-white to pinkish, fragrant
Distribution tropical Africa. **Habitat** always near water; stream banks; swamp forest

Economic and other uses
Food - fruit; alcoholic drink from fruit. **Timber.** **Other uses** dye from bark; medicinal

Nectar rating + honeybee species; blooms, nectar flow
N1 MOZ[tm](Cra/73); ZAM[tm](Sto/82)
Blooms viii-xi (southern Africa, Pag/77). **Nectar secretion** abundant (Pag/77)

Pollen
P ?ZAM. **Pollen grain** illustrated and described (Smt/56a)

Honey no data

395 Syzygium cuminii (L.) Skeels; Myrtaceae
syn Eugenia jambolana Lam.; Syzygium jambolanum (Lam.) DC.

jambolan, Java plum; jambul, jamun, naraidu, naval, neralumara (INI)
Tree, 13–30 m, evergreen; fls white with greenish tinge, pink
Distribution tropical Asia, Africa, C America, Caribbean, Oceania;
subtropical N America, Oceania; native to tropical Asia. **Habitat** widely
planted and often naturalized; orchards and forests; subtropical
evergreen forest; river banks and watery places; altitudes <1850 m;
predominant species of Western Ghats flora (INI/MAH)
Soil wide range. **Temperature** tropical and warm subtropical conditions
required. **Rainfall** heavy rainfall and prolonged flooding tolerated; not
drought resistant when young

Economic and other uses
Food – fruit; preserves, vinegar, wine and fruit juice. **Fodder** –
fruit. **Fuel. Timber. Land use** hedges, windbreak, shade, amenity.
Other uses tannin from bark; medicinal

Warning
Noxious weed in HAW where birds have spread the seeds; shades out more
desirable forage plants (Usa/80). Rotting fallen fruit is a nuisance in
urban areas (USA/FL, Usa/80)

Nectar rating + honeybee species; blooms, nectar flow; composition
N1 HAW(Eck/52; Ord/83); INI/BIH,ac(Nai/76); INI/KAR[ac](Kha/59;
Sat/75); INI/KER(Kha/59); INI/MAH,ac(Chu/65; Chu/80; Ded/53; Koh/58;
Sat/75); INI/PUN[ac](Chd/77); INI/TAM,ac(Chn/74); INI/UTT,ac,ad(Koh/58;
Rae/80); TAN/ZAN[tm](Cra/73)
N2 INI/KAR[ac](Diw/72)
Blooms ii–iv (INI/KAR); ii–iii (tropical America, Ord/83). **Nectar
secretion** erratic (Sat/75). **Sugar concentration** [high] up to 72.0%, but
may be as low as 9% (Sat/75)

Honey flow
Honey yield "second most important source after mesquite" in HAW (Mot/64);
30–40% of local harvest in Western Ghats (INI/MAH, Sat/75)

Pollen
P1 INI/BIH, TAM. **P2** INI/MAK. **P3** INI/KAR, PUN. **P** INI/KAR, KER.
Pollen grain illustrated and described (Smt/56a). **Reference slide**

Honey: chemical composition
Water [medium] 18.4% (8 samples, Naa/70); 18.85% (Raj/70)
Glucose [medium] 32.26% (Naa/70). **Fructose** [high] 43.30%. **Sucrose,
maltose, raffinose, melezitose** present (Pha/70). **Dextrin** 1.555% (Naa/70)
Ash [medium] 0.182% (Naa/70); see also Kal/64; Pha/70
Amino acids qualitative analysis (Kal/64). **Protein** 0.655% (Naa/70)
Fermentation on storage likely after a few mths (Koh/58)
Vitamins riboflavin present, also ascorbic acid, thiamine, niacin (Kal/65)

Honey: physical and other properties
Colour light burnt sienna (Chu/65); light reddish brown (Ded/53). **Pfund**
amber (Koh/58; Naa/70); light amber (Mot/64)

Relative density 1.407 (Naa/70).   **Viscosity** at 26° and 45°, approx 83
poise and 12 poise (Raj/70).   **Optical rotation** -2.05 deg (Naa/70)
Granulation [slow] none after storing for yrs (Ded/53; Mot/64)

396 Syzygium jambos (L.) Alston; Myrtaceae
syn Eugenia jambos L.

rose apple; pomarrosa (Es/CUB, DOR); pomo (Es/DOR); manzana rosa
(Es/HOD, NIA)
Tree, <10 m, evergreen, often forms thick groves; fl stamens greenish white
Distribution subtropical N America; tropical C America, Caribbean, Asia;
native to Indonesia.   **Habitat** river banks and wet savanna; widely grown
in CUB

Economic and other uses
Food - fruit, preserves, vinegar.   **Fodder** - lvs.   **Timber.**   Land use
hedges in coffee plantations, amenity

Nectar rating; blooms, nectar flow
N1 CUB(Ord/44)
N2 CUB(Ord/56; Ord/83); DOR(Ord/64); NIA(Ord/63a)
N HOD(Ord/63)
Blooms ii-iii (tropical America, Ord/83); spring (USA/FL)

Honey flow
Honey yield "a good harvest every yr" (CUB, Mot/64)

Pollen
P DOR

Honey: chemical composition
Water [medium] 18.15% (Pha/62)
Sugars, total 78.25%.   **Glucose** [medium] 36.09%.   **Fructose** [medium]
39.72%.   **Reducing sugars** 75.81%.   **Dextrin** 1.15%
Free acid [low] 13.1 meq/kg

Honey: physical and other properties
Pfund amber (Mot/64; Ord/83; Pha/62)
Relative density 1.415 (Pha/62).   **Optical rotation** -36 minutes
Flavour mild, like sorghum syrup (Mot/64)

397 Tamarindus indica L.; Leguminosae

tamarind; tamarindo (Es/DOR, HOD, NIA); tamarin (Fr/MAY); asem (In);
hunase, imli, puli, tentul (INI)
Tree, <25 m, evergreen, vigorous, hurricane-resistant; fls yellow spotted
red
Distribution tropical Asia, C America, Caribbean, Africa, Oceania;
subtropical N America; native to tropical Africa.   **Habitat** widely
cultivated in tropics; low altitudes; coastal and inland areas; open
sites preferred; dry savanna and monsoon regions; wasteland

Soil wide range; sandy; deep soil preferred but must be well drained.
Temperature damaged by frost. **Rainfall** >800 mm; extended period of dry
weather required; humid and sub-humid zones; drought resistant

Economic and other uses
Food - fruit for flavouring eg drinks, soups; lvs, fls and pods in
curries; oil from seed. **Fodder** - crushed seed. **Fuel. Timber. Land
use** shade, amenity, firebreak. **Other uses** medicinal; crushed seed for
making size; oil from seed

Nectar rating + honeybee species; blooms, nectar flow
N1 CHA[tm](Gad/80); ?HAI(Mul/78); INI/KAR,KER[ac](Kha/59); INI/TAM-
[ac](Ram/37); MAY(Bro/82; Cra/73); THA(Smt/83)
N2 DOR(Ord/64); INI/MAD[ac](Khn/48); INI/TAM[ac](Sig/62); INI/UTT[ac]-
(Koh/58); INO[ac](Bee/77); NIA(Ord/63a)
N3 INI/MAH[ac](Chu/80)
N HOD(Ord/63)
Blooms v-viii (central America, Ord/83); iv-vii (INI); xii-i (MAY)

Pollen
P1 INO. **P3** INI/MAH. **P** DOR; INI/KAR, KER. **Pollen grain** illustrated
and described (Smt/54a)

Honey: physical and other properties
Colour dark (Cra/75); rich golden (Koh/58)
Viscosity "thin" (Koh/58)
Flavour sour (Khn/48); "quite sweet, but when swallowed tastes slightly
acid like flower" (Koh/58)

398 Taraxacum officinale Weber; Compositae
syn Taraxacum dens-leonis Desf.

dandelion; diente de león (Es); lechuguilla (Es/CHL); amargon (Es/COS);
dent de lion, pissenlit (Fr); maelkebotte (Da); Frühlingslöwenzahn,
Kuhblume (De); soffione (It)
Herb, <50 cm, perennial, fls bright yellow
Distribution temperate N America, S America, Europe, Oceania, Asia, (Med)
Africa; native to Europe. **Habitat** pastures, lawns, wasteland
Soil damp areas; chalky soil preferred

Economic and other uses
Food - lvs for salad; fls for wine; roasted roots for coffee substitute. **Other uses** medicinal

Warning; alert to beekeepers
Warning persistent perennial weed. **Alert to beekeepers** it may be
necessary to mow flowering dandelions in orchards, to prevent bees foraging
on them instead of apple blossom. Blooms early in season so beekeepers
must make colonies strong to harvest honey (How/79)

Nectar rating; blooms, nectar flow; composition
N1 CAF(Cha/48; Cou/59); CAF/ALTA(Lov/62c); CHN(Tse/54); FRA(Lou/81);
IRN(Cra/73)

N2 AUS/SA(Pur/68); BEG(Grn/65); ?CAF/MAN(Mit/49); CAF/SASK(Mcc/58); COS(Ord/83); DEN(Jon/54); GFR(Gle/77); ITA(Ric/78); NEZ(Rob/56); USA/UT(Nye/71); USA(Lov/56)
N3 CAF/ALTA(Hen/77; Wes/49); CAF/NWT(Hat/81); CHL(Jav/38); UK(How/79); URS(Fed/55); USA/CO(Wio/58); USA(Pel/60)
N ALG(Ske/72); AUT(Maz/82); CAF/ALTA,MAN,SASK(Pel/76); EUR(Maz/82); FIN(Koc/74); NEZ(Wal/78); POL(Dem/79); SWI(Maz/82); URS(Glu/55)
Blooms iv-vi (CAF; EUR); early spring (NEZ); v-viii (POL); v-vi (USA/CO). **Alert to beekeepers** it may be necessary to mow flowering dandelions in orchards, to prevent bees foraging on them instead of apple blossom. Blooms early in season so beekeepers must make colonies strong to harvest honey (How/79). **Nectar secretion** 1.0-19.0 mg/fl/day (Dem/79); none if temperature >75°; fls close on cloudy days (USA, Lov/59). **Sugar concentration** [high] 11.60-72.74% (Dem/79); [medium] 18-51% (Maz/82); 26-56% (Shw/53). **Sugar value** [high] 0.43-9.79 mg/fl/day (Dem/79). **Sugar analysis** (Maz/59; Maz/82; Pec/61)

Honey flow
Honey yield (kg/colony/season) [high] 50-95 (CAF/ALTA, Lov/62c); 27, 9-13 (USA/CO, Lov/62c). **Honey potential** (kg/ha) [high] 800-1000 (GDR, Bec/67); [moderate] up to 25 (260 fls/sq m, POL, Dem/79); 200 (ROM, Apc/68; Cir/80)

Pollen
P1 CAF/ALTA, MAN; FRA; GFR; URS; ?USA/CO. **P2** AUS/SA; COS; ITA. **P** ALG; CAF; CAF/ALTA, MAN, SASK; EUR; NEZ; UK; USA/UT; USA. **Yield** 1.2 mg/fl/day (Maz/82); load 6.2 mg (Maz/82). **Pollen value** high (EUR, Sta/74). **Chemical analysis** (Gob/81; Maz/82; Sta/74). **Colour** of load dull yellow (Cir/80); load orange-yellow (Han/80); load orange (Hod/74; Ric/78). **Pollen grain** illustrated and described (Hod/74; How/79); under-represented in honey (Fea/74; Fea/74a). **Reference slide**

Honey: chemical composition
Sugars (as % of total, Maz/64): **glucose** 40.1% ; **fructose** 49.9%; **sucrose** 6.3%; **maltose** 2.0%; **fructomaltose** 1.7%; **melezitose** 0.27% (Hir/51)
pH 4.3 (Dus/72)
Peroxide number 243.7 µg/g/h (186.2 after 10 minutes in sunlight, Dus/72)
Amino acids - 12 identified (Kum/74)
Fermentation on storage unlikely, low yeast count (Maa/73); likely (Pia/81)

Honey: physical and other properties
Colour intense golden yellow (Cra/75); lemon yellow (Erb/83); bright yellow (Fos/56); deep yellow (Lov/56); dark (Pel/76); pale yellow (Ric/78). **Pfund** light amber to golden amber (Pia/81); medium amber (Wal/78); Lovibond-Pfund grade (Aub/83)
Viscosity "thick" (Glu/55). **Electrical conductivity** 0.000560/ohm cm (Vor/64)
Granulation rapid, coarse, hard (Cra/75); rapid, fine (Fos/56); rapid, fine-grained, regular (Pia/81; Ric/78)
Flavour sharp (Cra/75; Fos/56); sharp, strong, but can be mild (Lov/59); strong, characteristic and like animal glue (Pia/81). **Aroma** pronounced (Cra/75; Fos/56); like the flowers (Lov/56); initially repellent (Pia/81); distinctive (Ric/78)

399 Terminalia arjuna (Roxb.) Wight & Arn;   Combretaceae

mathi (INI);   arjan (PAK)
Tree, >16 m, deciduous; fls white/greenish
Distribution tropical Caribbean, Asia, Africa;   subtropical N America;
native to central India.   **Habitat** tropical moist deciduous forest (INI/MAH)
Soil damp.   **Rainfall** 1000-2500 mm (INI/MAH)

Economic and other uses
Land use shade, amenity.   **Other uses** medicinal

Nectar rating + honeybee species; blooms, nectar flow; composition
N1 INI/KAR,KER[ac](Kha/59);   PAK[ac](Sig/62)
N3 PAK,ac(Shr/48)
Blooms iv-vi (INI/KAR, KER); vi-vii (tropical America, Ord/83)

Pollen
P INI/KAR, KER

Honey no data

400 Terminalia bellerica (Gaertn.) Roxb.;   Combretaceae

behda, ghotinga, shanthi, thari (INI)
Tree, >16 m; fls white/yellow
Distribution tropical Asia.   **Habitat** semi-evergreen and moist deciduous
forest (INI/KAR)

Economic and other uses
Medicinal;   dyes and tannins

Nectar rating + honeybee species; blooms, nectar flow
N1 INI/KAR,KER[ac](Kha/59)
N3 INI/KAR,ac,ad, Trigona(Diw/64);   INO[ac](Bee/77)
Blooms iv-v (INI/KAR, KER)

Pollen
P INI/KAR, KER

Honey: physical and other properties
Flavour "nauseating" (Diw/64)

401 Terminalia chebula Retz.;   Combretaceae

inknut, orpiment;   anilekai, hirda (INI)
Tree, >16 m; fls pale yellow
Distribution tropical Asia.   **Habitat** evergreen forests of eastern edges
and slopes of Western Ghats (INI/MAH);   moist deciduous forest (INI/KAR)
Rainfall 3600-6000 mm (INI/MAH)

Economic and other uses
Medicinal; dyes and tannins

Nectar rating + honeybee species; blooms, nectar flow
N1 INI/KAR,KER[ac](Kha/59); INI/MAH[ac](Chu/65; Ded/53)
N3 INO[ac](Bee/77)
Blooms v-vi (INI/MAH); iv-v (INI/KAR, KER)

Honey flow
Honey yield contributes 20-25% of annual honey flow (INI/MAH, Chu/65)

Pollen
Pollen grain illustrated and described (Chu/65)

Recommended for planting to increase honey production
INI/MAH (Sub/79)

Honey: chemical composition
Water [medium] 17.2% (8 samples, Naa/70); 17.13% (Pha/62); 18.21% (Raj/60)
Sugars, total 79.06% (Pha/62). Glucose [medium] 35.69% (Naa/70); 34.06% (Pha/62). Fructose [medium] 40.23% (Naa/70); 41.30% (Pha/62). Sucrose, maltose, melezitose present (Pha/70). Reducing sugars 75.36% (Pha/62). Dextrin 1.80% (Naa/70); 1.49% (Pha/62)
Ash [medium, also low] 0.162% (Naa/70); 0.014% (Pha/62). Contents of Ca, P, Fe, Mg (Kal/64); Na, K, Ca, Mg, Fe, P, Si (Naa/70); see also Pha/70
Free acid [low] 7.7 meq/kg (Pha/62). Tartaric, citric, malic, succinic acids present (Naa/70; Pha/70)
Amino acid analysis (Kal/64; Pha/70). Protein 0.8% (Kal/64); 0.530% (Naa/70); 0.61% (Pha/62)
Vitamins niacin, riboflavin (Kal/65)

Honey: physical and other properties
Colour light yellow (Chu/65; Naa/70; Pha/62). Pfund amber to dark amber (Sig/62); Pfund-Lovibond grade (Pha/62)
Relative density 1.416 (Naa/70); 1.42 (Pha/62). Optical rotation -4 deg 4 minutes (Naa/70); -1 deg 48 minutes (Pha/62)
Granulation medium (Naa/70)
Flavour characteristic (Sig/62). Aroma pungent

402 Terminalia tomentosa Wight & Arn.; Combretaceae

ain, mathi (INI)
Tree, >16 m; fls white/yellow
Distribution tropical Asia. Habitat evergreen forest of Mahabaleshwar hills, INI; moist deciduous forest

Economic and other uses
Fuel. Timber

Nectar rating + honeybee species; blooms, nectar flow
N1 INI/KAR[ac](Diw/64; Kha/59); INI/KER[ac](Kha/59)
N3 ?INI/MAH[ac](Tha/76)

Blooms iv-vi (INI/KAR, KER).   **Nectar flow** "more fully worked by bees when monsoon delayed" (INI/KAR, Diw/64)

Pollen
P3 ?INI/MAH.   P INI/KAR, KER.   **Pollen grain** illustrated and described (Chu/65)

Honey no data

403 Thelepaepale ixiocephala (Benth.) Bremk.;   Acanthaceae
syn Strobilanthes ixiocephalus Benth.

darmori, tita karava, whayati (INI)
Shrub;   fls white, gregarious flowering every 8 yrs
Distribution tropical Asia.   **Habitat** moist deciduous and semi-evergreen forests (INI/KAR);   open forest and outskirts of forests;   altitudes >500 m on tablelands
Temperature 20-30° Mahabaleshwar (INI/MAH).   **Rainfall** 6500-7500 mm, Mahabaleshwar (INI/MAH)

Nectar rating + honeybee species;   blooms, nectar flow;   composition
N1 INI/KAR,ac,ad,Trigona(Diw/64);   INI/MAH[ac](Chu/64;   Pha/65)
Blooms xi-i (INI/KAR).   **Nectar flow** every 8 yrs.   **Sugar concentration** [high] 35% at 08.00 h, 46% at 15.00 h (Pha/65).   **Sugar analysis** (Wak/81)

Pollen
P1 INI/MAH.   P INI/KAR.   **Pollen grain** illustrated and described (Chu/65).   **Reference slide**

Recommended for planting to increase honey production
INI/MAH (Sub/79)

Honey: chemical composition
Water [medium] 18.30% (8 samples, Naa/70);   19.11% (Pha/62)
Glucose [medium] 38.31% (Naa/70);   38.32% (Pha/62).   **Fructose** [medium] 39.48% (Naa/70);   39.80% (Pha/62).   **Sucrose, maltose, raffinose, melezitose** present (Pha/70).   **Reducing sugars** 78.12% (Pha/62).   **Dextrin** 1.105% (Naa/70);   1.08% (Pha/62)
Ash [medium] 0.147% (Naa/70);   0.138% (Pha/62);   Na, K, Ca, Mg, Fe, P, Si contents (Naa/70;   Pha/70)
Free acid [medium] 23.3 meq/kg (Pha/62).   Citric, malic, succinic acids present (Pha/70)
Amino acid analysis (Pha/70).   **Protein** 0.522% (Naa/70);   0.579% (Pha/62)

Honey: physical and other properties
Colour colourless, transparent (Chu/65);   light yellow (Naa/70);   colourless to light (Sur/75).   **Lovibond** grade (Naa/70)
Relative density 1.409 (Naa/70);   1.405 (Pha/62).   **Optical rotation** -1 deg 30 minutes (Naa/70);   -1 deg 12 minutes (Pha/62)
Granulation rapid, uniform (Chu/65;   Sur/75);   rapid and complete soon after extraction (Pha/65);   medium, within 20-30 days (Naa/70)

404 Thunbergia grandiflora (Roxb. ex Rottl.) Roxb.; Acanthaceae

conejitos (PHI)
Shrub, climber, 1.5 m, perennial; fls blue/white
Distribution tropical Asia; native to Asia

Economic and other uses
Land use amenity

Nectar rating + honeybee species; composition
N1 PHI[ac,ad,af also am](Row/76)
Sugar concentration [low] 13% (Row/76). **Sugar analysis** (Row/76). **Nectar analysis** (AA117/66)

Honey no data

405 Thymus capitatus (L.) Hoffm. & Link; Labiatae
syn Coridothymus capitatus (L.) Reichenb. f.

Mediterranean wild thyme, mountain thyme, wild thyme; saghtar (MAQ)
Shrub, 20-50, exceptionally to 150 cm, v aromatic; fls purplish-pink
rarely white
Distribution temperate (Med) Europe and Africa; native to Europe.
Habitat dry sunny hills; bare parched hillsides (CYP)
Soil stony. **Temperature** on Malta, where shrub grows, temp is <40° but
rarely <0°. **Rainfall** drought resistant

Economic and other uses
Oil for medicine and perfumery

Nectar rating; blooms, nectar flow
N1 CRE(Adm/54; Nic/55); CYP(Adm/54); GRC(Mai/52; Nic/55); MAQ(Far/79)
Blooms v-viii (EUR); vi-vii (GRC); v-vii (MAQ). **Nectar flow** vii-viii
(MAQ). **Nectar secretion** total yield 0.0001 ml/fl, (Mcg/59); dependent on
RH (Adm/54); dry wind from Sahara desert stops flow (GRC, Nic/55)

Honey flow
"Hymettus honey" from GRC is mostly a mixture from Thymus spp, Satureia spp
and Origanum vulgare

Honey: chemical composition
Amino acids – contents of individual free amino acids; proline,
phenylamiline, tyrosine high (Mak/78)

406 Thymus serpyllum L.; Labiatae

mother-of-thyme, wild thyme; serpolet (Fr); Feldthymian, Quendel (De)
Herb, creeping, <10 cm, aromatic; fls purple

Distribution temperate Europe, N America; native to Europe. **Habitat** moorlands, dry pastures and chalk districts (UK); sunny hills, cliffs (EUR); dry slopes, grassland, dunes, bushy places
Soil chalky

Nectar rating; blooms, nectar flow; composition
N1 URS(Fed/55); USA/MA,NY(Pel/76)
N2 FRA(Lou/81); GFR(Gle/77); POL(Dem/64a); USA/MA,NY(Lov/56)
N3 ?UK(How/79)
Blooms iv–ix (EUR); vi–vii (UK); vii–frost (USA). **Nectar secretion** 0.16, 0.18 mg/fl/day (Dem/63a). **Sugar concentration** [medium] 26.9, 42.8% (Dem/63a); 27–45% (Han/80); 30.1% (Pek/78). **Sugar value** [low] 0.043– 0.077 (Dem/63a). **Sugar analysis** (Bat/73a)

Honey flow
Honey yield [high] 57 kg/colony/season (USA/NY, Pel/76); "especially important" in southern URS (Fed/55). **Honey potential** (kg/ha) [moderate] 40.8 (BUL, Pek/78); 48–161, together with T. pulegioides (EUR, Maz/82); 48, 149 (POL, Dem/63a); 100–200 (POL, Dem/64a); 80–120 (ROM, Cir/80)

Pollen
P2 FRA. **P3** GFR. **Colour** of load orange–yellow (Han/80). **Pollen grain** illustrated and described (Saw/82). **Reference slide**

Honey: chemical composition
Water [medium] 16.8% (20 mths, ?heated, Whi/62)
Glucose [medium] 31.20%. **Fructose** [medium] 37.13%. **Sucrose** [low] 0.85%. **Maltose** 8.83%. **Higher sugars** 1.70%. **Melezitose** 0.34%
Ash [medium] 0.384%
pH 4.80. **Total acid** 27.88 meq/kg. **Free acid** [medium] 22.41 meq/kg. **Lactone** 5.47 meq/kg
Nitrogen 0.035% dry wt (Bos/78); 0.057% (Whi/62). **Amino acids**, free 0.91, protein 0.103% dry wt (Bos/78)

Honey: physical and other properties
Pfund light amber (Cha/48); v light amber (Pel/76); 50–70 mm, light amber (Whi/62)
Viscosity "good body" (Pel/76)
Flavour and aroma strong, minty (Lov/56)

407 Thymus vulgaris L.; Labiatae

garden thyme, thyme; thym (Fr); Thymian (De)
Herb, sub-shrub, 10–30, exceptionally to 50 cm; fls whitish/pale purple
Distribution temperate (Med) Europe, Oceania; native to W Med. **Habitat** often cultivated; dry slopes, rocks or maquis
Soil limestone or clay essential; light warm soils preferred

Economic and other uses
Food – culinary herb. **Other uses** oil for medicine and perfumery

Nectar rating; blooms, nectar flow; composition
N1 CYP(Geo/52; Mel/30); FRA(Lou/81); URS(Glu/55)
N2 GFR(Gle/77); POL(Dem/64a)
N NEZ(Wal/78)
Blooms iv-vii (EUR); x-xi (NEZ). **Nectar secretion** 0.14, 0.13 mg/fl/day
(Dem/63a). **Sugar concentration** [medium] 36.4, 45.4% (Dem/63a). **Sugar
value** [low] 0.051, 0.059 mg/fl/day (Dem/63a). **Sugar analysis** (Wyk/52)

Honey flow
Honey yield [moderate] 19 kg/colony/season (NEZ, Bry/77). **Honey potential**
(kg/ha) [moderate] 100-200 (POL, Dem/64a); 150 (ROM, Cir/80)

Pollen
P1 FRA. P3 GFR. **Reference slide**

Honey: chemical composition
Water: refractive index at 20°, 1.5017 (Moh/82)
Glucose [low] 24.30%. **Fructose** [medium] 36.60%. **Sucrose** [medium] 1.55-
2.29% (3 samples, Moh/82). **Reducing sugars** 77.88% (Moh/82). **Maltose**
9.30%. **Raffinose** 11.10% [sic]
Ash [low] 0.06%. K, Na, Ca contents
pH 6.2. **Free acid** [low] 11.50 meq/kg
"Rich in enzymes" (Cra/75)
Other constituents carbonyl compounds identified (Hoo/63)

Honey: physical and other properties
Colour golden amber (Cra/75). **Pfund** amber (Erb/83); medium amber (Wal/78)
Viscosity at 20°, 205.80 poise (Moh/82). **Optical rotation** -10.53 deg
Granulation slow
Flavour fairly strong, distinctive (Bry/77); v strong (Mao/82a); rather
minty flavour which does not fade on storage (Wal/78). **Aroma** strong
(Cra/75); fragrant (Glu/55)

408 Tilia americana L.; Tiliaceae

lime, linden; basswood (En/CAF, USA); linn, whitewood (En/USA)
Tree, <36 m, deciduous; fls cream, fragrant
Distribution temperate N America; subtropical N America; native to N
America. **Habitat** deciduous forest region of southern ONT (CAF); good
locations no longer plentiful due to cutting of forests in USA (Pel/76);
grows poorly in EUR
Soil rich soil preferred; moderately damp but not too acid or peaty.
Temperature cooler regions preferred

Economic and other uses
Timber - soft fine-grained wood, used eg for comb honey sections and
separators (Lov/66; Pel/76). **Land use** shade, amenity. **Other uses** paper
from pulp, fibre from bark

Nectar rating; blooms, nectar flow; composition
N1 CAF/NB(Wil/59); CAF/ONT(Tow/76); ?CAF/QUE(Cha/48); USA/KS,MA-
(Pel/76); ?USA/MA(Shw/50); USA/MD(Die/71); USA/WI(Pel/76)

N2 CAF/QUE(Cou/59);   USA/IN,KY(Pel/76)
N CAF/ONT,QUE(Sel/49)
Blooms vi (USA);   vii (USA/MA).   **Nectar flow** 10-14 days, irregular - only
in 2-3 yrs out of 5 (Pel/76);   rain sometimes stops flow (Lov/56f).
Nectar secretion 5.39-6.73 mg/fl/day (Dem/60a);   only at temps >18°
(Rof/75).   **Sugar value** max when fl age is 4-5 days (AA535/80).   **Contents
of K, Na, Ca** in nectar (AA361/68)

Honey flow
Honey yield (kg/colony/season) [moderate] 9-22 (USA/VT, Lov/56f);   5-18
(USA/WI, Lov/77)

Pollen
Pollen grain illustrated and described (Ada/76)

Recommended for planting to increase honey production
USA (Lov/56f;   Stm/77)

Honey: chemical composition
Water [medium] 17.0% (1 sample, age 15 mths, Whi/62).
Glucose [low] 30.12%.   **Fructose** [medium] 36.99%.   **Sucrose** [low] 0.63%.
Maltose 8.02%.   **Higher sugars** 1.87%
Ash [low] 0.068%
pH 4.28.   **Total acid** 25.74 meq/kg.   **Free acid** [medium] 17.84 meq/kg.
Lactone 7.89 meq/kg
Nitrogen 0.024%
Other constituents methyl anthranilate 0.04 μg/g (Whi/66)

Honey: physical and other properties
Pfund white (Cha/48;   Pel76);   water white (Lov/56);   27-34 mm, white
(Whi/62)
Flavour rather strong (Lov/56;   Pel/76)

409 Tilia amurensis Rupr.;   Tiliaceae

Amur lime
Tree, deciduous
Distribution temperate Asia, Europe.   **Habitat** forests (URS/Far East and
Urals)
Soil moderately damp but not too acid or peaty

Nectar rating;   blooms, nectar flow;   composition
N1 URS(Ave/78;   Glu/55;   Kov/65;   Pem/65)
Nectar secretion 3.86-9.24 mg/fl/day (2 yrs, Dem 60a);   max at 24-26° and 60-
74% RH, trees aged 50-80 yrs;   little secretion from trees aged 15-20 yrs;
rain, wind, sun and aspect also affect secretion (Pem/65).   **Sugar value**
[medium] 0.71-1.07 mg/fl/day (Dem/60a).   **Sugar analysis** (Pem/65)

Honey flow
Honey yield [high] 45 kg/colony/season (mean for 19 yrs, 3 Tilia spp, URS,
Kov/65).   **Honey potential** [high] up to 1000 kg/ha (URS, AA852/81);   trees
50-80 yrs, 5.02 kg/tree (Pem/65);   together with T. mandschurica, T.
taqueti and Phellodendron amurensis, yields 75-80% of honey of URS/Far East
(Pem/65);   0.08 to 5 kg/tree, depending on age (URS, AA852/81)

Pollen
P URS

Honey no data

410 Tilia cordata Mill.; Tiliaceae
syn Tilia sylvestris Desf.

small-leaved lime, small-leaved linden; Winterlinde (De)
Tree, <35 m, deciduous; fls whitish, fragrant
Distribution temperate Europe, N America; native to Europe. **Habitat** low
altitudes up to and including the beechwood zone (ROM); cliffs and damp
woods (western UK); also planted in parkland?
Soil fertile damp soil (ROM); limestone (UK); not too acid or peaty

Economic and other uses
Timber.   **Land use** amenity.   **Other uses** fibre from bark

Alert to beekeepers
In dry yrs bees may be found dead, paralysed or "drunk" under trees; sugar
metabolism implicated (Cra/77)

Nectar rating; blooms, nectar flow; composition
N1 BEG(Cra/84);   FRA(Lou/81);   GFR(Gle/77);   ROM(Int/65);   URS (Ave/78;
Fed/55;   Glu/55)
N EUR(Maz/82)
Blooms vii (EUR).   **Nectar flow** 12 days (ROM).   **Alert to beekeepers** in dry
yrs bees may be found dead, paralysed or "drunk" under trees; sugar
metabolism implicated (Cra/77).   **Nectar secretion** (mg/fl/day) 1.64-8.02
(Dem/60a);   1.17 (Han/80);   1.09-2.4 (AA847/64);   2.09-4.94 (AA129/72);
1.94 (AA520/78).   **Sugar concentration** [medium] 45-56% (Cir/80);   39.7%
(Han/80);   24.97-34.65% (Pet/72);   17.1-42.6% (AA129/72);   [low] 17.0-19.3%
(AA847/64).   **Sugar value** (mg/fl/day) [medium] 0.148-0.344 (Dem/60a);   0.39
(Han/80);   2.60-3.02 (4 yrs, AA766/65).   **Sugar analysis** (Kay/78;   Maz/59;
Maz/82;   Wan/64;   AA678/66).   **Amino acid analysis** (Moo/65).   Nectar
contains calcium oxalate (Maz/82)

Honey flow
Honey yield [moderate] 15-20 kg/colony/season, 3 Tilia spp (ROM, Int/65).
Honey potential (kg/ha) [high] 600-1000 (ROM, Cir/80);   1000 (ROM,
Sad/60);   580-855 (ROM, AA766/65);   500 (C European URS, Stm/77);
[moderate] 90-125 (CZE, AA520/78);   125 (POL, AA847/64);   83-225 (POL,
AA129/72);   450 (ROM, AA846/64);   also 30 kg/tree (URS, Fed/55)

Pollen
P1 FRA.   **P3** GFR.   **P** EUR.   **Yield** low, 43 000 grains/fl (Maz/82).
Chemical analysis - contains calcium oxalate (Maz/82).   **Colour** dull yellow
(How/79);   load pale to strong yellow, small and compact (Maz/82).   **Pollen
grain** described (How/79);   under-represented in honey (EUR, Maz/82).
Reference slide

Honeydew

Honeydew produced and collected by bees from **Eucallipterus tiliae** (L.)
Callaphididae – rated **D1**, (mid EUR, Hag/66); flow may be intense in vi or
vii, viii, possibly ix, but bees can collect honeydew only early and late
in day ?and/or when RH is high, because it granulates rapidly on tree;
honey data below (mid Eur, Klo/65); honeydew analysis (ITA, Lom/77).
Honeydew produced in FRA (insect not specified, Lou/81). For Tilia spp:
honeydew analysis (from Eucallipterus tiliae, Klo/65); flow vi-vii in some
yrs, honey yield 6-12 kg/colony (ROM, Cir/80)

Honey: chemical composition (honeydew honey)

Glucose [low] 30.09, 30.28% (Tou/80). **Fructose** [medium] 40.01, 39.62%.
Sucrose [medium] 0.89, 1.00%. Contents of these sugars and of **maltose,
fructomaltose** as % of total sugars (Maz/64). **Melezitose** present
(Bel/79). Sugar analysis of honeydew honey (Bel/79; Lom/77); "lime"
honeydew honey contains up to 10% dextrin (How/79)
Malic acid 0.12, 0.13% (Tou/80). **Citric acid** 0.043%
Other constituents calcium oxalate crystals (Maz/82)

Honey: physical and other properties

Colour water clear (Maz/82); honeydew honey dark green (from E. tiliae,
Klo/65)
Electrical conductivity 0.000289/ohm cm (?honeydew honey, Vor/64)
Granulation - colour whitish, light yellow or greenish (Maz/82); honeydew
honey granulates v slowly although melezitose content high (Klo/65)
Flavour "pepperminty" (Maz/82); "lime honeydew honey tastes of treacle"
(How/79). **Aroma** of lime, strong

411 Tilia japonica (Miq.) Simonk.; Tiliaceae

Japanese lime, lime, linden; shina-no-ki (JAP)
Tree, large, deciduous; fls whitish yellow, fragrant
Distribution temperate Asia. **Habitat** mountainous districts (especially
Hokkaido, JAP)
Soil moderately damp but not too acid or peaty

Economic and other uses

Timber. **Other uses** fibre from bark

Nectar rating; blooms, nectar flow

N1 JAP(Inu/57; Sak/82)
Blooms vii-viii (JAP). **Nectar secretion** (mg/fl/day) 1.19-6.79 (Dem/60a);
irregular, highest on rich soil (Inu/57); reduced by low temps (AA563/65).
Sugar value [medium] 0.198-1.318 mg/fl/day (Dem/60a)

Honey flow

Honey yield [high] 60 kg/colony/season, every 2 yrs (JAP/Hokkaido,
Inu/57).

Pollen

P2 JAP

Honeydew produced (ROM, Cir/80)

Honey: chemical composition
Water [high] 21.7% (Ech/75)
Sugars, total 78.3% (Ech/75). As % of total (Ech/75): **glucose** 38.86%; **fructose** 47.83%; **sucrose** 3.36%; **maltose** 8.36%; **higher sugars** 0.56%.
Lactose present (Wan/64)
pH 3.9 (Ech/75)
Amylase 55 500 units/100 ml honey. **Glucose oxidase** 2100 units/100 ml
Protein 0.19%
Other constituents 27 compounds identified: alcohols (mainly 8-p-menthene-1,2-diol), ketones, esters and acids, and (in aroma) 4-isopropylidene-2-cyclo-hexene-1-one (Tsu/74)

412 Tilia koreana Nakai; Tiliaceae

Korean lime
Tree, deciduous
Distribution temperate Asia. **Habitat** mountainous regions (central KOS)
Soil moderately damp but not too acid or peaty

Nectar rating; blooms, nectar flow
N1 KOS(Bek/65)
Blooms vi-vii (KOS). **Nectar flow** 20 days (KOS, Bek/65)

Honey flow
Honey yield 6-10 kg/colony/day, together with T. mandschurica (KOS, Bek/65)

Honey no data

413 Tilia mandschurica Rupr. & Maxim.; Tiliaceae

Manchurian lime
Tree, deciduous
Distribution temperate Asia. **Habitat** forests (URS/Far East); mountainous regions (central KOS)
Soil moderately damp but not too acid or peaty

Nectar rating; blooms, nectar flow
N1 KOS(Bek/65); URS(Ave/78; Glu/55; Kov/65; Pem/65)
Blooms vi-vii (KOS); mid vii (URS, AA852/81). **Nectar secretion** max at 24-26° and 60-74% RH, trees aged 50-80 yrs; little secretion from trees aged 15-20 yrs; rain, wind, sun and aspect also affect secretion (Pem/65)

Honey flow
Honey yield [high] 45 kg/colony/season, mean for 19 yrs, 3 Tilia spp (URS, Kov/65); 6-10 kg/colony/day, together with T. koreana (KOS, Bek/65).
Honey potential [high] 680-900 kg/ha (URS, AA852/81); trees 50- 80 yrs, 1.692 kg/tree (URS,Far East, Pem/65); together with T. amurensis, T. taqueti and Phellodendron amurensis yields 75-80% of honey of URS/Far East (Pem/65)

Pollen
P URS

Honey: physical properties
Pfund white (Glu/55)

414 Tilia maximowicziana Shiras; Tiliaceae

lime, linden
Tree, deciduous; fls whitish yellow, fragrant
Distribution temperate Asia
Soil moderately damp but not too acid or peaty

Nectar rating; blooms, nectar flow; composition
N1 JAP(Sak/82)
Blooms vi-vii (JAP). **Nectar secretion** 5.84-15.41 mg/fl/day (Dem/60a).
Sugar value [medium] 0.67-0.92 mg/fl/day (Dem/60a)

Pollen
P2 JAP

Honey no data

415 Tilia platyphyllos Scop.; Tiliaceae

large-leaved lime; Sommerlinde (De); tiglio (It)
Tree, <33 m, deciduous; fls yellowish white, fragrant
Distribution temperate Europe, N America; native to Europe. **Habitat** low
altitudes up to and including beechwood zone (ROM); cliffs and woods
(western UK)
Soil fertile damp soil (ROM); limestone (UK); not too acid or peaty

Economic and other uses
Land use amenity. **Other uses** understock for grafts of rarer Tilia spp

Alert to beekeepers
In 1928 honeydew reported toxic to bees in one area of GFR (Cra/77)

Nectar rating; blooms, nectar flow; composition
N1 BEG(Cra/84); FRA(Lou/81); GFR(Gle/77); ROM(Int/65); UK(How/79);
URS(Fed/55)
N2 ITA(Ric/78)
N EUR(Maz/82)
Blooms vi (EUR; UK). **Nectar flow** 10 days (ROM); 2-3 wks but longer if
weather is cold or wet (UK). **Nectar secretion** (mg/fl/day) 4.31-13.20
(Dem/60a); 1.77 (Maz/82); 4.48 (AA520/78); higher with warm nights, warm
days, high RH than on hot bright sunny days; little or no secretion in
cold weather; optimum temp 66-70° (EUR, How/79). **Sugar concentration**
[medium] 36-48% (Cir/80); 33.9% (Maz/82); 28.4-47.2% (3 yrs, Pet/72);
21% (AA520/78). **Sugar value** (mg/fl/day) [high] 0.54 (Maz/82); 2.73-3.26
(AA766/65); [medium] 1.51-1.87 (Dem/60a). **Sugar analysis** (Maz/59; Maz/82)

Honey flow
Honey yield [moderate] 15-20 kg/colony/season, 3 Tilia spp (ROM, Int/65); "important in Ukraine and Crimea" (URS, Fed/55). **Honey potential** (kg/ha) [high] 400-800 (ROM, Cir/80); 800 (ROM, Sad/60); [moderate] 250-375 (ROM, AA766/65)

Pollen
P1 FRA. P3 GFR; ITA. P EUR. **Yield** low (Maz/82). **Colour** dull yellow (How/79); load grey (Ric/78). **Pollen grain** described (How/79; Sta/74); under-represented in honey; usually mixed with pollen of Castanea (ITA, Ric/78)

Honeydew
Honeydew produced, and collected by bees from **Eucallipterus tiliae** (L.), Callaphididae - rated **Dl** (mid EUR, Hag/66); flow may be intense vi or vii, viii, and possibly ix, but bees can collect honeydew only early and late in day ?and/or when RH is high because it granulates rapidly on tree; honeydew analysis for Tilia spp, also honey data below (S EUR, Klo/65). Honeydew produced (insect not specified): FRA (Lou/81); ITA (Ric/78); UK (How/79). **Alert to beekeepers** in 1928 honeydew reported toxic to bees (GFR, Cra/77)

Honey: chemical composition
Calcium oxalate crystals present in honeys of this and other Tilia spp (Dem/63)

Honey: physical and other properties
Colour honeydew honey dark green (from Eucallipterus tiliae, Klo/65)
Granulation honeydew honey v slow, although melezitose content high (Klo/65)

416 Tilia taqueti C. Schn.; Tiliaceae

take lime, taquet lime
Tree, deciduous
Distribution temperate Asia. **Habitat** forests (URS/Far East)
Soil moderately damp but not too acid or peaty

Nectar rating; blooms, nectar flow
N1 URS(Ave/78; Glu/55; Kov/65; Pem/65)
Nectar secretion max at 24-26°, 60-74% RH, trees aged 50-80 yrs; little secretion from trees aged 15-20 yrs; rain, wind, sun and aspect also affect secretion (Pem/65)

Honey flow
Honey yield [high] 45 kg/colony/season, mean for 19 yrs, 3 Tilia spp (URS, Kov/65). **Honey potential** together with T. amurensis, T. mandschurica and Phellodendron amurensis, yields 75-80% of honey of URS/Far East (Pem/65)

Pollen
P URS

Honey no data

417 Tilia tomentosa Moench; Tiliaceae

white lime, silver-leaved lime; Silberlinde (De)
Tree, <30 m, deciduous; fls dull white or yellow, fragrant
Distribution temperate Europe, N America; native to SE Europe and Asia
Minor. **Habitat** mixed deciduous woodland, valleys and mountain slopes at
low altitudes, occasionally forming small woods up to and including
beechwood zone (ROM); thrives in urban areas
Soil fertile moist soil (ROM); not too acid or peaty

Economic and other uses
Timber. Land use amenity

Alert to beekeepers
In dry yrs bees may be found dead, paralysed or "drunk" under trees; sugar
metabolism implicated (Cra/77)

Nectar rating; blooms, nectar flow; composition
N1 ROM(Cir/77)
N3 BEG(Cra/84)
N EUR(Maz/82)
Blooms vi (EUR); vi-vii (ROM, Cir/80); vii (UK). **Nectar flow** 8 days
(ROM). **Alert to beekeepers** in dry yrs bees may be found dead, paralysed
or "drunk" under trees; sugar metabolism implicated (Cra/77). **Nectar
secretion** (mg/fl/day) 2.4-7.26 (Dem/60a); 3.05 (Maz/82). **Sugar conce-
ntration** [medium] 46-62% (Cir/80); 25.9% (Maz/82); 48.05% (Sad/60).
Sugar value (mg/fl/day) [high] 3.05-3.38 (AA766/65); [medium] 0.629-1.033
(Dem/60a); 0.71 (Maz/82). **Sugar analysis** (Sad/60)

Honey flow
Honey potential (kg/ha) [high] 1200 (ROM, Cir/77; Sad/60); 560 (ROM,
AA846/64); 645-990 (ROM, AA766/65)

Pollen
P EUR; ROM. **Yield** low (Maz/82). **Chemical analysis** (Cir/80). **Pollen
grain** illustrated and described (Ayt/71)

Honeydew produced EUR (Maz/82); ROM (Cir/77; Cir/80)

Honey no data

418 Tilia x europaea L.; Tiliaceae

common lime, linden, lime
Tree, <46 m, deciduous, hybrid of T. cordata and T. platyphyllos; fls pale
yellow, fragrant
Distribution temperate Europe, N America
Soil moderately damp but not too acid or peaty

Economic and other uses
Land use amenity

Nectar rating; blooms, nectar flow; composition
N1 UK(How/79)
N EUR(Maz/82)
Blooms vi (UK). **Nectar flow** 2-3 wks but longer if weather cold or wet
(UK, How/79). **Nectar secretion** 2.72 mg/fl/day (Maz/82). **Sugar conc-
entration** [high] 52-71% (AA526/82); [moderate] 32.5% (Maz/82). **Sugar
value** [medium] 0.80 mg/fl/day (Maz/82). **Sugar analysis** (Maz/82; AA526/82)

Pollen
P EUR. **Yield** low (Maz/82). **Colour** dull yellow (Hod/74; How/79).
Pollen grain illustrated (Hod/74; How/79)

Honeydew produced UK (How/79)

Honey no data

419 **Tipuana tipu (Benth.) O. Kuntze; Leguminosae**
syn Tipuana speciosa Benth.

tipa (Es/BOL); tipu (Pt/BRA); tipu (SOU)
Tree, <24 m; fls yellow
Distribution tropical S America; subtropical Africa; native to Brazil

Economic and other uses
Timber

Nectar rating + honeybee species; blooms, nectar flow; composition
N1 BOL(Mun/53; Smt/60); BOL,tm(Kem/80)
N BRA,tm(Caa/72)
Blooms x-xii (BOL); ix-xi (BRA/RG, SP). **Sugar concentration** [medium] 18-
25% (Caa/72); 22% (Kem/80)

Honey flow
Honey yield [high] up to 80 kg/colony/season in wooded areas with this sp
as main source (BOL, Smt/60)

Pollen
Pollen grain illustrated and described (Mag/78; Sao/61). **Reference slide**

Honey no data

420 **Tithonia tubaeformis Cass.; Compositae**

acahual (Es/MEX)
Herb, erect, woody; fls orange or yellow
Distribution tropical C America. **Habitat** medium to high altitudes of
central states and Sierra Madre of Michoacán and Guerrero (MEX); disused
corn fields (MEX)

Nectar rating; blooms, nectar flow
Nl MEX(Brs/82; Ord/72; Ord/83; Wis/55)
Blooms x-xii (MEX). **Nectar secretion** varies with altitude and soil type
(Ord/83)

Honey flow
Honey yield [high] 50-60 kg/colony/season, with Lippia umbellata (MEX,
Brs/82); "main source in Morelos" (MEX, Ord/83)

Pollen
P MEX

Honey: physical and other properties
Pfund light amber (Ord/83); with Lippia umbellata this sp gives water-
white honey in the Sierra Madre, MEX (Brs/82)
Viscosity "v thick" (Ord/83)
Granulation rapid, 2-3 days after extraction

421 Toona ciliata M. Roem.; Meliaceae
syn Cedrela toona Roxb. ex Rottl. & Wild.

toon; tun (INI)
Tree, <25 m, deciduous; fls cream, fragrant
Distribution subtropical Asia. **Habitat** lower hills and plains in northern
and central INI, especially stream banks and marshy locations
Soil moist soil preferred. **Temperature** not affected by frost. **Rainfall**
not drought resistant

Economic and other uses
Fodder - chopped lvs. **Timber** - termite resistant; used for hives (INI,
Koh/58). **Land use** amenity. **Other uses** medicine from bark; dyes from fls

Nectar rating + honeybee species; blooms, nectar flow
Nl INI/HIM,ac(Goy/74); INI/KAS,ac(Sar/72); INI/PUN[ac](Chd/77);
INI/UTT[ac](Koh/58; Rae/80)
N INI/HIM,ac(Sig/48); INI[ac](Sig/62)
Blooms iii-iv (INI). **Nectar flow** 10-15 days (Sig/62)

Honey flow
Honey yield (kg/colony/season) [moderate] 9 (INI, Koh/58); 4-7 (INI,
Sig/62); usually with Dalbergia sissoo and Ehretia acuminata (INI, Koh/58)

Pollen
P3 INI/PUN. P INI/KAS; INI. **Colour** reddish (Sig/48). **Pollen grain**
illustrated and described (Nak/65; Smt/56a)

Recommended for planting to increase honey production
INI (Sig/62). Propagate by seed

Honey: physical and other properties
Pfund white to light amber (Sig/62)
Flavour pronounced

422 Tournefortia argentea L.f.;   Boraginaceae

tree heliotrope
Tree/shrub
Distribution tropical Oceania, Asia.   **Habitat** widespread coastal plant;
coral islands
Soil salt tolerant

Nectar rating; blooms, nectar flow
N1 WAK(Lar/72)
Blooms all yr (WAK)

Honey flow
Honey yield "10 kg/colony every 6-8 wks throughout yr" (WAK, Hit/76); "in
the 1 yr (1971/72) since honeybees arrived on Wake Island the 2 original
colonies and others reared from them have produced 680 kg of honey" (Lar/72)

Pollen
P WAK

Honey: chemical composition
Water [medium] 18.6, 16.3% (Hit/76)
Glucose [low] 28.6%.   **Fructose** [medium] 38.2%.   **Sucrose** [high] 5.1%
Fermentation on storage may sometimes occur

Honey: physical and other properties
Colour deep gold to red to almost black
Flavour strong

423 Trichilia havanensis Jacq.; Meliaceae

bastard lime (En/BEL); siguaraya (Es/CUB); barrehorno, limoncillo
(Es/HOD); cucharillo, garrapatillo (Es/MEX)
Tree/shrub, <12 m; fls whitish-green
Distribution tropical C America, Caribbean, S America.   **Habitat** rain
forests of Andes <1500 m; river banks, roadsides, foothills in Antilles
especially CUB

Economic and other uses
Timber

Nectar rating; blooms, nectar flow
N1 CUB(Ord/83);   HOD(Ord/63)
N2 BEL(Ord/83);   MEX(Ord/72)
Blooms i-v (tropical America, Ord/83)

Honey: physical properties
Pfund light amber (Ord/72;   Ord/83)

424 Trichostema lanceolatum Benth.; Labiatae

bastard-pennyroyal, blue curls, camphor weed, flea weed, turpentine weed,
vinegar weed (En/USA)
Herb, 10-40 cm, aromatic; fls blue
Distribution subtropical and temperate N America. **Habitat** valleys and
foothills; grain fields after cutting; waste places
Soil drier soil preferred.   **Temperature** damaged by frost

Nectar rating; blooms, nectar flow; composition
N1 USA/CA(Van/41)
N2 USA/CA(Lov/56; Pel/76)
Blooms viii-frost (USA/CA, Van/41).   **Sugar concentration** [medium] 27.1%
(Pel/76)

Honey flow
Honey yield [high] 45 kg/colony/season (USA/CA, Van/41)

Pollen
P2 USA/CA.   **Colour** of load green (Van/41)

Honey: physical and other properties
Pfund white (Lov/56; Pel/76; Van/41)
Granulation rapid, fine (Pel/76; Van/41)
Flavour mild (Lov/56)

425 Tridax procumbens L.; Compositae

dagad-phul (INI)
Herb, annual; fls yellow
Distribution tropical Asia, Africa, Caribbean.   **Habitat** wasteland

Nectar rating + honeybee species; blooms, nectar flow
INI/MAH[ac](Chu/80);   TAN/Zanzibar[tm](Cra/73)
Blooms i-xii (INI/MAH)

Pollen
P1 INI/MAH.   **Colour** grey

Honey no data

426 Trifolium alexandrinum L.; Leguminosae

berseem, Egyptian clover
Herb, 0.5-1.0 m, annual; fls yellowish
Distribution temperate Asia; subtropical Asia, Africa; native to Asia
Minor. **Habitat** cultivated crop plant
Soil heavy alkaline loam preferred; also light soil; salt tolerant.
Temperature v high temps not tolerated; damaged by frost, killed at temps
below -3°.   **Rainfall** >250 mm; winter dryland crop, or irrigated crop sown
in spring/summer; drought resistant

Economic and other uses
Fodder - forage, green fodder, pasture.   **Soil benefit** green manure, soil cover

Nectar rating + honeybee species; blooms, nectar flow;   composition
N1 EGY(Waf/51); INI/BIH,ac(Nai/76); INI/PUN, ac,ad also am(Atw/70; Atw/73; Chd/77); PAK,ac(Pak/77; Shr/48)
Blooms iv-vi (INI/PUN); iv-v (PAK).   **Sugar concentration** [medium] 32.4% (Pek/77)

Honey flow
Honey yield (kg/colony/season) [moderate] 9 (PAK, Pak/77);   27, together with T. resupinatum (PAK, Shr/48).   **Honey potential** [moderate] 165 kg/ha (BUL, Pek/77); "important in Baluchistan, Punjab, Sind, NWFP" (PAK, Pek/77)

Pollen
Pl INI/BIH; INI/PUN.   **P** EGY; INI/PUN; PAK

Honey: chemical composition
Water [low] 15.6-16.8% (39 samples, almost certainly from this sp, Els/79)
Sugars, total 73.4-83.0%.   **Glucose** [medium] 31.6-35.8% (Els/79);   30.18% (Moh/82).   **Fructose** [medium] 38.2-42.5% (Els/79);   38.80% (Moh/82).
Sucrose [medium] 3.6-4.7% (Els/79);   1.67-2.30% (3 samples, Moh/82).
Reducing sugars 71.82% (Moh/82).   **Maltose** 3.30%.   **Raffinose** 3.90%
Ash [low] 0.085-0.098% (Els/79);   0.06% (Moh/82).   Contents of K, Na, Ca, Mg, Fe, Cu, Mn, P (Els/79);   K, Na, Ca (Moh/82)
pH 5.0 (Moh/82).   **Free acid** [medium] 21.00 meq/kg
Nitrogen 0.340-0.470% (Els/79)

Honey: physical and other properties
Viscosity at 20°, 78.14 poise (Moh/82).   **Optical rotation** -8.74 deg
Granulation slow

427 Trifolium fragiferum L.; Leguminosae

strawberry clover (En/AUS, NEZ)
Herb, low creeping perennial; fls pinkish-white
Distribution temperate Europe, Oceania.   **Habitat** pastures; common in swampy ground in southern VIC but also in drier areas (AUS/VIC)
Soil moist alkaline soil; heavy swampy ground; prolonged flooding by salt water tolerated.   **Rainfall** regions with limited or no summer drought preferred

Economic and other uses
Fodder pasture.   **Soil benefit** - improves heavy swampy soil

Nectar rating; blooms, nectar flow
N1 AUS/SA(Pur/68)
N2 AUS/VIC(Gom/73);   ROM(Int/65)
N NEZ(Wal/78)
Blooms iv (NEZ); i to mid-autumn (AUS/VIC); xii-iii (AUS/SA).   **Nectar secretion** varies with soil moisture (AUS/SA, Pur/68)

Honey flow
Honey potential [moderate] 100 kg/ha (ROM, Cir/80)

Pollen
P2 AUS/SA, VIC

Honey: physical and other properties
Pfund white (Wal/78)
Viscosity "rather light body"
Flavour delicate

428 Trifolium hybridum L.; Leguminosae

alsike clover, hybrid clover, Swedish clover; trèfle hybride (Fr/ALG);
alsike trèfle (Fr/CAF), trevo-hibrido (Pt/BRA)
Herb, <60 cm, perennial but grown as annual/biennial; fls pinkish white
Distribution temperate N America, S America, Europe, Asia, Oceania; native
to Atlantic Europe. **Habitat** cultivated crop plant; low altitudes
Soil clay; marshy land; as for T. pratense but wet and acid conditions
better tolerated; not on black earth (URS). **Temperature** winter-hardy;
warm climate not suitable. **Rainfall** higher rainfall areas and under irrigation (AUS/VIC)

Economic and other uses
Fodder - hay, pasture; not liked by cows, harmful to horses (Maz/82).
Soil benefit green manure; improves soil; erosion control

Warning
Harmful to horses (Maz/82)

Nectar rating; blooms, nectar flow; composition
N1 BEG(Grn/65); CAF/ALTA(Hen/77; Wes/49); CAF/BC(Con/81); CAF/NB-
(Wil59); CAF/NWT(Hat/81); ?CAF/ONT(Tow/76); CAF/QUE(Cha/48);
CAF/SASK(Mcc/58); FIN(Enb/54; Kay/79); IRN(Cra/73); UK(How/79);
URS(Ave/78; Fed/55); USA/CO(Wio/58; Wio/65); USA/ID,IL,MA,MI,MN,-
NY,PA,WI(Pel/76); USA/MO(Has/55); USA/OH(Bai/55); ?USA/OK,WV(Pel/76);
USA/VT(Med/54)
N2 AUS/VIC(Gom/73); USA/CA(Jay/54); USA/CT,IN(Pel/76); USA/MD(Die/71);
USA/NH(Hol/75); USA/UT(Nye/71)
N3 USA/CA(Van/41); USA/KY(Pel/76)
N ALG(Ske/72); BRA[tm](Caa/72); CAF/NB,NS,ONT,PRI,QUE,SASK(Pel/76);
CAF/QUE(Cou/59); EUR(Maz/82); USA/MA(Shw/50)
Blooms v-ix (EUR); late vi (URS). **Nectar secretion** 0.026 mg/fl/day
(Maz/82); stops during drought (Lov/56e); increased by low N supply in
soil (AA83/56); and by high boron content (Mcg/76). **Sugar concentration**
[medium] 40-45% (Caa/72); 43% (Maz/82); 37.7% (Mog/58); 31% (Pek/80a);
43.3% (Pel/76); 26.5-46.5% (Shw/53). **Sugar value** [low] 0.011 mg/fl/day
(Maz/82). **Sugar analysis** (Maz/59; Maz/82; Pec/61)

Honey flow
Honey yield (kg/colony/season) [high] 90 (CAF/ALTA, Lov/56e); [moderate]
16 (CAF/ONT, Lov/56e). **Honey potential** (kg/ha) [moderate] 120 (GDR,
Bec/67; URS, Ave/78); 125 (URS, Fed/55)

Pollen
P1 URS.   P2 AUS/VIC.   P3 USA/CA.   P CAF/QUE;   FIN;   USA/MA.   **Yield** good
(Cir/80).   **Chemical analysis** (Sta/74).   **Colour** of load dull brown
(Cir/80);   load dark green-brown (Han/80);   load dark grey (Van/41).
Pollen grain illustrated (Hod/74).   **Reference slide**

Recommended for planting to increase honey production
URS (Ave/78).   See **Warning**

Honey: chemical composition
Water [medium] 16.6% (1 sample, age 12 mths, Whi/62)
Glucose [medium] 31.03%.   **Fructose** [medium] 38.37%.   **Sucrose** [medium]
1.53%.   **Maltose** 7.59%.   **Higher sugars** 1.58%
Ash [low] 0.090%
pH 3.86.   **Total acid** 27.97 meq/kg.   **Free acid** [medium] 17.91 meq/kg.
Lactone 10.06 meq/kg
Amylase 17.6
Nitrogen 0.032%

Honey: physical and other properties
Colour light (Fed/55).   **Pfund** white (Cra/75;   Lov/56e);   17-27 mm, white
(Whi/62);   white to extra light amber (Van/41)
Viscosity "heavy body" (Lov/56e)
Granulation rapid (Cra/75)
Flavour mild (Cra/75;   Pel/76);   delicate (Lov/56e)

429 Trifolium incarnatum L.;   Leguminosae

carnation clover, crimson clover, Italian clover;   trèfle incarnat (Fr/ALG)
Herb, annual/biennial;   fls deep crimson
Distribution temperate N America, S America, Europe;   subtropical N
America;   native to southern Europe.   **Habitat** cultivated crop plant;
grown as catch crop after cereals;   volunteer on roadsides (USA)
Soil wide range;   light;   sandy and clay soils;   acid conditions better
tolerated than by T. pratense and T. repens.   **Temperature** not completely
hardy.   **Rainfall** not drought resistant

Economic and other uses
Fodder - hay, pasture;   but old fl heads prickly, can cause digestive
trouble in livestock (How/79).   **Soil benefit** green manure, erosion control

Warning
Old fl heads prickly, can cause digestive trouble in livestock (How/79)

Nectar rating;   blooms, nectar flow;   composition
N1 USA/AL(Bas/67);   USA/TN(Lit/54)
N2 FRA(Lou/81);   NEZ(Rob/56);   USA/TN(Lov/56)
N3 URS(Glu/55);   USA/KY(Pel/76)
N ALG(Ske/72);   EUR(Maz/82);   NEZ(Wal/78);   UK(How/79)
Blooms v-vii (EUR);   i-ii (NEZ);   early summer (UK);   iv-v (USA/TN).
Nectar secretion 0.04-0.23 mg/fl/day (Maz/82).   **Sugar concentration**
[medium] mean 31-44%, max 60% (Maz/82);   39.4% (Pek/77);   47.7% (Pel/76);
35.5% (Pet/77);   58%, 38% (nectar from honey sacs of bees, Frj/70).   **Sugar
value** [low] 0.02-0.07 mg/fl/day (Maz/82).   **Sugar analysis** (Maz/59;   Maz/82)

Honey flow
Honey yield (kg/colony/season) [high] mean 34, max 95 (USA/TN, Lov/57c);
[moderate] 13-22 (USA/MS, Lov/57c); 25 (Frj/70). **Honey potential**
[moderate] 8.6 kg/ha (BUL, Pek/77)

Pollen
P2 FRA. **Yield** bees collect 50%, max 62%, of their pollen from this sp in
USA/GA (Frj/70). **Pollen value** ?high (Mcg/76). **Chemical analysis**
(Maz/82). **Pollen grain** illustrated (Lie/72; Hod/74). **Reference slide**

Honey: chemical composition
Water [medium] 15.8-19.1% (4 samples, age 10-22 mths, Whi/62)
Glucose [high] 29.31-32.81%. **Fructose** [medium] 36.72-39.66%. **Sucrose**
[low] 0.73-1.29%. **Maltose** 6.26-10.27%. **Higher sugars** 1.14-2.26%
Ash [low] 0.040-0.080%
pH 3.63-3.83. **Total acid** 17.19-28.13 meq/kg. **Free acid** [medium] 12.81-
20.64 meq/kg. **Lactone** 4.38-7.49 meq/kg
Amylase 16.7-31.9
Nitrogen 0.021-0.036%

Honey: physical and other properties
Colour yellowish tinge (Lov/56); v light yellow (Pel/76); pale (Ric/78);
light yellow (Ske/72). **Pfund** extra light amber (Lov/56); 4-27 mm, water
white to white (Whi/62; also Pia/81)
Viscosity "good body" (Lov/56; Lov/57c)
Granulation, whitish solid "paste" but grains redissolve readily (Pia/81)
Flavour delicate, nondescript (Pia/81). **Aroma** can be rather strong
(Lov/57c); v slight (Pia/81); almost none (Ric/78)

430 Trifolium pratense L.; Leguminosae
syn Trifolium sativum Crome

red clover; trébol morado (Es/ARG); trèfle violet (Fr); rødkløver (Da);
Rotklee, Wiesen-Rotklee (De); trifoglio violetto (It); trevo vermelho
(Pt/BRA)
Herb, 5-100 cm, perennial or biennial/annual in warmer climates, diploid
and tetraploid cvs; fls pink or reddish purple
Distribution temperate Europe, N America, S America, Asia, Oceania, (Med)
Africa; subtropical S America, N America; native to Europe. **Habitat**
cultivated crop plant; meadows and pastures; coastal NW Europe; high
altitudes in subtropics
Soil fertile, moist but well drained soil preferred; ?pH >5.5, average to
heavy neutral loam and clay; lime-deficiency not tolerated. **Rainfall**
irrigation needed in subtropics

Economic and other uses
Fodder - hay, silage, pasture. **Soil benefit** green manure

Alert to beekeepers
1st crop (of 2 or 3 in season) often cut for hay during early bloom (Mcg/76)

Nectar rating + honeybee species;   blooms, nectar flow;   composition
N1 CAF/NWT(Hat/81);   DEN(Jon/54);   FRA(Lou/81);   IRN
(Cra/73);   PAK,ac(Pak/77)
N2 CAF/ONT(Tow/76);   GFR(Gle/77);   ITA(Ric/78);   NEZ(Rob/56);   UK(How/79);
URS(Fed/55);   ?USA/OH(Bai/55);   USA(Pel/76)
N3 BEG(Cra/84);   JAP(Sak/82);   USA(Lov/56)
N ARG(Per/80);   BRA,tm(Caa/72);   EUR(Maz/82);   NEZ(Wal/78)
Blooms ix–xi (BRA/SC, SP);   vi–ix (EUR);   summer (JAP);   x–ii (NEZ);   vii–
viii (PAK);   viii–ix (USA).   **Nectar flow** early vii to mid–x (EUR/Maz/82).
Alert to beekeepers 1st crop (of 2 or 3 in season) often cut for hay during
early bloom (Mcg/76).   **Nectar secretion** (mg/fl/day) 0.08–0.90 (Han/80;
Maz/82);   diploid 0.052–0.141, tetraploid 0.071–0.328 (AA759/75);
secretion increased (also sugar concentration) by presence of boron and
ammonium molybdenate (Mcg/76;   AA574/81);   secretion higher in hot dry
summer (Frj/70).   **Sugar concentration** [high] mean 17–60%, max 70%
(Maz/82);   diploid 27–70% (AA758/75);   [medium] 35% (Caa/72);   27.24%
(Pek/77);   35.06% (Pet/77);   15–61% (during 1 day, AA635/66);   diploid 52.6–
55.8%, tetraploid 39.5–46.8% (AA759/75);   diploid 20.5–28.7%, tetraploid
18.2–26.6% (AA573/81).   **Sugar value** (mg/fl/day) [low] 0.02–0.3 (Maz/82);
diploid 0.029–0.075, tetraploid 0.033–0.130 (AA759/75).   **Sugar analysis**
(Bat/73a;   Kay/78;   Maz/59;   Pec/61;   Wyk/52;   Yak/73;   AA156/55)

Honey flow
Honey potential (kg/ha) [moderate] 28.9 (BUL, Pek/77);   6–25 (GDR,
Bec/67);   100 (POL, AA758/75);   25–50 (ROM, Cir/80);   10–20 (URS, Fed/55)

Pollen
P1 ITA;   USA/UT.   **P3** GFR;   JAP.   **P** BRA;   EUR;   NEZ;   USA.   **Pollen value**
"together with T. incarnatum one of the most important sources" in central
ITA (Ric/78).   **Chemical analysis** (Shp/80;   Sta/74).   **Colour** dark green
(Wal/78);   load khaki (Nye/71);   load brown (Hod/74);   load dark reddish
brown (Maz/82).   **Pollen grain** illustrated and described (Nye/71).
Reference slide

Honeydew produced USA (Pel/76)

Honey: chemical composition
Sugars, total 79.6% (Ver/65).   **Glucose** 49.0% of total sugars (Maz/59).
Fructose 50.1% of total.   **Sucrose** [medium] 2.28% (Ver/65)
pH 4.2 (Dus/72);   3.40 (Had/63)
Peroxide number 114.2 ug/g/h (72.1 after 10 minutes in sunlight, Dus/72)
Nitrogen 0.045% dry wt (Bos/78).   **Amino acids**, free 0.120%, protein 0.234%
dry wt (Bos/78);   12–14 individual acids (8 samples, Kum/74)
Other constituents – carbonyl compounds identified (Hoo/63)

Honey: physical and other properties
Colour reddish or pinkish tinge (Lov/56;   Pel/76).   **Pfund** water white
(Cra/75);   light amber (Lov/56)
Granulation rapid (Cra/75)
Flavour mild (Cra/75).   **Aroma** almost none (Ric/78).

431 Trifolium repens L.; Leguminosae

Dutch clover, white clover; trebol blanco (Es/ARG); trèfle blanc (Fr);
trevo branco de jardin (Pt/BRA); hvidkløver (Da); Weissklee (De)
Herb, perennial or annual; fls white/pinkish
Distribution temperate Europe, N America, S America, Asia, Oceania;
subtropical N America, S America; native to Europe. **Habitat** cultivated
crop plant; pasture plant; mountainous districts, forest, forest-steppe
and steppe zones (URS); common in lawns and on heavily grazed or mowed
areas; roadsides, wasteland
Soil heavier soil preferred; light soil if rainfall is adequate; optimum
pH 5-7; gravel and sand by roadsides; slag heaps; swampland (NEZ);
volcanic soil (NEZ). **Temperature** hot weather limits growth. **Rainfall**
dry conditions limit growth, irrigation needed in dry areas (AUS)

Economic and other uses
Fodder - pasture, silage, hay. **Soil benefit** enrichment

Nectar rating + honeybee species; blooms, nectar flow; composition
N1 ARG(Lut/63); BEG(Grn/65); CAF/BC(Con/81); CAF/MAN(Mit/49);
CAF/NB(Wil/59); CAF/NWT(Hat/81); ?CAF/ONT(Tow/76); CAF/QUE(Cha/48);
CHL(Kar/56; Kar/60); CHN(Tse/54); COL(Cor/76); DEN(Jon/54; Fre/57);
ECU(Cra/73); FIN(Enb/54; Kay/79); FRA(Lou/81); GRC(Adm/54; Nic/55);
IRN(Cra/73); IRR(Hil/68); LUX(Poo/65); NER(Mod/52); NEZ(Coo/67);
UK(How/79); URS(Ave/78; Fed/55); USA/AL(Bas/67); USA/IA,ID,IL,IN,KS,KY,
MA,ME,MN,NE,PA,WI(Pel/76); USA/LA(Lie/72); USA/MO(Has/55; Pel/76);
USA/MS(Tat/56); USA/OH(Bai/55); ?USA/OK,WV(Pel/76); ?USA/VA(Gra/50);
USA/VT(Med/54)
N2 AUS/NSW(Goo/47); AUS/QD(Bla/72); AUS/VIC(Gom/73); BRA/SC,tm(Wie/80);
CAF/ALTA(Wes/49); GFR(Gle/77); JAP(Sak/82); ROM(Int/65); ?USA/AR(War/65);
USA/CT(Pel/76); USA/MD(Die/71); USA/NH(Hol/75); USA/TN(Lit/54);
USA/UT(Nye/71)
N3 LEB(Fli/62); POL(Dem/64a); USA/CO(Wio/58); USA/KY(Pel/76)
N ALG(Ske/72); ARG,tm(Per/80); AZO(Hab/72); CAF/ALTA,NB,NS,PRI,SASK-
(Pel/76); CAF/QUE(Cou/59; Pel/76); NEZ(Wal/78); USA/CO(Wio/65)
Blooms late winter to early summer (AUS/QD); ix-xi (BRA/SC); v-x (EUR);
spring-summer (JAP); xii-ii (NEZ); vi and all summer (UK); vi-autumn,
peak vi-vii (URS). **Nectar flow** vi-viii (USA); heaviest in season
following a yr of excessive rain (Pel/76); starts 10 days after onset of
flowering (How/79). **Nectar secretion** (mg/fl/day) 0.05-0.40 (Maz/82);
0.070 (Sim/80); highest on limestone (Mot/64); and when weather hot and
soil moist (USA, Pel/76); and at temps <24° (UK, How/79); secretion low
on muck (deep, rich, organic soil), none on sandy soil (USA, Mot/64); none
on lime-deficient soil (UK, How/79); low during long periods of high
humidity (NEZ, Wal/78). **Sugar concentration** [high] mean 25-52%, max 64%
(Maz/82); [medium] 32% (Jul/72); 34.5-55.1% (Mog/58); 27.2% (Pek/77);
42.69% (Pet/77); 39.3% (Sim/80); 24% (Wie/80); 42.7% (AA664/80); 20-57%
(AA526/82); concentration increased by treatment with boron (AA402/70).
Sugar value (mg/fl/day) [low, also medium] 0.01-0.20 (Maz/82); 0.054
(Sim/80). **Sugar analysis** (Frj/70; Maz/59; Maz/82; Pec/61; Wan/64;
Wyk/52; AA156/55; AA678/66; AA654/70; AA526/82). **Amino acid analysis**
(Bak/77)

Honey flow
Honey yield (kg/colony/season) [high] 50 (NEZ, Wai/75); rarely <90 (USA/MN, Pel/76); [moderate] 27 (AUS/NSW, Goo/47). **Honey potential** (kg/ha) [moderate] 32.2 (BUL, Pek/77); 100 (GDR, Bec/67); 100-200 (ROM, Cir/80); 100 (URS, Ave/78); 16.2 (URS/Lithuania, AA664/80)

Pollen
P1 AUS/QD, VIC; FRA. **P2** JAP. **P3** GFR; URS. **P** ALG; CAF/BC, QUE; DEN; FIN; NEZ; UK. **Yield** 2980 grains/fl, 385 250 grains/pellet (Frj/70); good (Cir/80); wt of load 5.5-5.7 mg (Maz/82). **Chemical analysis** (Ave/78; Maz/82). **Colour** of load dull brown (Cir/80); load green brown (Han/80; Maz/82); load dull green (How/79); pale yellow (How/79); green, almost olive (Wal/78). **Pollen grain** illustrated (Hod/74; Lie/72). **Reference slide**

Recommended for planting to increase honey production
URS (Ave/78)

Honey: chemical composition
Water [medium, also high] 19.4-21.3% (Rav/75); 17.5% (Sci/81); 16.4-21.0% (5 samples, age 3-15 mths, Whi/62)
Glucose [low, also medium] 28.13-32.13% (Whi/62). **Fructose** [medium] 37.62-39.93%. **Sucrose** [medium, also low] 0.74-1.35%. **Maltose** 6.86-9.20%. **Higher sugars** 1.35-1.83%. Also contents as % of total sugars (Maz/59; Maz/64; Peo/72). **Lactose** present (Wan/64)
Ash [medium, also low] 0.087-0.618% (Whi/62); see also Peo/70; Ver/65; Zie/79
pH 4.0, 4.1 (Dus/72); 4.0 (Lan/66); 3.3-3.92 (Rav/75); 3.62-4.08 (Whi/62). **Total acid** 18.27-50.72 meq/kg (Whi/62). **Free acid** [medium] 13.69-31.35 meq/kg. **Lactone** 3.27-19.37 meq/kg
Amylase 23.8, 29.4 (Lan/66); 10.6-61.2 (Whi/62). **Sucrase** 30.2 (Dus/72a). **Peroxide number** 332.5 μg/g/h (44.1 after 10 minutes in sunlight, Dus/72). **HMF** 9.6 ppm (Sci/81)
Nitrogen 0.022% dry wt (Bos/78); 0.031-0.055% (Whi/62). **Amino acids**, free 0.123%, protein 0.073% dry wt (Bos/78); 0.101% (also contents of individual acids, Peo/74); 12 acids present (Moo/65); 16 acids present (Peo/71). **Protein** content before dialysis (Whi/62), and after (Whi/67)
Volatile constituents in sample 1, major: phenol, minor: furfuraldehyde + 9 others (trace); sample 2, major: 5-hydroxymethyl-2-furaldehyde, minor: methyl syringate + 6 others (trace) (Grd/79)

Honey: physical and other properties
Colour bright yellow (Dus/72); bright (How/79); clear (Wie/80). **Pfund** white (Fed/55; Lov/56; Pia/81; Rob/56); 8.1 mm, extra white (Lan/66); 32 mm, white (Peo/72); 4-59 mm, extra white to light amber, darkens on storage (Roc/68; also Bla/72); 27-42 mm, white to extra light amber (Whi/62)
Electrical conductivity (per ohm cm) 0.000127-0.000538 (Rav/75); 0.000192 (Vor/64)
Granulation rapid, transparent and slightly coarse grain (Bla/72); slow, uniform (Cra/75); slow, fine, smooth and white like lard (How/79); rapid (Lov/56); mixture of fine and coarse grains (Pia/81); fine, homogeneous (Sci/81)
Flavour characteristic, mild, sweet (Bla/72); mild (Cra/75; Lov/56; Rob/56); acrid (Mot/64); slight but persistent, with delicate aftertaste (Pia/81). **Aroma** sweet (Fed/55; Glu/55); delicate, like the flower (Pia/81); delicate (Rob/56)

432 Trifolium resupinatum L.; Leguminosae

annual strawberry clover, Persian clover; shaftal (INI)
Herb, 10-30 cm, annual; fls lavender-pink
Distribution temperate Europe, Asia, Oceania; subtropical N America,
Oceania; native to ?S Europe. **Habitat** cultivated crop plant; grassy
places, disturbed ground
Soil deep heavy soil; moist soil preferred; not soil of low fertility.
Temperature severe cold not tolerated

Economic and other uses
Fodder - hay, pasture. **Soil benefit** green manure, soil cover

Nectar rating + honeybee species; blooms, nectar flow; composition
Nl INI/PUN[ac](Chd/77); PAK,ac(Pak/77; Shr/48); URS(Fed/55; Glu/55)
Blooms iv-vi (INI/PUN); v-vi (PAK); iv-v (USA). **Nectar secretion** 0.013-
0.067 mg/fl/day (AA335/83). **Sugar concentration** [medium] 50% (Frj/70); 30-
40% (AA335/83). **Sugar value** [low] 0.006-0.021 mg/fl/day (AA335/83)

Honey flow
Honey potential (kg/ha) [moderate] 70-100 (URS, Fed/55); <150 (URS, Glu/55)

Pollen
Pl INI/PUN. **P** PAK. **Reference slide**

Honey: physical and other properties
Colour light (Cra/75)

433 Triplaris surinamensis Cham.; Polygonaceae

Christmas candle, long jack, long john
Tree/shrub; fls white turning pink
Distribution tropical S America. **Habitat** upper reaches of rivers (SUR)

Economic and other uses
Timber. **Land use** amenity

Nectar rating + honeybee species; blooms, nectar flow
Nl GUY(Cra/73; Smt/60); SUR[tm](Oti/79)
Blooms viii-ix (GUY)

Honey no data

434 Turbina corymbosa (L.) Raf.; Convolvulaceae
syn Rivea corymbosa (L.) Hall.

camotillo (Es/BOL); aguinaldo blanco (Es/CUB); aguinaldo de pascua,
bejuco de pascuas (Es/DOR), campanilla blanca, manto, pascua (Es/MEX)
Herb, climber, perennial; fls white

Distribution tropical S America, C America, Caribbean; subtropical N
America. **Habitat** climbs over fences, shrubs and tall trees; altitudes
<1000 m
Soil wide range, but clay or calcareous soil preferred

Nectar rating + honeybee species; blooms, nectar flow
N1 BOL,tm(Kem/80); CUB(Ord/66; Ord/83)
N2 DOR(Dyc/65; Ord/64)
N MEX(Ord/72)
Blooms v-vii (BOL); x-ii (MEX). **Nectar secretion** abundant (Ord/83)

Pollen
P DOR

Honey: chemical composition
Water [medium] 17.8% (Ord/83)
Sucrose [medium] 2.3%. **Reducing sugars** 71.24%
Ash [low] 0.0193%

Honey: physical and other properties
Colour transparent (Ord/83). **Pfund** water white, said to be the lightest
in the world (Cra/75)
Optical rotation -26 deg (Ord/83)
Flavour delicate (Cra/75)

435 Vaccinium uliginosum L.; Ericaceae

black whortleberry, bog whortleberry
Shrub, 75-100 cm, deciduous; fls white tinged pink
Distribution temperate Europe. **Habitat** moors, heaths, coniferous woods,
subalpine pastures and tundra
Soil nutrient-poor raw humus of subalpine and alpine regions; peaty;
calcifuge

Economic and other uses
Food – berries

Nectar rating; blooms, nectar flow; composition
N1 KOS(Bek/65); URS(Fed/55; Glu/55)
N EUR(Maz/82); NOR(Lun/71); POL(Maz/82)
Nectar flow "sometimes rich" (NOR, Lun/71). **Nectar secretion** 0.9-2.2
mg/fl/day (Szk/73). **Sugar concentration** [medium] 13-42%. **Sugar value**
(mg/fl/day) [medium] mean 0.3-0.14 [sic] (Maz/82); 0.25-0.44 (Szk/73)

Honey flow
Honey potential (kg/ha) [moderate] 4.4-6.4 in woods (POL, Maz/82); 140-270
in pinewoods (POL, Szk/73)

Pollen
P EUR. **Colour** of load whitish to silver grey, or dark grey to brownish if
mixed with spores (Maz/82). **Reference slide**

Honey no data

436 Vernonia polyanthes Less.; Compositae

suquinay (Es/GUM); assapeixe, mata-pasto (Pt/BRA)
Shrub/tree; fls whitish, fragrant
Distribution subtropical S America; native to Brazil. **Habitat** large
areas of plateau, tableland and marshy ground; upland forest 19-23 degrees
S (BRA)

Nectar rating + honeybee species; blooms, nectar flow; composition
N1 BRA,tm(Wie/80); BRA/SP[tm also am](Mun/54; Ord/83)
Blooms vi-viii (BRA). **Nectar secretion** abundant in warm weather
(Bar/68). **Sugar concentration** [medium] 56% (Bar/68); 43-56% (Ker/60); 34-
37% (Sao/54)

Pollen
Colour white (Ker/60). **Pollen grain** illustrated and described (Sao/54)

Honey: chemical composition
Water [low] 15% (Fle/63)
Ash [medium] 0.159%
pH 2.3

Honey: physical and other properties
Colour golden (Bar/68); light (Ord/83). **Pfund** 11.5 mm, extra white
(Fle/63)
Viscosity "thick" (Bar/68)

437 Vernonia poskeana Vatke & Hildebrandt; Compositae

Herb, <60 cm; fls deep reddish-purple
Distribution tropical Africa. **Habitat** plains, dunes and ridges of shrub
savanna zone (BOT); grasslands (MAI)
Soil sandy. **Rainfall** good rains needed in January for full development
(BOT, Cla/83)

Alert to beekeepers
Pollen sometimes turns rancid (BOT, Cla/83)

Nectar rating + honeybee species; blooms, nectar flow
N1 BOT,tm(Cla/83)
Blooms ii-vii (BOT). **Nectar flow** - optimum temps 21-29° (Cla/83).

Honey flow
Honey yield "the most important nectar herb" in Botswana (Cla/83)

Pollen
P BOT. **Colour** of load white. **Alert to beekeepers** pollen sometimes turns
rancid (BOT, Cla/83)

Honey no data. "Vernonia (species not stated) honey was v thick and had a
flavour like butterscotch" (KEN, Nih/83)

438 Vicia faba L.; Leguminosae

broad bean, faba bean, field bean, horse bean, mazagan bean, tick bean;
haba (Es/MEX); fève des champs (Fr); fava (It); fava (Pt/BRA)
Herb, <2 m, annual/biennial; fls white often with a black blotch,
fragrant; nectary in fl, also extrafloral nectaries on undersides of
stipules (Frj/70)
Distribution temperate Europe, (Med) N Africa; subtropical Africa, S
America, Asia; tropical Africa, S America, C America; native to N Africa
and SW Asia. **Habitat** cultivated crop plant; seed seldom sets in warm
regions therefore grown at altitudes >2000 m in tropics
Soil moderate to good fertility preferred; medium texture; plant
moderately salt tolerant. **Temperature** a cool-season crop grown in winter
period in subtropics and at altitudes >2000 m in tropics; high temps cause
fl drop; v hardy. **Rainfall** much water required at flowering time
followed by dry conditions and sun for seed ripening

Economic and other uses
Food - beans, both fresh and dried. **Fodder** - beans; whole plant for
silage. **Soil benefit** N-fixation

Nectar rating + honeybee species; blooms, nectar flow; composition
N1 BEG(Grn/65); ETH[tm](Cra/73); UK(How/79)
N2 FRA(Lou/81); ITA(Ric/78); URS(Glu/55)
N BRA/SP[tm](Caa/72); ZIM[tm](Pap/73)
Blooms vii-viii (BRA/SP); v-vii (EUR). **Nectar secretion** increased by
higher planting density (AA510/69); sometimes absent, depends on weather
(URS, south west, Glu/55). **Sugar concentration** [medium] 28% (Caa/72).
Sugar value higher in 2nd crop than 1st (AA304/69). **Sugar analysis**
(Wyk/52). **Amino acid analysis** of floral and extrafloral nectar (AA905/80)

Honey flow
Honey potential (kg/ha) [moderate] 20 (GDR, Bec/67); 30-60 (ROM, Cir/80)

Pollen
P2 FRA; ITA. **Chemical analysis** (Maz/82). **Colour** of load grey
(Han/80); load grey-green (Ric/78). **Reference slide**

Honeydew
Honeydew produced EUR (Maz/82); ROM (Cir/80); analysis of honeydew from
Megoura viciae Buckton, Aphididae (mid EUR, Klo/65)

Honey: physical and other properties
Colour light but dark if honeydew present (Cra/75). **Pfund** white
(Cha/48); light to dark amber (How/79)
Granulation often rapid, coarse (Cra/75); fairly rapid, coarse (How/79)
Flavour mild (Cra/75; How/79)

439 Vicia sativa L.; Leguminosae

vetch; vesce (Fr/ALG)
Herb, trailing vine, <1 m long, annual; fls violet-purple rarely white;

nectary in fl, also extrafloral nectaries on stipules at base of petiole
(Lov/59e; Mcg/76)
Distribution tropical C America; subtropical N America, Asia; temperate N
America, Europe, Asia, S America; native to Europe and Asia. **Habitat**
cultivated crop plant; often naturalized
Soil wide range, must contain lime; dry soil (UK). **Temperature** semi-hardy

Economic and other uses
Fodder hay; silage; green fodder but can cause bloat in animals
(AA142/73). **Soil benefit** green manure, N-fixation

Nectar rating; blooms, nectar flow; composition
N1 CHN(Mad/81); FRA(Lou/81)
N3 URS(Fed/55)
N ALG(Ske/72)
Blooms spring to early summer (EUR). **Sugar analysis** (Bat/73a; Pec/61).
"Thick" extrafloral nectar more attractive to bees than floral nectar
(Mcg/76)

Honey flow
Honey potential [moderate] 10-30 kg/ha (ROM, Apc/68; Cir/80)

Pollen
P1 FRA

Honey no data

440 Vicia villosa Roth; Leguminosae

vetch; hairy vetch (En/USA); Sandwicke, Zottelwicke (De)
Herb, vine, 30-200 cm, annual/biennial/perennial; fls violet/purple/blue,
sometimes with white/yellow wings
Distribution temperate Europe, N America; subtropical Africa. **Habitat**
cultivated crop plant; naturalized in N Europe
Soil sandy; acidity and alkalinity tolerated. **Temperature** v hardy.
Rainfall drought resistant

Economic and other uses
Fodder hay, green fodder. **Soil benefit** green manure, soil binder for
pasture land, cover crop in orchards

Nectar rating; blooms, nectar flow; composition
N1 CHN(Mad/81); USA/MS(Tat/56)
N2 HUN(Pet/77); USA/OK(Lov/56)
N3 URS(Fed/55)
Blooms vi-vii (CAF/BC); v (USA/OK). **Nectar secretion** 0.35-0.89 mg/fl/day
(AA792/74); "high" (Pet/77). **Sugar concentration** [medium] <55%
(Frj/70); 44% (Pel/76); 30.96 (Pet/77); 29.5-45.3% (AA792/74). **Sugar
value** [medium] 0.31 mg/fl/day (Pet/77)

Honey flow
Honey yield (kg/colony/season) [high] 18-45 (USA/OK, Lov/64);   [moderate]
<10 (HUN, Pet/77);   18 (USA/TX, Lov/64).   **Honey potential** [moderate] 30-
100 kg/ha (ROM, Apc/68;   Cir/80)

Honey: chemical composition
Water [medium] 15.8-17.2% (5 samples, age 12-19 mths, Whi/62)
Glucose [low] 25.51-32.86%.   **Fructose** [medium] 36.55-40.34%.   **Sucrose**
[medium, also high] 0.96-5.48%.   **Maltose** 6.12-10.95%.   **Melezitose** 0.56-
1.09 (3 samples).   **Higher sugars** 1.64-2.78%
Ash [low] 0.039-0.081% (5 samples)
pH 3.70-4.00.   **Total acid** 14.14-33.16 meq/kg.   **Free acid** [low, also
medium] 11.25-22.48 meq/kg.   **Lactone** 2.45-10.68 meq/kg
Amylase 6.1-11.3 (3 samples)
Nitrogen 0.017-0.044%

Honey: physical and other properties
Colour light (Cra/75).   **Pfund** water white (Lov/56);   4-42 mm, water white
to extra light amber (Whi/62)
Viscosity "heavy body" (Cra/75)
Granulation rapid
Flavour mild (Cra/75);   v mild (Lov/56)

441 Viguiera helianthoides Kunth;   Compositae

romerillo de costa (Es/CUB, MEX);   tah (MEX)
Herb, 1-1.5 m, forms large clumps;   fls yellow
Distribution tropical C America, Caribbean.   **Habitat** coastal areas,
wasteland, hills and roadsides (CUB);   rocky areas and fields where maize
and henequen have been cultivated (MEX/Yucatán)
Soil calcareous, gravelly;   rocky.   **Rainfall** low;   semi-arid areas of MEX

Nectar rating;   blooms, nectar flow
N1 MEX(Ord/83;   Saf/73;   Smt/60;   Wis/53)
N2 CUB(Ord/44;   Ord/56)
Blooms i-ii (CUB);   late xii to early i;   (MEX/Yucatán).   **Nectar flow**
intense but short (Ord/44);   fairly short (Saf/73)

Honey flow
Honey yield 30% of MEX/Yucatán honey crop (Saf/73)

Pollen
P1 CUB;   MEX

Honey: physical and other properties
Colour light (Cra/75).   **Pfund** dark amber (Ord/44);   light amber (Ord/83)
Granulation rapid (Cra/75)
Flavour and aroma pronounced

442 Vitex agnus-castus L.; Verbenaceae

chaste-tree
Tree, <6 m, deciduous, aromatic; fls pale blue/pink
Distribution temperate (Med) Europe and Africa; subtropical N America;
native to Mediterranean region. **Habitat** coastal (EUR); by streams;
naturalized in southern USA
Soil sandy (southern USA); damp (EUR)

Economic and other uses
Food - fruits as pepper substitute. **Other uses** medicinal; source of
yellow dye; twigs for basket-making

Nectar rating; blooms, nectar flow; composition
N1 GRC(Nic/55)
Blooms vi-ix (southern EUR). **Sugar concentration** [medium] 50.0%
(Fah/49). **Sugar value** [medium] 0.483 mg/fl/day (Sim/75)

Honey no data

443 Vitex cymosa Bert.; Verbenaceae

tarumá (Es/BOL)
Tree; fls lilac
Distribution tropical S America. **Habitat** woodland

Nectar rating + honeybee species; blooms, nectar flow; composition
N1 BOL,tm(Kem/80)
Blooms ix-x (BOL). **Nectar flow** reliable and prolonged (Kem/80). **Sugar
concentration** [medium] 33% (Kem/80)

Honey no data

444 Weinmannia racemosa L.f.; Cunoniaceae

birch (En/NEZ); kamahi (NEZ)
Tree, <25 m; fls white, small
Distribution temperate Oceania; native to NEZ. **Habitat** lowland to
montane forest from 37 degrees southwards (NEZ)

Nectar rating; blooms, nectar flow
N1 NEZ(Coo/67)
N NEZ(Wal/78)
Blooms xi-i (NEZ)

Pollen
P NEZ. **Colour** whitish (Wal/78). **Reference slide**

Recommended for planting to increase honey production
NEZ (Wal/78)

Honey: physical and other properties
Colour distinctive, green (Wai/75). **Pfund** extra light amber (Cra/75;
Wal/78); light amber (Rob/56)
Granulation coarse, uneven (Cra/75; Rob/56)
Flavour bitter, increasing with age and dominant in blends (Cra/75;
Wal/78); distinctive (Mao/82a); pronounced, bitter (Rob/56); not v sweet
(Wai/75)

445 Weinmannia trichosperma Cav.; Cunoniaceae

tineo (Es/CHL)
Tree; fls fragrant
Distribution temperate S America; native to Chile. **Habitat** forest

Nectar rating; blooms, nectar flow
N1 CHL(Car/38; Kar/56; Kar/60)
Blooms xi (CHL). **Nectar flow** less in yrs with good flow from Eucryphia
cordifolia (Car/38)

Pollen
Pollen grain illustrated and described (Heu/71; Mag/78). **Reference slide**

Honey no data

446 Wendlandia notoniana Wall.; Rubiaceae

renda, tiliya (INI)
Tree, <5 m; fls white
Distribution tropical Asia. **Habitat** wasteland/roadsides; open areas
where forest has been cleared; semi-evergreen forest

Nectar rating + honeybee species; blooms, nectar flow
N1 INI/KAR,ac,ad and Trigona(Diw/64)
N3 INI/MAH[ac](Tha/62)
Blooms ii (INI/KAR)

Honey flow
Honey yield "the most important bee plant of Castle-rock area" (INI/KAR,
Diw/64)

Pollen
P1 INI/KAR. **P3** INI/MAH

Honey no data

447 Ziziphus jujuba Mill.; Rhamnaceae

Chinese jujube; unab (PAK)
Tree/shrub, <8 m, deciduous, many cvs
Distribution temperate Asia, (Med) Europe; native from SE Europe to
China. **Habitat** cultivated crop plant; occasionally naturalized (EUR/Med)

Economic and other uses
Food - fruits stewed, dried or pickled

Nectar rating + honeybee species; blooms, nectar flow; composition
N1 CHN(Mad/81); PAK,ac(Pak/77)
Blooms vii-ix (PAK)

Honey flow
Honey yield "important" in several parts of PAK (Pak/77)

Pollen
P PAK.   Reference slide

Honey: physical and other properties
Pfund amber (Mad/81)
Granulation slow
Flavour extra sweet

448 Ziziphus mauritania Lam.; Rhamnaceae
syn Ziziphus jujuba (L.) Gaertn.

zyzyphus; Indian jujube, bor (INI)
Tree, 3-12 m, evergreen, spiny but spineless varieties in INI/Assam; fls
greenish, small
Distribution tropical Africa, Asia, Oceania, C America, Caribbean, S
America; native to S Asia. **Habitat** cultivated; sub-Himalayan hill
country but best below 600 m
Soil wide variety including oolitic limestone.   **Temperature** severe heat
and frost tolerated. **Rainfall** 300-500 mm; v drought resistant

Economic and other uses
Food - fruit, fresh, dried or for drinks. **Fodder** - lvs and fruit; lvs
for silkworms. **Fuel.** **Timber.** **Land use** living fence. **Other uses**
tannins; host plant for lac insects

Warning
Can form dense spiny clumps (Usa/80)

Nectar rating + honeybee species; blooms, nectar flow; composition
N1 CHA[tm](Gad/80); PAK,ac(Pak/77)
N3 INI/MAH[ac](Chu/80); INO[ac](Bee/77)
Blooms vii-x (INI/MAH). **Sugar concentration** [medium] usually >50% (Zma/80)

Pollen
P3 INI/MAH; INO. P PAK

Honey: chemical composition
Water [high] 23.0% (TAI, Lin/77)
Glucose [medium] 31.5%. **Fructose** [medium] 35.3% (Lin/77); "high"
(Mad/81). **Sucrose** [low] 0.1% (Lin/77)
Ash [medium] 0.63%
pH 5.9. **Free acid** [medium] 15.4 meq/kg (28.0 after 1 yr)

Honey: physical and other properties
Colour yellow brown (Lin/77). **Pfund** amber (Mad/81)
Relative density 1.38 (Lin/77)
Granulation slow (Mad/81)
Flavour extra sweet

449 Ziziphus mucronata Willd.; Rhamnaceae

buffalo thorn; blinkblaar-wag-'n-bietjie (Af); mokgalo (BOT)
Tree/shrub, <9 m, may be spiny; fls yellowish, inconspicuous
Distribution tropical Africa; subtropical Africa. **Habitat** open woodland,
along rivers, on termite mounds (southern Africa); dunes and dune valleys
(BOT)
Soil alluvial; sand, gravel (BOT)

Economic and other uses
Food – fruit, fresh or for alcoholic drink. **Fodder** – lvs and fruit.
Other uses medicinal, tannin from bark and lvs

Nectar rating + honeybee species; blooms, nectar flow
N1 BOT[tm](Cla/83; Cra/73)
N3 SOU,tm(And/73)
N ZAM[tm](Sto/82)
Blooms x-ii, irregular (BOT); xi-ii (southern Africa, Pag/77). **Nectar**
flow ii (BOT). **Nectar secretion** copious (Pag/77)

Pollen
P3 SOU. **P** ZAM. **Pollen grain** illustrated and described (Smt/56a)

Recommended for planting to increase honey production
ZAM (Sto/82)

Honey: physical properties
Pfund dark (And/73)

450 Ziziphus nummularia (Burm. f.) Wight & Arn.; Rhamnaceae

kokan ber (PAK)
Shrub, <4 m
Distribution subtropical Asia; native to Arabia, Iran, Afghanistan and
Pakistan. **Habitat** desert areas of NW India and Pakistan; wadis
Soil gravel/sand. **Rainfall** drought tolerant

Economic and other uses
Food – berries.   **Fodder** – browse plant for camels.   **Fuel.**   **Land use**
hedges.   **Other uses** medicinal

Nectar rating + honeybee species;   blooms, nectar flow
N1 PAK,ac(Pak/77)
Blooms iii-vi (PAK)

Honey flow
Honey yield "important" in several parts of PAK (Pak/77)

Pollen
P PAK

Honey no data

451 Ziziphus oxyphylla Edgew.;   Rhamnaceae

amlai (PAK)
Shrub
Distribution subtropical Asia;   tropical Asia

Nectar rating + honeybee species;   blooms, nectar flow
N1 PAK,ac(Pak/77)
N3 INO(Bee/77)
Blooms vi-ix (PAK)

Honey flow
Honey yield "important" in several parts of PAK (Pak/77)

Pollen
P3 INO.   **P** PAK

Honey no data

452 Ziziphus spina-christi (L.) Desf.;   Rhamnaceae

Christ's thorn;   elb (YEA)
Tree, 3-10 m, evergreen, spiny
Distribution tropical Africa, Asia;   subtropical Africa, Asia;   native to
Africa and (Med) Asia.   **Habitat** dry desert areas but wadis preferred;
altitudes <1500 m;   rocky hills (EUR/Med)
Soil deep (alluvial plains) preferred, with access to ground water.
Temperature v high temperatures tolerated.   **Rainfall** desert (100 mm
rainfall), also less arid areas;   v drought resistant

Economic and other uses
Food fruit.   **Fodder** – lvs, branches, fruit.   **Fuel.**   **Timber.**   **Land use**
hedges, windbreaks.   **Soil benefit** erosion control, dune stabilization

Warning
Forms spiny impenetrable thickets; planted only in v dry areas where few other species can survive (Usa/80)

Nectar rating + honeybee species; blooms, nectar flow; composition
N1 YEA(Fie/80; Fil/80)
N2 OMA(Dut/79)
N ?ISR(Eis/80); OMA,af(Dut/77)
Blooms iv-vi (Africa and (Med) Europe); viii-xi (ISR). **Nectar flow** xi (OMA). **Nectar secretion** 0.2-2.6 mg/fl/day (Eis/80). **Sugar concentration** [medium] 25.3, 51.5% (Eis/80). **Sugar value** [medium] 0.12-0.66 mg/fl/day (Eis/80)

Honey flow
Honey yield [moderate] 1 kg/colony/season (OMA, Dut/79)

Pollen
P ISR; YEA

Honey: physical properties
Pfund white (Dut/77)

01D Abies alba Miller; Pinaceae
syn Abies pectinata (Lam.) DC.

silver fir; Edeltanne, Weisstanne (De); abete bianco (It)
Tree, <50 m, evergreen
Distribution temperate Europe. **Habitat** forests in mountainous areas of
EUR; altitudes >500 m; widely planted in EUR; areas with atmospheric
pollution not suitable
Soil deep moist soil preferred. **Rainfall** not drought tolerant

Economic and other uses
Timber. **Land use** amenity

Alert to beekeepers
Honey from Cinara pectinatae granulates v rapidly in combs, difficult to
extract (FRA, Bab/56)

Honeydew
From **Cinara confinis** (Koch), previously Todolachnus abieticola (Cholodkovsky) and T. confinis (Koch), Lachnidae: flow viii-ix, "important to
beekeepers" in parts of Black Forest, Alps, and on North Sea coast,
honeydew analysis (mid EUR, Klo/65); flow (ROM, Cir/80); flow but not
every yr (YUG, Rih/77)
From **Cinara pectinatae** (Nördlinger), previously Buchneria pectinatae
Nördlinger, Lachnidae: rated **D1**, flow vii-ix (mid EUR, Hag/66); flow vii-
x, very heavy in some yrs (mid Eur, Klo/65); yield "v important" (FRA,
Bab/56); honey potential 40 kg/ha (ROM, Cir/80); yield 60-90 kg/colony/yr
(SWI, Scd/50); flow vii-viii, yield 19 kg/colony/yr (YUG/Slovenia,
Rih/77); honeydew analysis (Bab/56; Hag/66; Klo/65)
From **Cinara pilicornis** (Hartig), previously referred to as Cinara piceicola
(Cholodkovsky), Lachnidae: rate of flow at 25° is twice that at 15°, but
is reduced by direct sunlight and other factors (south-west GFR, Eck/72)
From **Mindarus abietinus** Koch, Thelaxidae: flow may be heavy, honeydew
"thin", honey yield reported May 1957 (?GFR, Klo/65). Some records of
honeydew "poisoning" of bees may relate to this insect
From **Physokermes piceae** (Schrank), Coccidae: flow mid-iv to early vii,
coincides with heavy nectar flow from other plants, so not fully used by
bees (mid EUR, Klo/65)
Insect not specified: mean honey yield 40-45 kg/colony/yr, mean honey
potential 58.4, max 96 kg/ha (1968-1974, AUT, Peh/77); yield unspecified
(GRE, Ric/80); "appreciable" (ITA/Apennines, Ric/78); honey potential 20
kg/ha (ROM, Apc/68); honey safe as winter food for bees (ROM/Muresh,
Magyar, Fra/65); yield unspecified (TUQ, Ric/80); "much in some yrs"
(YUG, Kul/59)

Honey: chemical composition
Water [low] 16.1%, 14.2% (insect not specified, AUT, GFR, Kir/61)
Sugars (insect not specified, ITA, Bat/73): **glucose** [medium] 36.90%;
fructose [low] 33.86%; **sucrose** [low] 0.40%; **maltose** 9.05%; **isomaltose**
1.45%; **trehalose** 4.88%; **gentiobiose** 0.64%; **raffinose** 1.57%; **melezitose**
8.10%. **Dextrin** 2.98%, 4.45% dry wt (Kir/61)
Ash 1.14%, 1.01% dry wt (Kir/61)
pH 4.78, 4.96 (Kir/61)
Sucrase high (Vor/68)
Amino acid analysis (GFR, Kum/74). **Colloids** 0.31%, 0.24% dry wt (Kir/61)

Honey: physical and other properties
Colour, from Cinara pectinatae: black-brown with greenish tinge (mid EUR,
Klo/65); white, yellow or brown (in combs, FRA, Bab/56); insect not
specified: dark green (AUT, Kir/61; YUG, Kul/59); khaki green (GFR,
Kir/61); v dark, often slightly greenish (ITA, Ric/78)
Optical rotation +2.89 deg (Bat/73); +9.55, +2.87 deg (insect not
specified, AUT, GFR, Kir/61)
Granulation (alert to beekeepers) v rapid, in combs, difficult to extract
(from Cinara pectinatae, FRA, Bab/56); rapid (YUG, Kul/59); none, or
irregular (ITA, Ric/78); medium, becoming more green (GFR, Vor/68)
Flavour of treacle (AUT, Kir/61); v sweet (GFR, Kir/61); mild to resinous
(aromatic) (GFR, Vor/68); "tonic" (ITA, Ric/78)

02D Abies borisii-regis Mattf.; Pinaceae

Tree, <60 m, evergreen; probably of hybrid origin, form variable
Distribution temperate (Med) Europe. **Habitat** mountains of Balkans

Honeydew
From **Eulecanium sericeum** (Lindinger), Coccidae: insect population small,
so honey yield low; flow x-vii, but bees collect honeydew only v-vii
(GRC/Tymphrystos, San/81)
From **Mindarus abietinus** Koch, Thelaxidae: of minor importance, flow late
spring to early summer (GRC/Iti, Tymphrystos, San/81). Some records of
honeydew "poisoning" may relate to this insect
From **Physokermes hemicryphus** Dalman, Coccidae: the main contributor,
giving "abundant" secretion for which hives are migrated to forests; 5-10%
of total honey production in Greece is from this insect on Abies borisii-
regis and A. cephalonica(03D); flow late v to early vii (GRC/Eperos,
Macedonia, Thessaly, San/81)

Honey no data

03D Abies cephalonica Loudon; Pinaceae

Greek fir
Tree, <30 m, evergreen

Distribution temperate (Med) Europe. **Habitat** mountainous areas of GRC; cultivated in ITA

Economic and other uses
Timber. **Land use** afforestation

Honeydew
Honeydew produced by the same insects as on Abies borisii-regis (02D). For A. cephalonica, insect distributions in Greece are: **Eulecanium sericeum** Parnis, Parnon; **Mindarus abietinus** central, Parnis, Iti, Tymphrystos; **Physokermes hemicryphus** central, Evia, Kephalonia, Peloponnessus (San/81)

Honey no data

04D Calocedrus decurrens (Torr.) Florin; Cupressaceae
syn Libocedrus decurrens Torr.

California incense cedar, white cedar (En/USA)
Tree, 21-30 m, evergreen
Distribution subtropical N America; temperate N America, Europe. **Habitat** higher altitudes, mountains of USA/CA,OR; atmospheric pollution not tolerated
Soil moist, well drained soil preferred. **Temperature** fully hardy

Economic and other uses
Timber. **Land use** amenity

Honeydew
From **Xylococcus macrocarpi** (Coleman), Coccoidea: 45-136 kg/colony/yr; flow abundant at times, early summer onwards; honeydew v "gummy", colour amber but white in xi (USA/CA,OR, Pel/76)

Honey: chemical composition
Water [low] 12.2, 15.2% (2 samples, age 9, 18 mths, insect not specified, USA/CA, Whi/62)
Glucose [low] 23.34, 27.94%. **Fructose** [low] 23.91, 26.22%. **Sucrose** [low] 0.83, 0.74%. **Maltose** 5.85, 6.08%. **Higher sugars** 11.50, 8.70%
Ash [high] 1.097, 1.047%
pH 4.42, 4.71. **Total acid** 76.49, 56.08 meq/kg. **Free acid** [high] 66.02, 49.91 meq/kg. **Lactone** 10.47, 6.16 meq/kg
Nitrogen 0.049, 0.047%

Honey: physical and other properties
Pfund 104 to >114 mm, amber to dark amber (insect not specified, USA/LA, Whi/62)
Viscosity "heavy body" (USA, Pel/76)
Granulation slow
Flavour bland

05D Fagus sylvatica L.; Fagaceae

beech, red beech; Rotbuche (De); faggio (It)
Tree, <30 m, deciduous; monoecious
Distribution temperate Europe; native to Europe. **Habitat** woodland;
mountains; widely planted
Soil well drained; lime tolerated. **Temperature** hardy

Economic and other uses
Fodder - nuts for pigs. **Timber.** **Land use** amenity

Honeydew
From **Lachnus pallipes** (Hartig), previously L. exsiccator Altum, and
Schizodryobius pallipes Hartig, Lachnidae: honeydew produced, amino acid
analysis reported (assumed to be F. sylvatica, GFR, Kum/74); honeydew
produced (ROM, Cir/80)
From **Phyllapis fagi** (L.), Callaphididae: rated **Dl**, flow v-vii (mid EUR,
Hag/66); flow heaviest in v, early vi (mid EUR, Klo/65); also reported
for ROM (Cir/80); honeydew v waxy, more attractive to bees after rain (mid
EUR, especially AUT, Klo/65)
Insect not specified: "yield obtained from time to time" (ITA/Central
Apennines, Ric/78)

Honey no data

06D Larix decidua Miller; Pinaceae
syn Larix europaea DC.

larch; europäische Lärche (De); larice (It)
Tree, <35 m, deciduous
Distribution temperate Europe, Asia. **Habitat** mountainous areas of EUR,
especially Alps and W Carpathians; widely planted
Soil well drained light or gravelly loam preferred; wet low-lying areas
not tolerated

Economic and other uses
Timber. **Land use** afforestation, amenity. **Other uses** tannin from bark;
medicinal; turpentine from resin

Alert to beekeepers
Honeydew and honey may granulate v rapidly, on tree or in combs (Klo/65;
Mal/79; Ric/78)

Honeydew
From **Cinara cuneomaculata** (del Guercio), previously C. boerneri Hille Ris
Lambers, and C. laricicola Börner, Lachnidae: rated **Dl**, flow v-viii (mid
EUR, Hag/66); "important", flow vi (or vii)-x, honeydew analysis (mid EUR,
Klo/65)
From **Cinara kochiana** (Börner) previously Laricaria kochiana (Borner),
Lachnidae: flow from late vi to autumn, visited by bees (mid EUR, Klo/65)

From **Cinara laricis** (Hartig), previously Lachnus muravensis Arnhart and Lachniella nigrotuberculata del Guercio, Lachnidae: honeydew analysis (CZE, Hag/63); "probably main honeydew producer on larch", flow heavy in vi, also late viii to x, honeydew may granulate on tree due to high melezitose content (mid EUR, Klo/65), then called manna; honey yield obtained, honeydew analysis (ITA, Mal/79)
Insect not specified: crystallized honeydew contained (% dry wt) - 53.36% melezitose, 53.36% [sic] sucrose, 13.94% invert sugar, 30.03% dextrin (AUT, Goa/52); honeydew produced (FRA, Lou/81); "a main honey source" (Germany, How/79); honey yield "rather scarce" (ITA/Alps, Apennines, Ric/78)

Honey: chemical composition
Glucose 15.94, 13.70% (dry wt, 2 samples, Cinara laricis, ITA/W Alps, Mal/79). **Fructose** 29.18, 24.33%. **Sucrose** 0.73, 4.16%. **Meso-inositol** 0.05,0.06%. **Turanose** 2.49, 1.29%. **Raffinose** 2.13, 2.00%. **Melezitose** 44.47, 42.76%; "high" (Cinara cuneomaculata, mid EUR, Klo/65)

Honey: physical and other properties
Granulation (alert to beekeepers) v rapid, in combs (from C. laricis, Klo/65; Mal/79; Ric/78); colour then white (Mal/79)

07D Nothofagus solandri var. cliffortioides (Hook. f.) Poole; Fagaceae

mountain beech (En/NEZ)
Tree, <15 m, shrub in subalpine belt, monoecious
Distribution temperate Oceania; native to NEZ, South Island.   **Habitat** montane and subalpine forests, scrub (NEZ)

Economic and other uses
Timber

Honeydew
From **Ultracoelostoma assimile** (Maskell), Margarodidae: estimated honey yield 60 kg/colony/yr (Bet/79); abundant flow in late summer and autumn, honey safe (and often used as) winter food for bees (Coo/81).
About 30 species of scale insects are found on Nothofagus spp, and many secrete honeydew which may contribute to beech honeydew honey (Wao/79)

Honey: chemical composition
Water [medium] 19% ("beech" honeydew honey, plant and insect not specified, NEZ, Dal/75)
Sugars, total 64%. **Fructose** [medium] 35.6%.   **Sucrose** [medium] 1%.
Reducing sugars [low] 63%
Ash [low] 0.79%
Free acid [medium] 32 meq/kg
Amylase 20. **HMF** 0 ppm

Honey: physical and other properties
Colour brown (beech honeydew honey, Dal/75)
Flavour distinctive (Coo/81). **Aroma** strong (Dal/75)

08D Picea abies (L.) Karsten; Pinaceae
syn Picea excelsa (Lam.) Link; Picea vulgaris Link

Norway spruce; Fichte, Rottanne (De); abete rosso (It)
Tree, <60 m, evergreen
Distribution temperate Europe; native to Europe. **Habitat** widely planted
as forest tree in N, W and C Europe; often in mountainous districts

Economic and other uses
Food - alcoholic drink from fermented shoots, lvs. **Timber.** **Land use**
afforestation, amenity. **Other uses** resin purified for pitch; turpentine
from shoots, lvs; tannin from bark; when young, sold for Christmas trees

Honeydew
From **Cinara costata** (Zetterstedt), previously Lachniella costata
(Zetterstedt), Lachnidae: heavy flow in some yrs, but only one report of
bees flying directly to this source (mid EUR, Klo/65)
From **Cinara piceae** (Panzer), previously Mecinaria piceae (Panzer),
Lachnidae: up to 15 kg/colony/yr; with Physokermes piceae reported to be
"most important source" on P. abies in Alps; flow 3-6 days in vii or viii,
almost every yr (mid EUR, Klo/65)
From **Cinara pilicornis** (Hartig), previously referred to as C. piceicola
(Cholodkovsky), Lachnidae: rated **D1**, flow vi-viii (mid EUR, Hag/66);
"important" in S, mid and N GFR, flow from late v to vii, honeydew analysis
(Klo/65); factors affecting secretion (Ecl/72)
From **Cinara pruinosa** (Hartig), previously C. bogdanowi (Mordvilko),
Lachnidae: rated **D1**, flow vi-viii (mid EUR, Hag/66)
From **Physokermes hemicryphus** Dalman, Coccidae: honey yield 3.2-42.8
kg/colony/yr, "the most important honeydew producer of mid EUR", flow from
early vi to early vii (AUT, Peh/76); rated **D1**, flow vi-vii (mid EUR,
Hag/66); honey yield up to 39 kg/colony/yr, flow late v to mid vii,
honeydew analysis (mid EUR, Klo/65); honeydew analysis (Hag/63)
From **Physokermes piceae** Schrank, Coccidae: flow mid-iv to mid-vi coincides
with main nectar flow, so not fully used by bees; honeydew analysis (mid
EUR, Klo/65); but important in Alps (see under Cinara piceae, Klo/65);
heavy flow, collected by bees, gains of up to 4 kg/day per hive, water
content of honeydew 14% (GFR, Got/51); honeydew analysis (Hag/63)
Insect not specified: honey potential 100-500 kg/ha (EUR, Nee/78);
honeydew produced: (FRA, Lou/81); mean honey yield 40 kg/colony/yr
(Germany, Pel/76); "rather scarce" (ITA/Alps, Ric/78); flow vii-viii
(NOW, Lun/71); honey potential: 20 kg/ha (ROM, Apc/68); 50 kg/ha (ROM,
Cir/80). Honey safe as winter food for bees (ROM, Fra/65)

Honey: chemical composition
Water [medium] 17.4% (insect not specified, AUT, Kir/61)
Reducing sugars 76.9% dry wt. **Dextrin** 3.57% dry wt
Ash 0.85% dry wt
pH 4.74
Colloids 0.24% dry wt. **Amino acid** analysis (insect not specified, GFR,
Kum/74)

Honey: physical and other properties
Colour dark red brown (from Cinara piceae), reddish (from C. pilicornis),
reddish brown (from Physokermes hemicryphus) (mid EUR, Klo/65); buff
(insect not specified, AUT, Kir/61); greenish black (insect not specified)
(SWI, Pel/76)

Optical rotation +3.48 deg (AUT, Kir/61)
Granulation slow (from C. piceae), slow (from C. pilicornis) (Klo/65)
Flavour fairly sweet (AUT, Kir/61)

09D Pinus halepensis Miller; Pinaceae

Aleppo pine, Jerusalem pine; pin d'Alep (Fr/ALG)
Tree, 20 m, evergreen
Distribution temperate (Med) Europe, Asia, Oceania; native to Med area
Soil shallow soil preferred; limestone; heavy clay better tolerated than
by other Pinus spp; not waterlogged or saline soils. **Temperature** brief
occasional cold spells -18° to -20° tolerated, also high temperatures.
Rainfall 355-400 mm (native range); drought tolerant, 250-800 mm annually
but young seedlings not drought tolerant

Economic and other uses
Fuel.   Timber.   Land use windbreak, shade, afforestation in poor dry
conditions, amenity.   **Other uses** resin

Honeydew
From **Marchalina hellenica** (Gennadius), previously Monophlebus hellenicus
(Gennadius), Margarodidae: "high" honey yield, mainly in Chalkidiki (GRC,
Mai/52); 5-10 kg/colony/yr, abundant flow in autumn (GRC/Chalkidiki,
Thasos, Nic/55); heavy flow from late vi to following spring, bees collect
honeydew viii-ix (GRC/Med basin, San/81); main source of pine honeydew at
altitudes <900 m, 60% of Greek honey is from this insect on pine trees,
mainly P. halepensis (San/81)
Insect not specified: much honeydew secreted, mainly in Oranie and
southern mountains (ALG, Ske/77); main honey crop on Thasos and Ikaria
(GRC, Adm/64); 60 000 colonies brought to forests by end vii for abundant
flow viii (TUQ/west coast, Adm/77)

Honey: chemical composition
Nitrogen 14 mg/100 g dry wt (insect not specified, ?ITA, Bos/78).   **Amino
acids** free 54, protein 64 mg/100 g dry wt (Bos/78)

Honey: physical and other properties
Colour dark (insect not specified, ALG, Ske/72).   **Pfund** light amber (from
Marchalina hellenica, GRC, Nic/55)
Viscosity "good body"
Granulation [slow] does not granulate
Flavour characteristic (Ske/72)

10D Pinus sylvestris L.; Pinaceae

Scotch fir, Scots pine; gemeine Kiefer, Rotkiefer (De); furu (No)
Tree, <40 m, evergreen
Distribution temperate Asia, Europe; native to Europe.   **Habitat** open
woodland; widely planted forest sp

Economic and other uses
Timber. **Land use** afforestation, amenity. **Other uses** rosin and turpentine from resin; tar, pitch and pine oil by distillation

Honeydew
From **Cinara cembrae** (Seitner), Lachnidae: rated **D1**, flow vii-ix (mid EUR, Hag/66)
From **Cinara nuda** (Mordvilko), Lachnidae: flow may be fairly heavy, honeydew collected by bees (mid EUR, Klo/65)
From **Cinara pinea** (Mordvilko), Lachnidae: flow vi-vii, honeydew collected by bees and in some areas hives are moved to forest for this flow, honeydew analysis (mid EUR, Klo/65); "good" flow vii-viii (NOW, Lun/71); honeydew honey potential on Pinus spp 10 kg/ha (ROM, Cir/80)
From **Marchalina hellenica** (Gennadius), Margarodidae: heavy flow late vi to following spring; 60% of Greek honey is from this insect on pine trees - P. sylvestris less important than P. halepensis (GRC, San/81)
From **Schizolachnus pineti** (Fabricius), Lachnidae: "considerable" flow visited by bees, but importance not known (EUR/Med, Klo/65)

Honey: chemical compositon
Water [medium] 20.2, 20.5% (insect not specified, Dus/67)
Glucose [medium] 30.05% (presumed P. sylvestris, CZE, Svo/56). **Fructose** [medium] 38.25%. **Dextrin** 3.11%
Ash [medium] 0.44%
Invertase 46, 47.3. **Peroxide number** 418.7, 662.5 ug/g/h
Amino acid analysis (Kum/74)

Honey: physical properties
Optical rotation laevorotatory (presumed P.sylvestris, CZE, Svo/56).
Electrical conductivity 0.00112, 0.0012/ohm cm (Dus/67)

11D Populus spp; Salicaceae

poplar; Pappel (De)
Tree, deciduous, usually dioecious
Distribution temperate Europe, Asia, N America, (Med) Africa; subtropical N America, Asia, C America

Economic and other uses
Timber - many hybrids planted for pulp-wood (EUR)

Honeydew
From **Chaitophorus populeti** (Panzer), Chaitophoridae, on Populus alba, P. nigra and P. tremula: heavy flow collected by bees (AUT, Klo/65)
From **Chaitophorus populeti**, together with **Pterocomma salicis** (L.), Aphididae: intense flow from late v to mid vi, colonies gained 10-20 kg/colony in 10 days, of which 60% attributed to poplar honeydew; honey potential for P. alba 20 kg/ha, for P. nigra 20 kg/ha (ROM, Cir/80)
From **Pachypappa vesicalis** Koch, Pemphigidae: flow v to early vi, honeydew analysis (URS/Ukraine, Blz/79)
Insect not specified: crystallized honeydew (manna) contained 40% melezitose (AUT, Goa/52); honeydew produced, NOW (Lun/71); USA (Pel/76); honeydew shortened life span of bees (URS/Voronezh, Orz/58)

Honey no data

12D Quercus robur L.; Fagaceae
syn Quercus pedunculata Ehrh.

common oak, English oak, pedunculate oak; farnia (It)
Tree, <45 m, deciduous, monoecious
Distribution temperate Europe, (Med) Africa, Asia; native to Europe.
Habitat woodland, where it is often the dominant sp
Soil wide range but brown-earth soils preferred

Economic and other uses
Timber v hard, used for ship-building etc

Alert to beekeepers
Honey from Quercus spp not suitable as winter food for bees (ROM, Fra/65)

Pollen
Pl ITA.   P3 FRA.   P URS.   **Pollen value** 80% of spring harvest (Ric/78).
Colour of load yellow green (Han/80)

Honeydew
From **Kermes quercus** (L.), Kermesidae: on Quercus robur and Q. pubescens -
rated **D1**, flow ?iv-v (mid Eur, Hag/66); "considerable" flow from end iv to
mid v or vi, honeydew tastes only slightly sweet (?to man) but bees
collect it actively in the afternoon (mid EUR, Klo/65)
From **Lachnus iliciphilus** (del Guercio), previously Schizodryobius
longirostris (Mordvilko), Lachnidae: collected by bees in vi (mid EUR,
Klo/65)
From **Lachnus roboris** (L.), Lachnidae: rated **D1**, flow v-vii (mid EUR,
Hag/66); importance to bees not certain (Klo/65)
From **Thelaxes dryophila** (Schrank), Thelaxidae: heavy flow with peak v or
vi, continuing till autumn, collected by bees (mid EUR, Klo/65)
From **Tuberculatus annulatus** (Hartig), previously Tuberculoides annulatus
(Hartig), Calliphididae : flow peak mid vi to vii, " most important
producer on oak", honeydew granulates rapidly, bees collect it while liquid
in the morning (mid EUR, Klo/65)
Insect not specified: flow every 4 yrs (ITA/Umbria, Marche, Abruzzo,
Ric/78); FRA (Lou/81); URS (Fed/55); honeydew honey potential from
Quercus spp 20 kg/ha (ROM, Cir/80). **Alert to beekeepers** honey from
Quercus spp not suitable as winter food for bees (ROM, Fra/65)

Honey: physical and other properties
Colour "less dark than other honeydew honeys" (insect not specified, ITA,
Ric/78)
Flavour sweet, also slightly sharp.   **Aroma** intense

13D Quercus suber L.; Fagaceae

cork oak; chêne liège (Fr)
Tree, <20 m, evergreen, monoecious
Distribution temperate (Med) Europe, (Med) Africa, Oceania
Temperature fairly hardy

Economic and other uses
Land use shade, amenity.   **Other uses** cork from bark for making hives

Alert to beekeepers
Honey from Quercus spp not suitable as winter food for bees (ROM, Fra/65);
honeydew shortened life span of bees (URS/Voronezh, Orz/58)

Pollen
P ALG

Honeydew
Insect not stated: flow sometimes abundant in Dellys forest (ALG,
Ske/77); "important" honey source in Morocco (Cra/73).   **Alert to
beekeepers** honey from Quercus spp not suitable as winter food for bees
(ROM, Fra/65); honeydew shortened life span of bees (URS/Voronezh, Orz/58)

Honey no data

14D Quercus virginiana Mill.; Fagaceae

live oak (En/USA)
Tree, <18 m, evergreen, monoecious
Distribution temperate N America; subtropical N America; tropical C America

Economic and other uses
Timber

Honeydew
From "live oak gall" (specific name not stated): yield 11 kg/colony/yr,
flow viii to late autumn, "v useful source during drought" (USA/TX, Pel/76)

Honey: chemical composition
Water [medium] 18.2, 16.2% (2 samples, age 8,9 mths, USA/FL, Whi/62, who
says "Quercus fagaceae", presumably Quercus (family Fagaceae) and we assume
it to be Q. virginiana)
Glucose [low] 29.51, 26.61%.   **Fructose** [medium] 38.12, 34.59%.   **Sucrose**
[medium, also low] 1.14, 0.63%.   **Maltose** 8.67, 10.59%.   **Higher sugars**
1.28, 2.47%.   **Melezitose** (1 sample) 0.38%
Ash [medium] 0.212, 0.799%
pH 3.90, 4.70.   **Total acid** 50.71, 67.27 meq/kg.   **Free acid** [medium, also
high] 36.62, 64.57 meq/kg.   **Lactone** 14.09, 2.58 meq/kg
Amylase 6.7, 41.4
Nitrogen 0.053, 0.223%

Honey: physical and other properties
Pfund 85 to >114 mm, amber to dark amber (Whi/62, see note under **Water**).
Colour dark (from live oak gall, USA/TX, Pel/76)
Viscosity "heavy" (Pel/76)

15D Zea mays L.; Gramineae

maize, sweetcorn; Indian corn (En/USA)
Herb, 1-8 m, annual, monoecious
Distribution temperate N America, Asia, (Med) Africa, Europe and (Med)
Europe; subtropical S America, N America, Asia, Africa; tropical S
America, C America, Caribbean, Asia, Africa; native to N America.
Habitat cultivated crop plant
Soil deep well drained fertile soil preferred; waterlogging not
tolerated. **Rainfall** – rain during growth essential

Economic and other uses
Food – fresh seeds as vegetables, dried for flour/oil. **Fodder** – lvs and
seeds for forage, silage. **Fuel.** **Other uses** spathes for paper-making

Pollen
Pl ALG (Ske/72); URS/South (Glu/55); USA/TX (Pel/76)

Honeydew
Insect not specified: occasional flow (ALG, Ske/72); honey yield reported
as 45 kg/colony/yr (USA/LA, Ord/83); honeydew honey also reported in
USA/IA, TX (Pel/76); USA/MA (Shw/50); ZIM (Pap/73). Bees also collect
sap from split stems (URS, Glu/55; USA, Pel/76)

Honey: physical and other properties
Colour yellow (USA/IA, Pel/76); dark (USA/TX, Pel/76)
Granulation v coarse (USA/IA, Pel/76)
Flavour "peculiar", like corn silk

Some nectar plants (MAIN ENTRIES 001-452) also produce honeydew:

013 Acer platanoides L.
014 Acer pseudoplatanus L.
015 Acer tataricum L.
080 Castanea sativa Mill.
209 Gossypium barbadense L.
221 Helianthus annuus L.
267 Liriodendron tulipifera L.
282 Malus domestica Borkh.
283 Mangifera indica L.
290 Medicago sativa L.
354 Robinia pseudacacia L.
357 Rubus spp [R. fruticosus L.]

359 Rubus ulmifolius Schott.
362 Saccharum officinarum L.
363 Salix alba L.
364 Salix caprea L.
410 Tilia cordata Mill.
411 Tilia japonica (Miq.) Simonk.
415 Tilia platyphyllos Scop.
417 Tilia tomentosa Moench
418 Tilia x europaea L.
430 Trifolium pratense L.
438 Vicia faba L.

6. BIBLIOGRAPHY

Publications cited in the MAIN ENTRIES are coded by the first three letters of the first author's surname, and the two final digits of the year. Thus Ada/72 represents Adams, R.J. and Morton, J.K. (1972). Two or more of an author's publications from the same year are distinguished by adding letters a, b, ..., e.g., Bat/73 and Bat/73a. Where the names of different authors have the same first three letters, a letter beyond the third is used for one code, for example, Brother Adam's 1954 publication cannot be coded Ada/54, and is Adm/54. The same three letters are used throughout for the same author. The Bibliography is arranged in alphabetical order of author <u>code</u> (not of name), Ada/72 preceding Adm/54. Users should, however, not have too much difficulty in locating a publication without knowing its code.

Publications marked °° were used extensively in compiling the botanical parts of MAIN ENTRIES.

For data on **Nectar** and **Honey flow**, some references are given as, e.g., <u>AA</u>932/81. This refers to a publication (not in the Bibliography) which provides data only for these sections; full details can be found in the journal Apicultural Abstracts, <u>AA</u>932/81 representing entry 932 in the 1981 volume.

Ada/72   ADAMS, R.J.; MORTON, J.K. (1972)   An atlas of pollen of the trees and shrubs of eastern Canada and the adjacent United States. I. Gymnospermae to Fagaceae.   Univ. Waterloo Biol. Ser. No. 8 : 52 pages
Bd, AA365/77
Ada/74   ADAMS, R.J.; MORTON, J.K. (1974)   An atlas of pollen of the trees and shrubs of eastern Canada and the adjacent United States. II. Ulmaceae to Rosaceae.   Univ. Waterloo Biol. Ser. No. 9 : 53 pages   Bd, AA366/77
Ada/76   ADAMS, R.J.; MORTON, J.K. (1976)   An atlas of pollen of the trees and shrubs of eastern Canada and the adjacent United States. III. Leguminosae to Cornaceae.   Univ. Waterloo Biol. Ser. No. 10 : 37 pages
Bd, AA737/77
Ada/79   ADAMS, R.J.; SMITH, M.V.; TOWNSEND, G.F. (1979)   Identification of honey sources by pollen analysis of nectar from the hive.   J. apic. Res. 18(4) : 292-297   Bb, AA1280/80
Adm/54   ADAM, BROTHER (1954)   In search of the best strains of bee: second journey.   Bee Wld 35(10) : 193-203; (12) : 233-244   Ba, AA126/57
Adm/64   ADAM, BROTHER (1964)   In search of the best strains of bee: concluding journeys.   Bee Wld 45(2) : 70-83; (3) : 104-118 Ba, AA539/65
Adm/77   ADAM, BROTHER (1977)   In search of the best strains of bees: supplementary journey to Asia Minor, 1973.   Bee Wld 58(2) : 57-66
Bj, AA195/78

Aga/71   AGANIN, A.V. (1971)   [Electrical conductivity of several unifloral
   honeys.]   Trudy saratov. zootekh. vet. Inst. 21 : 137-144   In Russian
                                                              Ba, AA421/73
Alb/78   ALBISETTI, J. (1978)   Le robinier - faux-acacia - et son miel.
   Bull. tech. apic. 5(4) : 33-40   Fiche technique, CIDA No. 75   In
   French   Bc, AA210/80
Ama/55   AMARAL, E.; KERR, W.E. (1955)   Práticas apícolas regionais.
   Estado de S. Paulo, Supl. agric. 1(27) : 12   In Portuguese   Ba, AA404/57
Ama/57   AMARAL, E. (1957)   Honey bee activities and honey plants in
   Brazil.   Am. Bee J. 97(10) : 394-395   Bj, AA291/58
Ama/79   AMARAL, E.; ALVES, S.B. (1979)   Insetos uteis.   Piracicaba,
   Brazil : Livroceres   192 pages   In Portuguese   Bd, AA31L/81
Amb/81   AMBROSE, J.T. (1981)   Prospects and recommendations for the
   development of a modern beekeeping industry in the Sudan.   Unpublished
   report : 27 pages   Bs
And/56°° ANDERSON, R.H. (1956)   The trees of New South Wales.   Sydney,
   Australia : N.S.W. Department of Agriculture   471 pages   3rd ed.
                                                              Bd, AA90L/62
And/73   ANDERSON, R.H.; BUYS, B.; JOHANNSMEIER, M.F. (1973)   Beekeeping
   in South Africa.   Bull. Dep. agric. tech. Serv. S. Afr. No. 394 : 191
   pages   Bd, AA91/74
Anr/74   ANDRAG, H.R. (1974)   Sugar gum Eucalyptus cladocalyx Muell.   S.
   Afr. Bee J. 46(2) : 6-7   Bj
Aoy/68   AOYAGI, S. ET AL. (1968)   [Honeybees and honey. II. Chemical
   composition of honey.]   Bull. Fac. Agric. Tamagawa Univ. No. 7/8 : 181-
   202   In Japanese; English summary   Bb, AA199/72
Apc/68   APICULTURA (1968)   Caracterizarea apicolă a celor mai răspîndite
   plante nectaro-polenifere din ţara noastră.   Apicultura, Bucureşti 21(9)
   : 13-16;   (10) : 9-12;   (11) : 9-11;   (12) : 14-15   In Romanian
                                                              Bj, AA880L/73
Apd/83   API DIVULGAÇÕES (1983)   A bracatinga (Bracatinga scabrella.)   Api
   Divulg. 1(1) : 7   In Portuguese   Bj
Api/38   APICULTURA CHILENA (1938)   Valdivia.   Apicultura chil. (5) : 154-
   155   In Spanish   Bj
Api/39   APICULTURA CHILENA (1939)   La flor del quisco.   Apicultura chil.
   (17) : cover   In Spanish   Bj
Aso/60   ASO, K.; WATANABE, T.; YAMAO, K. (1960)   Studies on honey. I. On
   the sugar composition of honey.   Tohoku J. agric. Res. 11(1) : 101-108;
   originally published in Japanese in Hakko Kogaku Zasshi 36(2) : 39-44
   (1958)   Bb
Atw/70   ATWAL, A.S.; BAINS, S.S.; SINGH, B. (1970)   Bee flora for four
   species of Apis at Ludhiana.   Indian J. Ent. 32(4) : 330-334   Bb, AA286/73
Atw/73   ATWAL, A.S.; GOYAL, N.P. (1973)   Introduction of Apis mellifera
   in Punjab plains.   Indian Bee J. 35(1/4) : 1-9   Bb, AA512/77
Aub/83   AUBERT, S.; GONNET, M. (1983)   Mesure de la couleur des miels.
   Apidologie 14(2) : 105-118   In French; English, German summaries   Bj
Aus/83   AUSTRALIAN BEE JOURNAL (1983)   Honey flora for reafforestation.
   Part II. Trees recommended for honey production.   Aust. Bee J. 64(11) :
   9-11   Bj
Ave/78   AVETISYAN, G.A. (1978)   Apiculture.   Bucharest, Romania : Api-
   mondia Publishing House   270 pages   Bd, AA143/80
Ayt/71   AYTUĞ, B.; AYKUT, S.; MEREV, N.; EDIS'IN, G. (1971)   İstanbul
   çevresi bitkilerinin polen atlası. [Atlas des pollens des environs d'Istanbul.] Istanbul, Turkey : İstanbul Üniversitesi Orman Fakültesi   330
   pages   In Turkish and French   Bd, AA34/L/70

Bab/56   BARBIER, E.C. (1956)   Récolte exceptionnelle de manne de Briançon.
Revue fr. Apic. 3(126) : 1775-1780   In French   Ba, AA67/58
Bab/61   BARBIER, E.C.; PANGAUD, C.-Y. (1961)   Origine botanique et carac-
téristiques physicochimiques des miels.   Annls Abeille 4(1) : 51-65   In
French   Bj, AA509/62
Bab/63   BARBIER, E.[C.] (1963)   Les lavandes et l'apiculture dans le sud-
est de la France.   Annls Abeille 6(2) : 85-159   In French   Bj, AA336/64
Bab/69   BARBIER, E.[C.] (1969)   Present state of apiculture in Morocco and
its development programme.   Proc. XXII int. Beekeep. Congr. : 372-374   Bd
Bac/60   BACULINSCHI, H. (1960)   Cercetări privind valoarea meliferă a
principalelor plante spontane şi cultivate din zona de stepă.   Lucr.
ştiinţ Staţ. cent. Seri. Apic. 2 : 219-239   In Romanian; English,
French, German, Russian summaries   Bj, AA133/64
Bac/61   BACULINSCHI, H. (1961)   Constantele fizico-chimice şi biologice la
unele sorturi de miere de albine din ţara noastră.   Apicultura, Bucu-
reşti 14(5) : 18-20   In Romanian   Bj, AA832/65
Bac/65   BACULINSCHI, H. (1965)   Constantele fizico-chimice şi biologice la
unele sorturi de miere.   Lucr. ştiinţ. Staţ. cent. Seri. Apic. 5 : 65-
70   In Romanian; English, French, German, Russian summaries Bj, AA170/69
Bae/49°°   BAILEY, L.H. (1949)   Manual of cultivated plants.   New York, NY,
USA : Macmillan Co.   1116 pages   rev. ed.   Reprinted 1971   Bd
Bag/65   BAGA, A.M. (1965)   Phacelia tanacetifolia Benth. as a melliferous
plant and as a means of controlling Bruchus pisorum.   Proc. XX int.
Beekeep. Congr. : 317-320   Bd
Bah/73   BARTH, O.M. (1973)   Rasterelektronenmikroskopische Beobachtungen
an Pollenkörnern wichtiger Brasilianischer Bienenpflanzen.   Apidologie
4(4) : 317-329   In German; English, French summaries   Bj, AA205/75
Bai/55   BAILEY, S.E. (1955)   Beekeeping in Ohio.   Glean. Bee Cult. 83 :
723-725   Bj, see AA70/56
Bak/77   BAKER, H.G.; BAKER, I. (1977)   Intraspecific constancy of floral
nectar amino acid complements.   Bot. Gaz. 138(2) : 183-191   Bc, AA605/79
Bal/76   BALLY, P. (1976)   La Martinique: sa géographie, son climat, sa
flore, son apiculture.   Revue fr. Apic. (342) : 230-233   In French
                                                            Bj, AA503L/77
Bar/68   BARROSO, A.A. (1968)   Apicultural panorama of Brazil.   Apiacta
3(4) : 10-16   Bj, AA681L/69
Bas/67   BASKIN, C.C.; BLAKE, G.H. [1967]   Beekeeping in Alabama.   Circ.
Auburn Univ. co-op. Ext. Serv. No. P-64 : 35 pages   Ba, AA258L/69
Bat/72   BATTAGLINI, M.; BOSI, G. (1972)   Ricerche comparate sulla natura
dei glucidi di alcuni mieli monoflora e dei rispettivi nettari.   Pages
123-129 from "Problemi di flora mellifera e impollinazione: Simposio
Internazionale di Apicoltura, Torino, 2-6 ottobre 1972", Bucharest,
Romania : Apimondia Publishing House   In Italian; English, French
summaries   Ba, AA476/74
Bat/73   BATTAGLINI, M.; BOSI, G. (1973)   Caratterizzazione chimico-fisica
dei mieli monoflora sulla base dello spettro glucidico e del potere
rotatorio specificio.   Scienza Tecnol. Aliment. 3(4) : 217-221   In
Italian; English summary   Bc, AA280/77
Bat/73a   BATTAGLINI, M.; BOSI, G.; GRANDI, A. (1973a)   Considerazioni
sulla frazione glucidica del nettare di 57 specie botaniche dell'Italia
centrale.   Annali Fac. Agr. Univ. Perugia 28 : 233-242   In Italian;
English summary   Bb, AA581/77

Bau/66   BAUDIN, R. (1966)   Étude du milieu naturel apicole pour l'implanta-
tion des ruchers au Bugesera-Mayaga. Prévue par convention C.E.E. No.
315/RM. Projet No. 215.14.02.   Kigala : Ministère de l'Agriculture et de
l'Élevage République Rwandaise   29 pages   In French   Bd, AA654/72

Bea/79   BERKLEY, A. (1979)   Beekeeping in Dhofar.   Unpublished report :
11 pages + 1 figure   Bs

Bec/67   B[ECKER], H. (1967)   Der Honigwert verschiedener Bienennähr-
pflanzen. Luxemb. Bienenztg 82(3) : 18-21   In German   Bj, AA360L/68

Bee/76   BEETSMA, J. (1976)   Improving honey production and disposal in
Guyana and Surinam.   Pages 81-83 from "Apiculture in tropical climates"
ed. E. Crane, London, UK : International Bee Research Association
                                                          Bc, AA1254L/77

Bee/77   BEETSMA, J. (1977)   Beekeeping in Indonesia.   Unpublished report
: 29 pages   Bs

Bei/40°° BEIJERINCK, W. (1940)   Calluna - a monograph on the Scotch heather.
Verh. K. ned. Akad. Wet. 38(4) : 180 pages + 30 plates   Bd

Bek/65   BEEKEEPERS' ASSOCIATION OF PHENIAN (1965)   A test on 177.6 kg of
honey harvesting from each bee colony of the Liphen Production Co-opera-
tive in the Province of Diagan.   Proc. XX int. Beekeep. Congr. : 670-675
                                                                       Bd

Bel/79   BELLIARDO, F.; BUFFA, M.; PATETTA, A.; MANINO, A. (1979)
Identification of melezitose and erlose in floral and honeydew honeys.
Carbohyd. Res. 71 : 335-338   Ba, AA270/81

Bep/65   BEEKEEPERS' ASSOCIATION OF KOREAN PEOPLE'S DEMOCRATIC REPUBLIC
(1965)   Melliferous plants on Peaktu plateau of Korea.   Proc. XX int.
Beekeep. Congr. : 286-293   Bd

Ber/75   BERGNER, K.-G.; DIEMAIR, S. (1975)   Proteine des Bienenhonigs.
II. Gelchromatographie, enzymatische Aktivität und Herkunft von Bienen-
honig-Proteinen.   Z. Lebensmittelunters. u. -Forsch. 157(1) : 7-13   In
German; English summary   Ba, AA711/79

Bes/81   BEES AND HONEY (1981)   Rubber honey. Bees and Honey (April) : 3 Bj

Bes/81a BEES AND HONEY (1981a)   Test of a new nectar source in New Zealand. Bees and Honey (April) : 2   Bj

Bet/79   BELTON, M. (1979)   Beech scale insect: the place of the beech
scale insect (Ultracoelostoma assimile) in the ecology of mountain beech
forest.   Apiarist No. 8 : 4-6   Bj, AA914/80

Beu/49   BEUTLER, R. (1949)   Ergiebigkeit der Trachtquellen.   Imkerfreund
4(11) : 207-208   In German   Bb, AA67/50

Bew/52   BEE WORLD (1952)   Science and practice.   Bee Wld 33(9) : 152-153
                                                                       Bj

Bey/67   BEYLEVELD, G.P. (1967)   Maculate aloes as nectar producers in the
Transvaal and its implications.   S. Afr. Bee J. 39(5) : 10-12
                                                          Ba, AA362/68

Bey/68   BEYLEVELD, G.P. (1968)   Nectar and pollen producing trees and
plants of the Transvaal.   S. Afr. Bee J. 40(4) : 10-11, (6) : 13-14
                                                     Bj, AA650L, 651L/70

Bia/79   BIAN JI (1979)   Beekeeping buzzes to new high.   Source unknown :
1 page   Bs

Bla/72   BLAKE, S.T.; ROFF, C. (1972)   The honey flora of Queensland.
Brisbane, Australia : Department of Primary Industries   234 pages
                                                          Bd, AA630/73

Bla/79   BLAEDEL, N. (1979)   Et år med bier - og egen honning på bordet.
Copenhagen, Denmark : Rhodos   258 pages   In Danish   Bd, AA840L/81

Bls/80   BLASCO, F. ET AL. (1980)   Les rivages tropicaux. Mangroves
   d'Afrique et d'Asie.   Talence, France : Centre d'Études de Géographie
   Tropicale   246 pages   In French   Bd, AA733L/81
Blz/79   BLAZHIEVSKAYA, A.P. (1979)   [Honeydew from the white poplar.]
   Pchelovodstvo (8) : 17   In Russian   Bj, AA1287/80
Bod/42   BODENHEIMER, F.S. (1942)   Türkiye'de bal arisi ve aricilik hakk-
   inda etüdler. [Studies on the honey bee and beekeeping in Turkey.]
   Istanbul, Turkey : Numune Matbaasi   179 pages   In Turkish and English
                                                                          Bd
Bon/65   BORNUS, L. (1965)   On organizational methods and measures.   Proc.
   XX int. Beekeep. Congr. : 601-607   Bd
Bon/66   BORNUS, L.; KALINOWSKI, J.; ZALEWSKI, W. (1966)   Produkcja i
   skład chemiczny miodu akacjowego w m. Cigacice.   Pszczel. Zesz. nauk.
   10(1/4) : 113-122   In Polish; English, Russian summaries   Bj, AA787/67
Boo/72°° BOOMSMA, C.D. (1972)   Native trees of South Australia.   Bull.
   Woods Forests Dep. S. Aust. No. 19 : 224 pages   Bd, AA597/79
Bor/59   BORNECK, R. (1959)   Facts about beekeeping in France.   Bee Wld
   40(2) : 29-37   Ba, AA401/59
Bor/76   BORNECK, R. (1976)   Apiculture moderne: possibilités d'implanta-
   tion en Côte d'Ivoire. Rapport préliminaire.   Unpublished report (con-
   fidential) : 36 pages + 13 pages with data on the morphology of bees   In
   French   Bs
Bor/79   BORNECK, R. (1979)   Coopération technique avec les pays les moins
   avancés pour le développement des exportations des produits de la ruche:
   République Centrafricaine.   Unpublished report (confidential) : 45
   pages   In French   Bs
Bor/79a BORNECK, R. (1979a)   Résumé de la situation apicole au Togo.
   Unpublished report (confidential) : 13 pages   In French   Bs
Bor/80   BORNECK, R. (1980)   Technical co-operation with least developed
   countries for the development of exports of bee products: Ethiopia.
   Unpublished report (confidential) : 17 pages   Bs
Bos/78   BOSI, G.; BATTAGLINI, M. (1978)   Gas chromatographic analysis of
   free and protein amino acids in some unifloral honeys.   J. apic. Res.
   17(3) : 152-166   Bj, AA1075/79
Bra/54   BRAUN, W. (1954)   Beekeeping notes from Paraguay.   Bee Wld 35(8)
   : 155-156   Bj, AA268/55
Bra/54a BRAUN, W. (1954a)   Katalog der bedeutenden Nektar- und Pollen-
   pflanzen in Primavera, Alto Paraguay, Departamento de San Pedro, Republik
   Paraguay.   Dusenia 5(2) : 61-67   In German   Ba, AA276/55
Bra/59   BRAUN, W. (1959)   [Correspondent's report.]   Bee Wld 40(4) :
   100   Bj
Bri/55   BRITTAN, O. (1955/56)   Introduction of modern beekeeping to
   Cyrenaica (Libya).   Bee Craft 37(12) : 145-146; 38(1) : 4-5   Bj, AA214/58
Bro/82   BROUARD, J. (1982)   Honey plants in Mauritius.   Unpublished
   report in collaboration with J. Gueho : 3 pages   Bs
Brs/82   BRASHEAR, R. (1982)   Visiting a Mexican bee paradise.   Am. Bee J.
   122(5) : 337-340   Bj
Brv/71   BRAVO, L.C.; VINUESA, A.G. DE; GIL, M.S. (1971)   La apicultura
   en España.   Mundo Abejas (15) : 28-30   In Spanish   Bj
Bry/77   BRYANT, T.G. (1977)   Thyme honey - liquid gold.   N.Z. Jl Agric.
   135(2) : 19, 21   Bc, AA985/79
Buk/76°° BURKART, A. (1976)   A monograph of the genus Prosopis (Leguminosae
   subfam. Mimosoideae).   J. Arnold Arbor. 57(3) : 217-249; (4) : 450-
   525   Spanish summary   Bd

Bur/82   BURGETT, M. (1982)   Sabbatical leave report – apiculture in
   Thailand.   Unpublished report : 10 pages   Bs
But/72   BUTLER, G.D., JR; LOPER, G.M.; MCGREGOR, S.E.; WEBSTER, J.L.;
   MARGOLIS, H. (1972)   Amounts and kinds of sugars in the nectars of
   cotton (Gossypium spp.) and the time of their secretion.   Agron. J.
   64(3) : 364–368   Bc, AA888/73
Caa/72   CAMARGO, J.M.F. DE (EDITOR) (1972)   Manual de apicultura.   São
   Paulo, Brazil : Editora Agronômica Ceres   252 pages   In Portuguese
                                                          Bd, AA322/73
Cab/81   CAMBRIDGE AGRICULTURAL PUBLISHING (1981, reprinted 1982)   Oilseed
   rape book: a manual for growers, farmers and advisers.   Cornish Hall
   End, Finchingfield, UK : Cambridge Agricultural Publishing   160 pages +
   4–page insert   Bd
Cal/69   CALKINS, C.F. (1969)   Notes on beekeeping in Costa Rica.   Glean.
   Bee Cult. 97(9) : 526–531   Bj, AA607L/70
Cam/53   CAMPBELL, D.J. (1953)   Beekeeping in the Netherlands.   Bee Wld
   34(2) : 25–29   Ba, AA231/55
Cao/79   CARON, D.M. (1979)   Honey plants: honey from tulip trees.   Glean.
   Bee Cult. 107(6) : 305–306   Bj, AA901L/80
Car/38   CÁRDENUS, B., L. (1938/39)   Flora melífera: monografía de la flora
   melífera de San Juan de la Costa.   Apicultura chil. (1) : 16–17, (2) :
   55–57, (3) : 72–74, (4) : 127–128, (5) : 135–138; (7) : 215–217, (8) :
   239–241, (9) : 280–283, (10) : 313–315, (11) : 328–330, (12) : 375   In
   Spanish   Bj
Cas/74   CASTRO DE LEPRATTI, M. DE (1974)   Honey flora in Uruguay.
   Apiacta 9(1) : 45–47; also in Mundo Abejas (26) : 30–31   In Spanish
                                                          Bj, AA390L/75
Cer/64   ČERMAGIČ, C.; JANKOVIČ, A. (1964)   An analysis of Yugoslav
   honey.   Proc. XIX int. Beekeep. Congr. : 127–133   Bd
Cha/48   CHABOT, J.N. (1948)   Plantes mellifères dans Québec.   Quebec,
   Canada : Ministère de l'Agriculture   4 pages   In French   Bc
Chd/77   CHAUDHARI, R.K. (1977)   Bee forage in Punjab Plains (India). 1.
   Pathankot and adjacent villages.   Indian Bee J. 39(1/4) : 5–20
                                                          Bj, AA527L/80
Che/74   CHANDLER, B.V.; FENWICK, D.; ORLOVA, T.; REYNOLDS, T. (1974)
   Composition of Australian honeys.   Tech. Pap. Div. Fd Res. C.S.I.R.O.
   Aust. No. 38 : 39 pages   Bb, AA845/75
Chi/65   CHAIM, K. (1965)   Beekeeping in Israel.   Proc. XX int. Beekeep.
   Congr. : 710   Bd
Chk/72   CHAKRABARTI, K.; CHAUDHURI, A.B. (1972)   Wild life biology of the
   Sundarbans forests: honey production and behaviour pattern of the honey
   bee.   Sci. Cult. 38(6) : 269–276   Bc, AA1130/78
Chl/75   CHANDLER, M.T. (1975)   Apiculture in Madagascar.   Bee Wld 56(4) :
   149–153   Bj, AA406L/76
Chl/80   CHANDLER, [M.] T. (1980)   Apiculture development plan for Mozam-
   bique – including detailed technical recommendations.   Unpublished
   report (confidential) : iii + 73 pages   Bs
Chn/74   CHANDRAN, K.; SHAH, F.A. (1974)   Beekeeping in Kodai Hills
   (Tamilnadu, India).   Indian Bee J. 36(1/4) : 1–8   Bj, AA780/78
Cho/40   CHOUDHURI, M.M. (1940)   Bee-keeping in Orissa.   Indian Bee J.
   2(11/12) : 136–137   Bj
Chp/70   CHAPMAN, G.P.; FRANKSON, C.D.; JAY, S.C. (1970)   Bees and
   beekeeping in Jamaica.   Bee Wld 51(4) : 173–181   Ba, AA356/72

Cht/69   CHATURVEDI, P.L. (1969)   A tip for migratory bee-keepers of Kumaon
   region.   Indian Bee J. 31 : 48-51   Bj
Cht/77   CHATURVEDI, M. (1977)   Further investigation on the pollen analysis of bee loads from Banthra, India.   New Botanist 4(1/4) : 41-47
                                                              Bb, AA1316/81
Chu/65   CHAUBAL, P.D.; DEODIKAR, G.B. (1965)   Morphological characteriza-
   tion of pollen grains of some major honey yielding plants of the Western
   Ghats (India).   Indian Bee J. 27(1) : 1-28 + 2 plates   Bb, AA136/70
Chu/80   CHAUBAL, P.D.; KOTMIRE, S.Y. (1980)   Floral calendar of bee
   forage plants at Sagarmal (India).   Indian Bee J. 42(3) : 65-68
                                                              Bj, AA877L/82
Cir/77   CÎRNU, I.; HARNAJ, A.; LUCESCU, A.; FOTA, G.; GROSU, E. (1977)
   New criteria and elements to classify and estimate beekeeping economic
   contribution of honey plants.   Pages 191-194 from "Honey plants - basis
   of apiculture: International Symposium on Melliferous Flora, Budapest,
   1976", Bucharest, Romania : Apimondia Publishing House   Bd, AA1220L/78
Cir/80   CÎRNU, I.V. (1980)   Flora meliferǎ.   Bucharest, Romania : Editura
   Ceres   202 pages   In Romanian   Bd, AA154/82
Cla/78   CLAUSS, B. (1978)   Bees and beekeeping at Kagcae in the western
   central Kalahari.   Unpublished report : 23 pages   Bd
Cla/83   CLAUSS, B. (1983)   Bees and beekeeping in Botswana.   Botswana :
   Ministry of Agriculture   v + 122 pages   Bd
Clr/44   CLARIDGE, W.I. (1944)   Beekeeping in the Bahamas, B.W.I.   Glean.
   Bee Cult. 72(1) : 5   Bj
Coc/81   CONCHOUSO FERNANDEZ, J.; CONCHOUSO, P.P. DE (1981)   Honey produc-
   tion/hive in two different zones of the Mexican Republic.   Proc. XXVIII
   int. Beekeep. Congr. : 146-148   Bd
Cod/83   CODEX ALIMENTARIUS COMMISSION (1983)   Proposed draft codex stan-
   dard for honey (world-wide standard).   Rome, Joint FAO/WHO Food Stan-
   dards Programme (CX/PFV 84/13) : 9 pages   Bc
Coe/67   COLEMAN, L. (1967)   Honey flows of Zululand.   S. Afr. Bee J.
   39(4) : 6-7   Bj
Coi/69   COLLINS, T. (1969)   Abroad for the day.   Br. Bee J. 97(4196) : 173
                                                                           Bj
Cok/63   COOKE-YARBOROUGH, R.E. (1963)   The honey industry in New South
   Wales.   Rev. Mktg agric. Econ., Sydney 31(1) : 337   Ba
Col/62   COLEMAN, R.S. (1962)   Commercial bee-keeping. 2. Honey flora of
   Western Australia.   Bull. Dep. Agric. West. Aust. No. 3038 : 15 pages   Bb
Con/81   CORNER, J. (1981)   Personal communication
Coo/67   COOK, V.A. (1967)   Facts about beekeeping in New Zealand.   Bee
   Wld 48(3) : 88-100   Bj, AA320/68
Coo/81   COOK, V.A. (1981)   New Zealand honeydew from beech [Nothofagus].
   Bee Wld 62(1) : 20-22   Bj, AA1073L/81
Cor/70   CORNEJO, L.G.; TOMASEVICH, R. (1970)   Estudio sumario de la
   calidad de las mieles de algunas zonas del estado de Rio Grande do Sul,
   Brasil.   1 Congr. bras. Apic. : 241-245   In Portuguese   Bd, AA880/74
Cor/76   CORNEJO, L.G. (1976)   Informe final sobre diagnostico de la
   situacion actual de la apicultura colombiana y bases para su desarrollo.
   Bogatá, Colombia : Centro Interamericano de Promoción de Exportaciones
   ii + 332 pages   In Spanish   Bd, AA1267/79
Cos/63   COSCIA, A.A. (1963)   Posibilidades económicas de la apicultura en
   la Argentina (mercado de miel).   Infme téc. Estac. exp. agropec. Perga-
   mino No. 13 : 82 pages   In Spanish   Bd, AA259L/69

Cou/59   COUTURE, J.M. (1959)   Beekeeping in Quebec.   Glean. Bee Cult.
   87(8) : 462-467   Bj, AA115/61
Cra/66   CRANE, E. (1966)   Canadian bee journey.   Bee Wld 47(2) : 55-65;
   (4) : 132-148   Bj, AA485/67
Cra/73   CRANE, E. (1973)   Honey sources in some tropical and subtropical
   countries.   Bee Wld 54(4) : 177-186   IBRA Reprint M73   Ba, AA875/73
Cra/73a CRANE, E. (1973a)   Personal communication
Cra/75   CRANE, E. (EDITOR) (1975)   Honey: a comprehensive survey.
   London, UK : William Heinemann in co-operation with the International Bee
   Research Association   xvi + 608 pages + 30 plates   Reprinted 1976 with
   corrections   Bd, AA542/76
Cra/77   CRANE, E. (1977/78)   Dead bees under lime trees. Sugars poisonous
   to bees.   Bee Wld 58(3) : 129-130; 59(1) : 37-38   Bj, AA232/79
Cra/81   CRANE, E. (1981)   When important honey plants are invasive weeds.
   Bee Wld 62(1) : 28-30   Bj, AA936L/81
Cra/82   CRANE, E. (1982)   Report on apiculture in Mauritius.   Unpublished
   report (confidential) : 47 pages + 8 figures   Bs
Cra/83   CRANE, E.; WALKER, P. (1983)   The impact of pest management on
   bees and pollination.   London, UK : Tropical Development and Research
   Institute   232 pages   Bd, AA347/84
Cra/83a CRANE, E. (1983a)   Surveying the world's honey plants.   Lect.
   cent. Ass. Beekprs : 12 pages   Ba
Cra/84   CRANE, E. (1984)   Personal communication
Cri/57   CRISP, W.E. (1957)   About South African honey flora.   S. Afr. Bee
   J. 32(1) : 1-3; (3) : 12-13; (5) : 7   Bj, AA87/62
Cul/81   CULVENOR, C.C.J.; EDGAR, J.A.; SMITH, L.W. (1981)   Pyrrolizidine
   alkaloids in honey from Echium plantagineum L.   J. agric. Fd Chem. 29(5)
   : 958-960   Bc, AA657/83
Cur/66   CURTI, R.; RIGANTI, V. (1966)   Ricerche sugli aminoacidi del
   miele.   Rass. chim. 18(6) : 278-282   In Italian   Bc, AA215/68
Dae/61°° DALE, I.R.; GREENWAY, P.J. (1961)   Kenya trees and shrubs.
   Nairobi, Kenya : Buchanan's Kenya Estates Limited in association with
   Hatchards, London   654 pages + 112 plates   Bd, AA843/64
Dai/70   DAVIDSON, V.R. (1970)   Trees for beekeeping.   S. Afr. Bee J.
   42(2) : 12-14   Bj
Dal/75   DALZELL, K.W.; SINGERS, W.A. (1975)   A survey of some South
   Island honeydew honeys.   N.Z. Jl Sci. 18(3) : 329-332   Bb, AA278/78
Dan/75   DANILENKO, P.L. (1975)   Beekeeping in the far east and its further
   development.   Proc. XXV int. Beekeep. Congr. : 177-179   Bd
Dav/69   DAVIDSON, J. [1969]   Native flowers for bees.   List Univ. Br.
   Columb. : 4 pages   Ba, AA652L/70
Ded/53   DEODIKAR, G.B.; THAKAR, C.V. (1953)   A pollen study of major
   honey yielding plants of Mahabaleshwar hills.   Poona, India : Village
   Industries Committee, Bombay State   8 pages   Ba, AA185/55
Ded/57   DEODIKAR, G.B.; THAKAR, C.V.; PHADKE, R.P. (1957)   High nectar
   concentration in floral nectaries of silver oak, Grevillea robusta.
   Indian Bee J. 19(7/8) : 84-85   Ba, AA711L/72
Dem/60   DEMIANOWICZ, Z. ET AL. (1960)   Wydajność miodowa ważniejszych
   roślin miododajnych w warunkach Polski.   Pszczel. Zesz. nauk. 4(2) : 87-
   104   In Polish; English, Russian summaries   Bj, AA853/64
Dem/60a DEMIANOWICZ, Z.; HŁYŃ, M. (1960a)   Porównawcze badania nad
   nektarowaniem 17 gatunkow lip.   Pszczel. Zesz. nauk 4(3/4) : 133-152
   In Polish; German, Russian summaries   Bj, AA91/62

Dem/63   DEMIANOWICZ, Z. (1963)   Sur l'origine des macles d'oxalate de calcium contenues dans les miels de tilleul.   Annls Abeille 6(4) : 249-255   In French;   English summary   Bj, AA575/66

Dem/63a DEMIANOWICZ, Z.; JABŁOŃSKI, B.; OSTROWSKA, W.; SZYBOWSKI, S. (1963a)   Wydajność miodowa ważniejszych roślin miododajnych w warunkach Polski.   Pszczel. Zesz. nauk. 7(2) : 95-111   In Polish;   German, Russian summaries   Bj, AA854/64

Dem/64   DEMIANOWICZ, Z. (1964)   Charakteristik der Einartenhonige.   Annls Abeille 7(4) : 273-288   In German;   French, English summaries
                                                        Bj, AA167/67

Dem/64a DEMIANOWICZ, Z. (1964a)   The nectar secretion and the honey yield of the most important nectariferous plants in Poland.   Proc. XIX int. Beekeep. Congr. : 173-176   Bd

Dem/77   DEMIANOWICZ, Z.; WARAKOMSKA, Z. (1977)   Influence of hive population on pollen spectrum of rape honey.   Pages 183-187 from "Honey plants - basis of apiculture: International Symposium on Melliferous Flora, Budapest, September 1976", Bucharest, Romania : Apimondia Publishing House
                                                        Bd

Dem/79   DEMIANOWICZ, Z. (1979)   Nektarowanie i wydajność miodowa Taraxacum officinale Web.   Pszczel. Zesz. nauk. 23 : 97-103   In Polish; English, Russian summaries   Bj, AA162/82

Den/74   DENING, R.C. (1974)   The development of beekeeping in Malawi. Lilongwe, Malawi : Ministry of Agriculture and Natural Resources   46 pages   Bd

Deo/72   DEODASAN, A. (1972)   Rubber plantation and bee-keeping.   Indian Bee J. 34(1/2) : 38-39   Bj, AA156/74

Des/62   DESHUSSES, J.; GABBAI, A. (1962)   Recherche de l'anthranilate de méthyle dans les miels espagnols de fleur d'oranger par chromatographie sur couche mince.   Mitt. Geb. Lebensmittelunters. Hyg. 53(5) : 408-411 In French   Ba, AA727/66

Dev/71   DEVADASON, A. (1971)   Migratory beekeeping in Kerala.   Indian Bee J. 33(3/4) : 35-38   Bj, AA817L/73

Dew/79   DEWAN, S.A.L. (1979)   Apiculture programme progress report 1978-1979.   Unpublished report : 26 pages   Bs

Die/71   DIETZ, A.; CARON, D.M. (1971)   Beekeeping in Maryland.   Ext. Bull. Univ. Md coop. Ext. Serv. No. 223 : 40 pages   rev.   Bc, AA96L/73

Diw/64   DIWAN, V.V.; SURYANARAYANA, M.C.; THAKAR, C.V. (1964)   Beekeeping potentialities of the Castle Rock area. 1. Preliminary surveys.   Indian Bee J. 26 : 4-15   Bb, AA356L/66

Diw/69   DIWAN, V.V.; RAO, S.K. (1969)   Foraging behaviour of Apis indica F. on the flowers of Synadenium grantii Hook f.   Indian Bee J. 31(1) : 22
                                                        Ba, AA331/72

Diw/72   DIWAN, V.V.; RAO, G.V. (1972)   Altitudinal variation in bee forage of West Coorg.   Indian Bee J. 33(3/4) : 39-50   Bj, AA884/73

Dol/61   DOULL, K.M. (1961)   Pollen-induced disease of honeybees in Australia.   Bee Wld 42(1) : 3-5   Bj

Dou/65   DOUHET, M. (1965)   Apicultural methods and the distribution of apicultural production in Madagascar.   Proc. XX int. Beekeep. Congr. : 634-638   Bd

Dou/70   DOUHET, M. (1970)   L'apiculture sénégalaise: situation et perspectives.   Nice, France : Laboratoire de Recherches Apicoles   64 pages In French   Bd

Dou/79   DOUHET, M. (1979)   L'apiculture en Empire Centrafricaine: situation et perspectives.   Maisons-Alfort, France : Institut d'Élevage et de
Médecine Vétérinaire des Pays Tropicaux   78 pages   In French   Bd

Dou/80   DOUHET, M. (1980)   L'apiculture en Côte-d'Ivoire: régions nord et
centre.   Maisons-Alfort, France : Institut d'Élevage et de Médecine
Vétérinaire des Pays Tropicaux   84 pages   In French   Bd

Dre/74   DRESCHER, W. (1974)   Evaluation of the present beekeeping situation in the Tanga region, including recommendations for possible future
improvement.   Frankfurt am Main, German Federal Republic : Bundesstelle
für Entwicklungshilfe   27 pages   Bd

Dre/80   DRESCHER, W. (1980)   The present situation of honey production in
the Binwa-Catchment (Himachal Pradesh) and recommendations for the
improvement of beekeeping. Project No. 78.2131.7.   Unpublished report :
12 pages   Bs

Dua/73   DUTAUT, P. (1973)   La récolte du miel sur la côte est en 1973.
Revue agric. Nouv. Caléd. (24) : 14-15   In French   Bc

Dub/50   DUBOIS, L.; COLLART, E. (1950)   L'apiculture au Congo Belge et au
Ruanda-Urundi.   Belgium : Ministère des Colonies, Direction de l'Agriculture, de l'Élevage et de la Colonisation   230 pages   In French
                    Bd, AA9/58

Duf/02   DUFOUR, L. (1902)   L'apiculture aux colonies françaises.   Apiculteur : 18 pages   In French   Ba

Dul/68   DUBBS, A.L. (1968)   Sainfoin as a honey crop.   Bull. Mont. agric.
Exp. Stn No. 627 : 108-109   Bc, AA660/71

Dus/67   DUSTMANN, J.H. (1967)   Messung von Wasserstoffperoxid und Enzymak-
tivität in Mitteleuropaischen Honigen.   Z. Bienenforsch. 9(2) : 66-73
In German   Bj, AA784/67

Dus/67a DUSTMANN, J.H. (1967a)   Messungen von Wasserstoffperoxid in Bienenhonig aus Edelkastanientracht (Castanea sativa M.).   Z. Lebensmittelunters. u. -Forsch. 134(1) : 20   In German   Ba, AA350/69

Dus/69   DUSTMANN, J.H. (1969)   Die Kornblume honigt nicht nur zur Blüte-
zeit.   Nordwestdt. Imkerztg 21(12) : 330-332   In German   Bj, AA929/71

Dus/71   DUSTMANN, J.H. (1971)   Über die Katalaseaktivität in Bienenhonig
aus der Tracht der Heidekrautgewächse (Ericaceae).   Z. Lebensmittelunters. u. Forsch. 145(5) : 294-295   In German; English summary
                    Ba, AA1055/72

Dus/72   DUSTMANN, J.H. (1972)   Über den Einfluss des Lichtes auf den
Peroxid-Wert (Inhibin) des Honigs.   Z. Lebensmittelunters. u. -Forsch.
148(5) : 263-268   In German; English summary   Ba, AA472/74

Dus/72a DUSTMANN, J.H. (1972a)   Einfluss der Dialyse bei der Bestimmung
der Sacharase-Aktivität in Honig.   Lebensm.-Wiss. Technol. 5(2) : 70-
71   In German   Bc, AA946/73

Dut/77   DUTTON, R.[W.]; SIMPSON, J. (1977)   Producing honey with Apis
florea in Oman.   Bee Wld 58(2) : 71-76   Ba, AA118/78

Dut/79   DUTTON, R.W.; FREE, J.B. (1979)   The present status of beekeeping
in Oman.   Bee Wld 60(4) : 176-185, 175   Ba, AA467/80

Dyc/65   DYCE, E.J. (1965)   Beekeeping developments in the Dominican Republic.   Glean. Bee Cult. 93(9) : 550-552, 567   Bj, AA427L/66

Ech/75   ECHIGO, T.; TAKENAKA, T.; EZAWA, M. (1975)   [Studies on quality
of honey. II. Changes in quality caused by heating.]   Nihon Shokuhin
Kôgyô Gakkaishi 22(4) : 148-153   In Japanese; English summary
                    Bb, AA1347/77

Ech/77   ECHIGO, T. (1977)   [Food chemical and biochemical studies of
   honey.]   Bull. Fac. Agric. Tamagawa Univ. No. 17(2) : 1-77   In Japanese;
   English summary   Bd, AA603/80
Eck/51   ECKERT, J.E. (1951)   Rehabilitation of the beekeeping industry in
   Hawaii.   Hawaii : Final Report I.R.A.C. Grant No. 19   29 pages
                                                             Bc, AA158/53
Eck/52   ECKERT, J.E.; BESS, H.A. (1952)   Fundamentals of beekeeping in
   Hawaii.   Ext. Bull. Univ. Hawaii No. 55 : 59 pages   Ba
Eck/60   ECKERT, J.E; SHAW, F.R. (1960)   Beekeeping: successor to "Beekeep-
   ing" by Everett F. Phillips.   New York, USA : Macmillan   536 pages
                                                             Bd, AA116/61
Ecl/72   ECKLOFF, W. (1972)   Beitrag zur Ökologie und forstlichen Bedeutung
   bienenwirtschaftlich wichtiger Rindenläuse.   Z. angew. Ent. 70(2) : 134-
   157   In German; English summary   Bb, AA425/74
Edw/75   EDWARDS, R.A.; FARAJI-HAREMI, R.; WOOTTON, M. (1975)   A rapid
   method for determining the diastase activity in honey.   J. apic. Res.
   14(1) : 47-50   Bb, AA461/75
Ehr/77   EHRBAR, A. (1977)   Personal communication
Ehr/78   EHRBAR, A. (1978)   Swaziland: Note.   Page 22 from "Bibliography
   of tropical apiculture. Part 2. Beekeeping in Africa south of the Sahara"
   ed. E. Crane, London, UK : International Bee Research Association
                                                             Bc, see AA149L/80
Eis/80   EISIKOWITCH, D.; MASAD, Y. (1980)   Nectar-yielding plants during
   the dearth season in Israel.   Bee Wld 61(1) : 11-18   Ba, AA1281/80
Eis/82   EISIKOWITCH, D.; MASAD, Y. (1982)   Preference of honeybees for
   different ornamental nectar-yielding plants during the dearth period in
   Israel.   Bee Wld 63(2) : 77-82   Ba, AA1276/82
Elm/52   ELMENHORST, C.W. (1952)   Beekeeping in Guatemala.   Bee Wld 33(6)
   : 93-96   Bj, AA261/53
Els/79   EL-SHERBINY, G.A.; RIZK, S.S. (1979)   Chemical composition of
   both clover and cotton honey produced in A.R.E.   Egypt. J. Fd Sci.
   7(1/2) : 69-75   Bb, AA1313/83
Enb/54   ENBOM, N.-J. (1954)   Beekeeping in Finland.   Glean. Bee Cult.
   82(1) : 24-26   Bj
Erb/83   ERBORISTERIA DOMANI (1983)   Tabella di fioritura di alcune impor-
   tanti specie mellifere.   Erboristeria Domani No. 2, Suppl. : 152-155
   In Italian   Bc, AA220L/84
Erd/50   ERDA, S. (1950)   L'apiculture en Turquie.   Gaz. apic. (518) : 12-
   14   In French   Bj
Eri/75   ERICKSON, E.H. (1975)   Honey bees and soybeans.   Am. Bee J.
   115(9) : 351-353, 372   Bj, AA1083/76
Erm/50   ERMIN, R. (1950)   Untersuchungen zur Honigtau- und Tannenhonig-
   frage in der Türkei.   İstanb. Üniv. Fen Fak. Mecm. Ser. B 15(3) : 185-
   225   In German   Bc, AA28/52
Ert/61   ERDTMAN, G.; BERGLUND, B.; PRAGLOWSKI, J. (1961)   An introduc-
   tion to a Scandinavian pollen flora.   Stockholm, Sweden : Almqvist &
   Wiksell   92 pages + 74 plates   Bd, AA7/63
Ert/63   ERDTMAN, G.; PRAGLOWSKI, J.; NILSSON, S. (1963)   An introduction
   to a Scandinavian pollen flora. Volume II.   Stockholm, Sweden : Almqvist
   & Wiksell   89 pages + 58 plates   Bd, AA454/64
Ert/69   ERDTMAN, G. (1969)   Handbook of palynology. Morphology - taxonomy
   - ecology. An introduction to the study of pollen grains and spores.
   Copenhagen, Denmark : Munksgaard   486 pages + 1 chart   Bd, AA11/71

Esa/81   ESPADA HERRERO, T. (1981)   Composition of two Catalonian (Spain)
   honeys.   Proc. XXVIII int. Beekeep. Congr. : 427   Bd
Esb/80   ESBENSHADE, H.W. (1980)   Kiawe: a tree crop in Hawaii.   Int. Tree
   Crops J. 1 : 125-130   Ba, AA192/83
Esp/64   ESPINA [PÉREZ], D. (1964)   Plan de fomento apícola.   El Salvador
   : Banco Central de Reserva de El Salvador   10 pages   In Spanish
                                                             Bc, AA55L/65
Esp/81   ESPINA PÉREZ, D. (1981)   Guidelines for the study of the bee-
   keeping potential of a region.   Proc. XXVIII int. Beekeep. Congr. : 148-
   152   Bd
Esp/83 [Ord/83, Ord/82]   ESPINA PÉREZ, D.; ORDETX ROS, G.S. (1983)   Flora
   apícola tropical.   Cartago, Costa Rica : Editorial Tecnológica de Costa
   Rica   406 pages   In Spanish   Bd
Eur/74   EUROPEAN COMMUNITIES (1974)   Council Directive 74/409/EEC of 22
   July 1974 on the harmonization of the laws of the Member States relating
   to honey.   Off. J. Eur. Communities 17(L221) : 10-14   Bc, AA286L/77
Ewa/56   EWART, W.H.; METCALF, R.L. (1956)   Preliminary studies of sugars
   and amino acids in the honeydews of five species of coccids feeding on
   citrus in California.   Ann. ent. Soc. Am. 49(5) : 441-447   Bb, AA56/58
Fag/69   FAGG, P. (1969)   Introduction to beekeeping in Uganda.   Unpub-
   lished report : 12 pages   Ba
Fah/49   FAHN, A. (1949)   Studies in the ecology of nectar secretion.
   Palest. J. Bot. Jerusalem Ser. 4 : 207-224   Bb, AA164/53
Fal/74   FALCONER, D. (1974)   Producing honey from sweet oranges (Citrus
   sinen[s]is).   S. Afr. Bee J. 46(3) : 4-5   Bj
Fan/52   FAN TSUNG DEH (1952)   Beekeeping in Formosa.   Bee Wld 33(9) : 150-
   151   Bj, AA260/53
Far/79   FARRUGIA, V. (1979)   Beekeeping in the Maltese Islands.   Pages 41-
   46, 190 from "Beekeeping in rural development: unexploited beekeeping
   potential in the tropics with particular reference to the Commonwealth"
   ed. Commonwealth Secretariat and International Bee Research Association,
   London, UK : Commonwealth Secretariat   Bd, AA1189L/80
Fea/74   FERRAZZI, P. (1974)   Mieli di "tarassaco" di Piemonte e Lombardia.
   1. Analisi mellisso-palinologica.   Apicolt. mod. 65(1/2) : 21-26   In
   Italian; English summary   Bc, AA265/76
Fea/74a FERRAZZI, P. (1974a)   Indagini sulla scarsità di polline di Tarax-
   acum officinale Weber in miele di "tarassaco" di Piemonte e Lombardia.
   Apicolt. mod. 65(3/4) : 58-64   In Italian; English summary Bj, AA266/76
Fea/83   FERRAZZI, P. (1983)   Insetti fitomizi ed api: incidenza di melata
   in mieli dell'Italia settentrionale.   Atti XIII Congr. naz. ital. Ent. :
   729-736   In Italian; English summary   Bd, AA230/84
Fed/55   FEDOSOV, N.F. (EDITOR) (1955)   [Beekeeper's encyclopaedia.]
   Moscow, USSR : State Publishing House for Agricultural Literature   419
   pages   In Russian   Bd, AA79/58
Feo/62   FEDOROWSKA, Z. (1962)   Effects of various purifying reagents on
   results of honey tests.   Acta. Pol. pharm. 19(1) : 910-922 Ba, AA749/70
Feo/71   FEDOROWSKA, Z.; ZBOROWSKY, J. (1971)   Comparative studies of
   Polish heath honey from the crops 1965-1967.   Acta Pol. pharm. 28(1) :
   106-116   Ba, AA806/72
Fer/78   FERNANDO, E.F.W. (1978)   Studies on apiculture in Sri Lanka. 1.
   Characteristics of some honeys.   J. apic. Res. 17(1) : 44-46
                                                             Bb, AA1329/78

Fer/79   FERNANDO, E.F.W. (1979)   The ecology of honey production in Sri
    Lanka.   Pages 115-125, 191-192 from "Beekeeping in rural development:
    unexploited beekeeping potential in the tropics with particular reference
    to the Commonwealth" ed. Commonwealth Secretariat and International Bee
    Research Association, London, UK : Commonwealth Secretariat Bd, AA1202/80
Fie/80   FIELD, O. (1980)   Beekeeping and honey production in Yemen Arab
    Republic.   Personal communication
Fil/80   FIELD HONEY FARMS (1980)   Bee keeping and honey production in
    Yemen Arab Republic : report of an investigation carried out by Field
    Honey Farms.   Report No. FHF/YAR/1 : 15 pages   Bs
Fin/74   FINI, M.A.; SABATINI, A.G. (1974)   Osservazioni sulla compo-
    sizione di alcuni tipi di miele della Sicilia.   Scienza Tecnol. Aliment.
    4(6) : 349-355   In Italian   Bc, AA279/78
Fis/38   FISSORE, C. (1938)   La industria apícola en la provincia de
    Valdivia.   Apicultura chil. 1(4) : 113-114   In Spanish   Bj
Fle/63   FLECHTMANN, C.H.W.; CALDAS FILHO, C.F.; AMARAL, E.; ARZOLLA,
    J.D.P. (1963)   Análise de méis do estado de São Paulo.   Bolm Ind. anim.
    21 : 65-73   In Portuguese; English summary   Bb, AA197/65
Fli/62   FLEISCH, R.P.H. (1962/63)   Apiculture libanaise.   Apiculteur 106
    : 33-40, 123-127, 175-181, 199-202; 107 : 40-45   In French Bj, AA78L/63
Foo/79   FOOD AND AGRICULTURE ORGANIZATION [1979?]   L'apiculture en Tunisie.
    Rome, Italy : FAO   64 pages   In French   Bd
Fos/56   FOSSEL, A. (1956)   Steirische Honige.   Bienenvater 77(5) : 156-163
    In German   Ba, AA303/56
Fra/65   FRANCISC, S.; EMERIC, T. (1965)   Wintering on candy of bee
    colonies which have supplies containing honeydew honey.   Proc. XX int.
    Beekeep. Congr. : 194-196   Bd
Fre/57   FREDSKILD, B. (1957)   Nogle faenologiske studier over de vigtigste
    af biernes traekplanter.   Tidsskr. PlAvl 61(1) : 133-148   In Danish;
    English summary   Ba, AA290/58
Frj/70   FREE, J.B. (1970)   Insect pollination of crops.   London, UK :
    Academic Press Inc. (London) Ltd   544 pages   Bd, AA478/71
Fur/75   FURUTA, T.; OKIMOTO, Y. (1975)   Study on yeasts in honey.   Bull.
    Fac. Agric. Tamagawa Univ. No. 15 : 37-43 Japanese summary Bb, AA1369/77
Gad/76   GADBIN, C. (1976)   Plant sources of honey in Chad.   Personal
    communication
Gad/80   GADBIN, C. (1980)   Les plantes utilisées par les abeilles au Tchad
    méridional.   Apidologie 11(3) : 217-254   In French; English, German
    summaries   Bj, AA930/81
Gen/67   GENSITSKII, I.P.; SEREDA, A.G. (1967)   [Honey protein.]   Pchelo-
    vodstvo (11) : 26-27   In Russian   IBRA translation E984   Bj, AA172/69
Geo/52   GEORGHIOU, G.P. (1952)   Beekeeping in Cyprus.   Glean. Bee Cult.
    80(1) : 12-15, 61   Bj
Gia/37   GIAVARINI, I. (1937)   Notizie sulle api e sull'apicoltura Abissina.
    5 Congr. naz. Apic. ital. : 12 pages   In Italian   IBRA translation E1374
                                                                            Ba
Gil/80   GILLIAM, M.; MCCAUGHEY, W.F.; WINTERMUTE, B. (1980)   Amino acids
    in pollens and nectar of citrus cultivars and in stored pollen and honey
    from honeybee colonies in citrus groves.   J. apic. Res. 19(1) : 64-72
                                                              Bb, AA570/81
Gle/77   GLEIM, K.H. (1977)   Nahrungsquellen des Bienenvolkes.   St.
    Augustine, German Federal Republic : Delta Verlag   159 pages   In German
                                                              Bd, AA895/78

Glu/55   GLUKHOV, M.M. (1955)   [Honey plants.]   Moscow, USSR : State
   Publishing House of Agricultural Literature   512 pages   6th ed.   In
   Russian   Bd, AA415/57

Goa/52   GORBACH, G. (1952)   Biene und Blatthonig.   Südwestdt. Imker 4(5)
   : 138-142   In German   Bj, AA319/57

Gob/81   GOROBETS, A.V.; BANDYUKOVA, V.A.; SHAPIRO, D.K.; ANIKHIMOVSKAYA,
   L.V.; NARIZHNAYA, T.I. (1981)   [Amino acid composition of the pollens
   of some nectar plants.]   Khimiya prirod. Soed. No. 5 : 672-673   In
   Russian   Bb, AA909/83

God/52   GODDARD, R. (1952)   Beekeeping in Tauranga County.   N.Z. Jl
   Agric. 84(3) : 223-224   Bb, AA155/54

Gom/73   GOODMAN, R.D. (1973)   Honey flora of Victoria.   Victoria, Australia : Department of Agriculture   175 pages   Bd, AA833/74

Gon/65   GONNET, M. (1965)   Les modifications de la composition chimique
   des miels au cours de la conservation.   Annls Abeille 8(2) : 129-146
   In French   Ba, AA573/66

Gon/79   GONNET, M. (1979)   Application au miel d'une méthode de dosage par
   voie enzymatique des monosaccharides réducteurs.   Apidologie 10(4) :
   395-401   In French; English, German summaries   Bj, AA667/81

Goo/47   GOODACRE, W.A. (1947)   The honey and pollen flora of New South
   Wales.   Sydney, Australia : New South Wales Department of Agriculture
   195 pages   Bd

Gor/64   GORENZ, A.M. (1964)   A start in bee-keeping in Ghana.   Ghana Fmr
   8(4) : 108-114   Bb, AA57L/66

Got/51   GONTARSKI, H. (1951)   Zur Analyse der Formbestandteile des Wald-
   honigs.   Z. Bienenforsch. 1 : 33-37   In German   Ba, AA162/51

Goy/74   GOYAL, N.P. (1974)   Apis cerana indica and Apis mellifera as
   complementary to each other for the development of apiculture.   Bee Wld
   55(3) : 98-101   IBRA Reprint M76   Bj, AA771/74

Gra/50   GRAYSON, J.M.; ROWELL, J.O. (1950)   Beekeeping in Virginia.
   Bull. Va agric. Ext. Serv. No. 178 : 29 pages   Ba, AA112/53

Grd/79   GRADDON, A.D.; MORRISON, J.D; SMITH, J.F. (1979)   Volatile
   constituents of some unifloral Australian honeys.   J. agric. Fd Chem.
   27(4) : 832-837   Bc, AA669/81

Gre/61   GREGORY, G.B. (1961)   Rural industry: bee keeping.   Agric. J.
   Dep. Agric. Fiji 31 : 42-43   Ba, AA548L/64

Gri/77   GRIGORENKO, V.N. (1977)   An important honey plant in the woods of
   the southern taiga.   Pages 76-79 from "Honey plants - basis of apiculture: International Symposium on Melliferous Flora, Budapest, September
   1976", Bucharest, Romania : Apimondia Publishing House   Bd, AA1208/78

Grn/65   GRANDJEAN, X. (1965)   Some remarks on the Belgian melliferous
   flora; achievement of an experimental botanical chequerboard; the country's melliferous map; multiplication and acclimatization of the best
   plants.   Proc. XX int. Beekeep. Congr. : 294-297   Bd

Gun/79   GUNN, C.R. (1979)   Genus Vicia with notes about tribe Vicieae
   (Fabaceae) in Mexico and Central America.   Tech. Bull. U.S. Dep. Agric.
   No. 1601 : ii + 41 pages   Bd

Guy/71   GUY, R.D. (1971)   South African bee plants. No. 2. Eucalyptus
   paniculata Sm. - grey ironbark.   S. Afr. Bee J. 43(5) : 7   Bj, AA750L/75

Guy/71a [GUY, R.D.] (1971a)   South African bee plants. No. 1. Faurea
   saligna Harv. Transvaal boekenhout.   S. Afr. Bee J. 43(4) : 3
                                                                Bj, AA750L/75

Guy/72   GUY, R.D. (1972)   South African bee plants. No. 6. Eucalyptus
   maculata Hook. Spotted gun.   S. Afr. Bee J. 44(3) : 11   Bj, AA750L/75

Guy/72a GUY, R.D. (1972a) South African bee plants. No. 8. Citrus spp. S. Afr. Bee J. 44(6) : 5-6 Bj, AA750L/75

Guy/74 GUY, R. (1974) Eucalyptus grandis Hill ex Maiden, as a honey plant in South Africa. S. Afr. Bee J. 46(1) : 6-7, 9 Bj, AA387L/75

Haa/60 HARAGSIMOVÁ-NEPRAŠOVÁ, L. (1960) Zjišťování nektarodárnosti rostlin. Ved. Pr. výzk. Úst. včelár. v Dole u Libčic In Czech; English, German, Russian summaries Bj, AA132/64

Hab/72 HABBIE (1972/73) Azorean idyll. Scott. Beekpr 49(12) : 210; 50(1) : 6, (2) : 22-23 Bj

Had/63 HADORN, H.; ZÜRCHER, K. (1963) Formolzahl von Honig. Gleichzeitige Bestimmung von Formolzahl, pH, freier Säuer und Lactongehalt in Honig. Mitt. Geb. Lebensmittelunters. Hyg. 54(1) : 304-321 In German; English, French summaries Ba, AA590/68

Hae/53 HAMMER, O. (1953) Observations sur la sécrétion nectarifère chez le romarin (Rosmarinus officinalis). Revue fr. Apic. 3(86) : 397-403 In French Ba, AA247/54

Hag/63 HARAGSIM, O. (1963) Medovice a její včelařské využití. Ved. Pr. výsk. Úst. včelař. v Dole u Libčic 3 : 277-318 In Czech; English, French, German, Russian summaries Bj, AA120/66

Hag/66 HARAGSIM, O. (1966) Medovice a včely. Prague, Czechoslovakia : Státni zemědělské nakladatelství 200 pages In Czech; also in Polish (Spadži i pszczoly. Warzaw, Poland : Państwowe Wydawnictwo Rolnicze i Leśen 252 pages (1970)) Bd

Hag/76 HARAGSIM, O. (1976) Académie d'Agriculture Tchécoslovaque. Abeille Fr. Apic. (595) : 221-228 In French Bj, AA215/77

Hai/69 HALMÁGYI, L. (1969) Die honigtauerzeugenden Blattläuse (Aphidoidea) der Eichen in Ungarn. Opusc. zool. Bpest 9(1) : 97-105 In German; English summary Ba, AA937/71

Hal/77 HALMÁGYI, C. (1977) Nectar secretion of acacia tree. Pages 46-53 from "Honey plants - basis of apiculture: International Symposium on Melliferous Flora, Budapest, September 1976", Bucharest, Romania : Apimondia Publishing House Bd, AA1227/78

Ham/73 HAMEED, S.F.; ADLAKHA, R.L. (1973) Preliminary comparative studies on the brood rearing activities of Apis mellifera L. and Apis cerana indica F. in Kulu Valley. Indian Bee J. 35(1/4) : 27-35 Bb, AA479/77

Han/80 HANSSON, A. (1980) Bin och biodling. Stockholm, Sweden : LTs Förlag 586 pages In Swedish Bd, AA839/81

Hao/75 HAMMER, B. (1975) Risiko ved drivfodring? Tidsskr. Biavl 109(10) : 168-169 In Danish Bj, AA1154/76

Har/75 HARGROVE, R.G. (1975) How many bees in Bermuda? Glean. Bee Cult. 103(1) : 7-8, 27 Bj, AA301L/75

Has/55 HASEMAN, L. (1955) Bee pasturage in Missouri. Glean. Bee Cult. 83 : 212-213 Bj, see AA70/56

Hat/81 HARTOP, B.V. (1981) Personal communication

Hay/81 HAYES, B. (1981) The horse chestnut has nectar potential. Am. Bee J. 121(5) : 320, 327 Bj

Haz/55 HAZSLINSZKY, B. (1955) A gesztenye méhészeti jelentösége. Méhészet 3(6) : 109-110 In Hungarian Bj, AA171/57

Hel/53 HELVEY, T.C. (1953) Colloidal constituents of dark buckwheat honey. Fd Res. 18(2) : 197-205 Bb, AA35/54

Hen/77 HENN, G.D.; NELSON, D.L. (1977) An analysis of Beaverlodge nectar flow records. Publ. Beaverlodge Res. Stn No. WRG 77-6 : 8 pages Bc, AA164/79

Her/83   HERRMANN, J.M. (1983)   Fruit flies and the eucalypt nectar flow.
    S. Afr. Bee J. 55(2) : 25, 27-33   Bj
Heu/71   HEUSSER, C.J. (1971)   Pollen and spores of Chile. Modern types of
    the Pteridophyta, Gymnospermae, and Angiospermae.   Tucson, AZ, USA :
    University of Arizona Press   167 pages   Bd, AA923/71
Hic/69   HICHERI, K.; BOUDERBALA, M. (1969)   Tunisian apiculture.   Proc.
    XXII int. Beekeep. Congr. : 440-443   Bd
Hil/68   HILLYARD, T.N.; MARKHAM, J. (1968)   A survey of beekeeping in
    Ireland.   Dublin : An Foras Taluntais, Horticulture and Forestry Divi-
    sion   49 pages   Bd, AA309L/70
Hir/51   HIRSCHFELDER, H. (1951)   Schwer schleuderbare sogenannte "Lärchen-
    honige".   Imkerfreund 6(10) : 337-339   In German   Bj, AA48/53
Hit/76   HITCHCOCK, L.E. (1976, 1979)   Personal communications
Hla/81   HLAING, C. (1981)   Development of modern beekeeping in Burma.
    Proc. XXVIII int. Beekeep. Congr. : 166-169   Bd
Hod/74   HODGES, D. (1974)   The pollen loads of the honeybe: a guide to
    their identification by colour and form.   London : Bee Research Associa-
    tion   114 pages (including 42 plates)   Bd, AA584/76
Hof/72   HOFFMAN, C. (1972)   Practical ideas for Afghan beekeepers.
    Kabul, Afghanistan : AID   106 pages   Bd, AA563L/73
Hol/65   HOLMES, F.O. (1965)   An apiary in India.   Glean. Bee Cult. 93(2)
    : 86-87   Bj
Hol/70   HOLMES, F.O. (1970)   Nectar sources to avoid.   Glean. Bee Cult.
    98(2) : 117, 120   Bj, AA1039L/71
Hol/75   HOLMES, F.O. (1975)   Nectar sources of New Hampshire.   Am. Bee J.
    115(10) : 395-397   Bj, AA1081L/76
Hoo/63   HOOPEN, H.J.G. TEN (1963)   Flüchtige Carbonylverbindungen in
    Honig.   Z. Lebensmittelunters. u. -Forsch. 119(6) : 478-482   In German
                                                          Ba, AA195/64
Hor/81°° HORA, B. (CONSULTANT EDITOR) (1981)   The Oxford encyclopedia of
    trees of the world.   Oxford, UK : Oxford University Press   288 pages
                                                          Bd, AA921L/81
How/79   HOWES, F.N. (1979)   Plants and beekeeping.   London, UK : Faber &
    Faber   236 pages   As 1945 ed., with minor amendments   Bd, AA202/80
Ich/81   ICHIKAWA, M. (1981)   Ecological and sociological importance of
    honey to the Mbuti net hunters, Eastern Zaire.   Afr. Study Monogr. 1 :
    55-68   Bc, AA1311/82
Ili/80   ILIE, B.; CÎRNU, I.; FOTA, G.; GROSU, E. (1980)   Donnés concer-
    nant la valeur mellifère et le degré d'attractivité envers les abeilles
    chez les variétés et hybrides de tournesol (Helianthus annuus L.) de
    Roumaine.   Proc. 9 int. Sunflower Conf. : 418-423   In French   Ba
Ind/73   INDIA, INDIAN STANDARDS INSTITUTION (1973)   Code for conservation
    and maintenance of honeybees.   Indian Standard No. IS:6695-1972 : 28
    pages   Ba, AA122/77
Ind/77   INDIA, INDIAN STANDARDS INSTITUTION (1977)   Specification for
    Carvia callosa honey.   Indian Standard No. IS:8512-1977 : 6 pages
                                                          Ba, AA674L/81
Ine/82   INTERNATIONAL AGRICULTURAL DEVELOPMENT (1982)   Rekindling of
    interest in neem tree insecticide.   Int. Agric. Dev. 2(8) : 14   Bj
Ini/73   INDIAN BEE JOURNAL (1973)   Mysore migrates bee colonies.   Indian
    Bee J. 35(1/4) : 71   Bj
Ini/81   INDIAN BEE JOURNAL (1981)   Mahua flowers.   Indian Bee J. 43(4) :
    137   Bj

Ino/77 INDONESIA (1977) Liste der Trachtpflanzen Indonesiens. Personal
  communication via H. Duisberg In German   Bs
Int/65 INTERNATIONAL BEEKEEPING CONGRESS (1965) Beekeeping in Romania.
  Proc. XX int. Beekeep. Congr. : 58-60   Bd
Int/75 INTERNATIONAL BEEKEEPING CONGRESS (1975) Characteristics of some
  French monofloral and honeydew honeys. Proc. XXV int. Beekeep. Congr. :
  505-512   Bd
Inu/52 INOUL, T. [INOUE, T.] (1952) Japanese beekeeping today. Bee Wld
  33 : 10   Bj
Inu/57 INOVE, T. [INOUE, T.] (1957) Important honey plants in Japan.
  Glean. Bee Cult. 85 : 726-730   Bj, AA158/58
Iva/77 IVANOV, TS. (1977) [Changes in acacia [Robinia] honey composition
  during storage.] Zhivot. Nauki 14(2) : 133-138   In Bulgarian; English,
  Russian summaries   Ba, AA1068/79
Iva/78 IVANOV, TS. (1978) [Composition and properties of Bulgarian
  honey.] Dissertation, Beekeeping Experimental Station, Sofia, Bulgaria
  : 34 pages. In Bulgarian; English, Russian summaries IBRA translation E1547 (tables only)   Ba, AA1327/78
Ivn/50 IVANOVA-PAROISKAYA, M.I. (1950) [Honey-yielding cotton in Uzbe-
  kistan.] Pchelovodstvo (9) : 479-483   In Russian   IBRA translation
  E301   Bj
Jab/68 JABŁOŃSKI, B. (1968) Wydajność miodowa ważniejszych roślin
  miododajnych w warunkach Polski. Część IV. Pszczel. Zesz. nauk. 12(3) :
  117-126 In Polish; German, Russian summaries   Bj, AA404/72
Jac/77 JACHIMOWICZ, T. (1977) Plantations for soil protection and
  improvement of the honeyflow. Pages 29-30 from "Honey plants - basis of
  apiculture: International Symposium on Melliferous Flora, Budapest,
  September 1976", Bucharest, Romania : Apimondia Publishing House   Bd
Jam/58 JAMIESON, C.A. (1958) Facts about beekeeping in Canada. Bee Wld
  39(9) : 232-236   Ba, AA324/59
Jar/60 JAROSZYŃSKA, T. (1960) Biologia kwitnienia i nektarowania klonu
  tatarskiego Acer tataricum L. Pszczel. Zesz. nauk. 4(3/4) : 153-165
  In Polish; German, Russian summaries   Bj, AA273/62
Jas/52 JASIM, K. (1952) Beekeeping in Iraq. Mod. Beekeep. 36(12) :
  405-407, 414   Bj, AA185/53
Jav/38 JAVET, G. (1938) Conversando con los principiantes: junio.
  Apicultura chil. (1) : 20-21   In Spanish   Bj
Jay/54 JAYCOX, E.R. (1954) Beekeeping regions of California. Glean.
  Bee Cult. 82 : 18-19   Bj, see AA70/56
Jay/70 JAYCOX, E.R. (1970) Ecological relationships between honey bees
  and soybeans. Am. Bee J. 110(8) : 306-307, (9) : 343-345, (10) : 383-
  385   Bc, AA712/72
Joh/73 JOHANNSMEIER, M.F. (1973) Beeplants of the northern Cape and
  southern half of South West Africa. Unpublished report : 6 pages   Bs
Joh/75 JOHANNSMEIER, M.F. (1975) Commercial beekeeping in the Transvaal
  in relation to honeyplants. Proc. 1 Congr. ent. Soc. S. Afr. : 181-184
                                                           Bb, AA204L/77
Joh/75a JOHANNSMEIER, M.F. (1975a) Personal communication
Joh/75b JOHANNSMEIER, M.F. (1975b) Personal communication
Joh/76 JOHANNSMEIER, M.F. (1976) Personal communication
Joh/76a JOHANNSMEIER, M.F. (1976a) Aloe marlothii in South Africa. S.
  Afr. Bee J. 48(2) : 10-14   Bj, AA1237/77
Joh/77 JOHANNSMEIER, M.F. (1977) Can nectar flows from saligna gum be
  predicted? S. Afr. Bee J. 49(4) : 9, 11-15   Bj, AA1253L/78

Jon/52 JOHNSEN, P. (1952) Vanskabte bier og hestekastanieforgiftning.
Nord. Bitidskr. 4(2) : 44-47 In Danish   Ba, AA104/57
Jon/54 JOHNSEN, P. (1954) Facts about beekeeping in Denmark. Bee Wld
35(6) : 105-110   Bj, AA201/55
Jos/59 JOHNSON, D.A. (1959) [Beekeeping in New Guinea.] Bee Wld 40(7)
: 180   Bj
Jua/64 JULA, F.; ILLYÉS, G.; PIRVU, E. (1964) Cercetări privind
cantitatea și calitatea nectarului la unele plante melifere din jurul
orașului Cluj. Lucr. științ. Inst. agron. Cluj 19 : 43-51 In Romanian;
English, Russian summaries   Ba, AA939/7C
Jul/70 JULIANO, J.C. (1970) Contribuição ao conhecimento da flora
apícola do Rio Grande do Sul. 1 Congr. bras. Apic. : 73-79 In Portu-
guese; English summary   Bd, AA622/74
Jul/72 JULIANO, J.C. (1972) Identificação de espécies de interesse
apícola da flora do Rio Grande do Sul. 2 Congr. bras. Apic. : 85-118
In Portuguese   Bd, AA136/75
Jur/65 JURUKOV, D. (1965) Beekeeping and its organization in Bulgaria.
Proc. XX int. Beekeep. Congr. : 664-667   Bd
Kai/64 KALIMI, M.Y.; SOHONIE, K. (1964) Mahabaleshwar honey. I. Proxi-
mate and mineral analysis and paper chromatographic detection of amino
acids and sugars. J. Nutr. Dietet. 1(4) : 261, 264
                                           Ba(abstract), AA753/7C
Kai/65 KALIMI, M.Y.; SOHONIE, K. (1965) Mahabaleshwar honey. III.
Vitamin contents (ascorbic acid, thiamine, riboflavin, and niacin) and
effect of storage on these vitamins. J. Nutr. Dietet. 2(1) : 9-11
                                           Ba(abstract), AA755/7C
Kal/77 KALMAN, C. (1977) The factors influencing the sucrose content of
honeys. Proc. XXVI int. Beekeep. Congr. : 221-222   Bd
Kan/40 KANNANGARA, A.W. (1940) Some bee plants of Ceylon. Bee Wld 21 :
94-96   Bj
Kap/57 KAPIL, R.P. (1957) The length of life and the brood-rearing cycle
of the Indian bee. Bee Wld 38(10) : 258-263   Ba, AA70/59
Kar/56 KARL-HEINZ, F. (1956) Die Bienenzucht in Chile. Öst. Imker
6(10) : 229-233 In German   Bj, AA343/57
Kar/60 KARL-HEINZ, F. (1960) Einiges über südchilenische Verhältnisse
und Betriebsmethoden. Bienenvater 81(2) : 47-49 In German Bj, AA162/61
Kat/68 KATZENELSON, M. (1968) Argentine beekeeping. Apiacta 3(1) :
6-8   Bj
Kat/77 KATZENELSON, M. (1977) Efforts to instal stock centers of Apis
mellifera in isolated valleys in Patagonia. Proc. XXVI int. Beekeep.
Congr. : 302-305   Bd, AA1170/78
Kau/76 KAUFFELD, N.M.; KNUTSON, H. (1976) Bee culture. Bull. Kans.
agric. Exp. Stn No. 357 : 62 pages   Bd, AA863L/77
Kay/78 KÄPYLÄ, M. (1978) Amount and type of nectar sugar in some wild
flowers in Finland. Ann. bot. fenn. 15 : 85-88   Bb, AA1354/79
Kay/79 KÄPYLÄ, M.; NIEMELÄ, P. (1979) Flowers visited by honey bee in
southern Finland. Maataloust. Alkakausk. 51 : 17-24   Ba, AA525/8C
Kee/77 KERESZTESI, B. (1977) Acacia (Robinia pseudacacia L.) basic
source of marketable honey production. Pages 35-42 from "Honey plants -
basis of apiculture: International Symposium on Melliferous Flora,
Budapest, September 1976", Bucharest, Romania : Apimondia Publishing House
                                                         Bd, AA1205L/78

Kee/77a KERESZTESI, B. (1977a)   Robinia pseudoacacia: the basis of com-
mercial honey production in Hungary.   Bee Wld 58(4) : 144-150
                                                              Bj, AA155/79
Kee/83°° KERESZTESI, B. (1983)   Breeding and cultivation of black locust,
Robinia pseudoacacia, in Hungary.   For. Ecol. Mgmt 6 : 217-244   Ba
Kee/83a KERESZTESI, B. (1983a)   Black locust forests: the basis of the
Hungarian beekeeping.   Proc. XXIX int. Beekeep. Congr. : 20 (typed) pages
                                                              Bc
Kel/68   KELLOGG, C.R. (1968)   Entomological excerpts from southeastern
China (Fukien Province). Aborigines: silkworms, honeybees and other
insects.   Claremont, CA, USA : published by the author   89 pages
                                                              Bd, AA476/78
Kem/71   KEMPFF MERCADO, N. (1971)   Potencial melífero de Amazonia boli-
viana.   Gac. Colmen. 33(10) : 304, 306-308   In Spanish   Bj, AA159L/74
Kem/80   KEMPFF MERCADO, N. (1980)   Flora apícola subtropical de Bolivia.
Santa Cruz, Bolivia : Departmento de Publicaciones de la UBGRM   ii + 96
pages   In Spanish   Bd, AA1351/81
Ken/76   KENT, R.B. (1976)   Beekeeping regions and the beekeeping industry
in Colombia.   Bee Wld 57(4) : 151-158   Bj, AA865/77
Ker/57   KERR, W.E.;   AMARAL, E. (1957)   Factores para o aumento da pro-
dução de mel no Estado de S. Paulo.   Solo 49(1) : 61-69   In Portuguese
                                                              Ba, AA228/62
Ker/60   KERR, W.E.;   AMARAL, E. (1960)   Apicultura científica e prática.
São Paulo, Brazil : Secretaria da Agricultura do Estado de São Paulo
148 pages   In Portuguese   Bd, AA227/62
Ket/76   KERKVLIET, J.D. (1976)   Enzymarme honingsoorten.   Warenchemicus 6
: 184-186   In Dutch   Bc, AA640/77
Ket/81   KERKVLIET, J.D. (1981)   Analysis of a toxic rhododendron honey.
J. apic. Res. 20(4) : 249-253   Bj, AA623/82
Kev/82   KEVAN, P.G. (1982)   Apiculture development potential in Malaysia.
Unpublished report : 3 pages   Bs
Key/77°° KELLY, S.;   CHIPPENDALE, G.M.;   JOHNSTON, R.D. (1977)   Eucalypts.
Volume 1.   Melbourne, Australia : Thomas Nelson   xiv + 82 pages
                                                              Bd, AA899/78
Kha/59   KHADI & VILLAGE INDUSTRIES COMMISSION (1959)   Bee flora of Karna-
tak and Kerala.   Indian Bee J. 21(7/8) : 90-92   Bj, AA85/62
Khn/48   KHAN, M.S.A. (1948)   Some important nectariferous plants and
pollen sources of Bhopal State, Central India.   Indian Bee J. 10(11/12)
: 107-108   Bj
Kia/54   KIAT, L.C. (1954)   Beekeeping in Singapore.   Glean. Bee Cult.
82(11) : 649-650, 657   Bj
Kil/80   KILON, E. (1980)   Beekeeping technical report for 1978-1980.
Unpublished report : 18 pages   Bs
Kin/79   KIANG, Y.T. (1979)   Beekeeping in Taiwan.   Am. Bee J. 119(5) :
363, 366-367   Bj
Kir/60   KIRKWOOD, K.C.;   MITCHELL, T.J.;   SMITH, D. (1960)   An examination
of the occurrence of honeydew in honey.   Analyst 85(1011) : 412-416
                                                              Ba, AA316/62
Kir/61   KIRKWOOD, K.C.;   MITCHELL, T.J.;   ROSS, I.C. (1961)   An examination of the occurrence of honeydew in honey. Part II.   Analyst 86(1020)
: 164-165 + 1 table   Ba, AA390/65
Klo/63   KLOFT, W. (1963)   Problems of practical importance in honeydew
research.   Bee Wld 44(1) : 13-18, 24-29   Ba, AA822/63

Klo/65   KLOFT, W.; MAURIZIO, A.; KAESER, W. (1965)   Das Waldhonigbuch.
   Munich, German Federal Republic : Ehrenwirth Verlag   218 pages + 4
   plates   In German   IBRA translation E1140 (extracts from pages 159-180)
                                                            Bd, AA114/67
Koc/65   KOCH, H.G. (1965)   Regional distribution of nectar flow on the
   whole territory of the German Democratic Republic based on studies of the
   scale hives.   Proc. XX int. Beekeep. Congr. : 374-379   Bd
Koc/74   KOCH, H.G.; VILJANEN, T. (1974)   Climate and honey flow in
   Finland.   Apiacta 9(3) : 99-101   Bj
Koc/77   KOCH, H.G. (1977)   A remarkable movement of the periods with a
   summer flow in these past 13 years in Central Europe.   Pages 70-76 from
   "Honey plants - basis of apiculture: International Symposium on Melli-
   ferous Flora, Budapest, September 1976", Bucharest, Romania : Apimondia
   Publishing House   Bd, AA1246L/78
Koh/58   KOHLI, N. (1958/59)   Bee flora of Northern India.   Indian Bee J.
   20 : 113-118, 132-134, 150-151, 178-179, 192-193;   21 : 7-8, 31-32, 61-
   62, 83-85, 106-107, 127-128   Bj, AA84/62
Koi/74   KOIVULEHTO, K. (1974)   Beekeeping in Finland.   Apiacta 9(3) : 141-
   144   Bj
Kok/78   KOCH, C.K.; CAMPOS, S., L. (1978)   Biocenosis del tamarugo
   (Prosopis tamarugo Philippi) con especial referencia de los artrópodos
   fitófagos y sus enemigos naturales.   Z. angew Ent. 85(1) : 86-108   In
   Spanish   Ba, AA1065/80
Kol/72   KOLB, H. (1972)   Herman Kolb in Guatemala.   Am. Bee J. 112(10) :
   371   Bj
Kon/65   KONSTANTINOVIĆ, B. (1965)   Biological and economic justification
   of some measures connected with the queen during the main honey flow.
   Proc. XX int. Beekeep. Congr. : 679-680   Bd
Kon/77   KONSTANTINOVIĆ, B. (1977)   Economic effects of application of
   Farrar's method under the conditions of Serbia (Yugoslavia).   Proc. XXVI
   int. Beekeep. Congr. : 525-526   Bd
Kon/81   KONSTANTINOVIĆ, B. (1981)   Comparative investigations of two-queen
   method in beekeeping under the conditions in Yugoslavia.   Proc. XXVIII
   int. Beekeep. Congr. : 177-178   Bd
Kos/77   KOCSIS, S. (1977)   Hungarian apiculture and its prospects.   Pages
   30-34 from "Honey plants - basis of apiculture: International Symposium
   on Melliferous Flora, Budapest, September 1976", Bucharest, Romania :
   Apimondia Publishing House   Bd, AA1139L/78
Kot/38   KOTESKY, A. (1938)   El ano apicola 1937-1938 en la provincia de
   Malleco.   Apicultura chil. (2) : 51-52   In Spanish   Bj
Kot/38a KOTESKY, A. (1938a)   Agosto en la provincia de Malleco.   Apicul-
   tura chil. (3) : 90-91   In Spanish   Bj
Kot/38b KOTESKY, A. (1938b)   Octubre en la provincia de Malleco.   Apicul-
   tura chil. (5) : 151-153   In Spanish   Bj
Kov/65   KOVALEV, A.M. (1965)   Orientations in bee production in the
   principal areas of USSR.   Proc. XX int. Beekeep. Congr. : 614-617   Bd
Kri/70   KRISHNASWAMY, S.V. (1970)   Soapnut trees: a nectar source.
   Indian Bee J. 32(3/4) : 83   Bj
Kri/80   KRISHNASWAMY, S.V. (1980)   Introduction and propagation of bee
   plants in Chikmagalur District (Karnataka).   Indian Bee J. 42(3) : 72-73
                                                            Bj, AA878L/82
Kud/81   KUDAGAMMANA, S M P. [1981?]   Bee keeping in Sri Lanka.   Unpub-
   lished report : 13 pages + 6 figures   Bc

Kul/59   KULINČEVIĆ, J. (1959)   Facts about beekeeping in Yugoslavia.   Bee
    Wld 40(10) : 241-250   Bj, AA258/60
Kum/74   KULLMANN, E. (1974)   Freie Aminosäuren, qualitativ bestimmt in
    einigen Blüten-, Wald- und Mischhonigen.   Apidologie 5(1) : 21-38   In
    German; English, French summaries   Bj, AA257/76
Kwe/78°° KWEI, T.; ESMONDS, T. (1978) Landscape plants in the United Arab
    Emirates.   New York, USA : Teresa Kwei   133 pages   Bd
Lan/66   LANGRIDGE, D.F. (1966)   An investigation into some quality aspects
    of Victorian honey.   J. Agric. Vict. Dep. Agric. 64 : 81-90, 119-126, 139
                                                                  Bb, AA573/67
Lar/72   LARSON, C.J.F. (1972)   Apis mellifera discovers Pacific paradise.
    Am. Bee J. 112(10) : 378-379   Bj, AA364L/74
Las/72   LASZLO, S. (1972)   Undreamed success in Africa.   Glean. Bee Cult.
    100(11) : 332, 345   Bj, AA805L/73
Lat/54   LATIQUE, P. (1954)   La "Mere du Miel", abeille de Madagascar,
    fournit 300 tonnes de cire par an.   Climats, France et Outremer : 1
    page   In French   Ba, AA155/57
Lau/73   LAURENCE, G.A. (1973)   Some bee plants of Trinidad.   J. agric.
    Soc. Trin. 73(1) : 100-101   Ba
Lau/76   LAURENCE, G.A.; MOHAMMED, I. (1976)   Beekeeping in Trinidad and
    Tobago.   J. agric. Soc. Trin. 76(4) : 342-354   Ba, AA434/78
Lau/76a   LAURENCE, G.A. (1976a)   Common bee weeds of Trinidad and Tobago.
    J. agric. Soc. Trin. 76(June) : 16 pages   Ba, AA1238L/77
Lav/71   LAVIE, P. (1971)   L'apiculture en Algérie.   Rapport technique No.
    4, Projet Algérie 30 : 40 pages + 30 pages appendices   In French   Bd
Lav/76   LAVIE, P. (1976)   Le romarin, Rosmarinus officinalis L.   Bull.
    tech. apic. 3(1) : 15-26   In French   Bj, AA516/78
Led/66   LEDENT, G. (1966)   Coup d'oeil sur l'apiculture libanaise.   Belg.
    apic. 30(1/2) : 10-13   In French   Bj, AA246L/66
Lei/72   LEIGH, J.H. (1972)   Honey and beeswax production in semi-arid and
    arid Australia.   Pages 264-283 from "The use of trees and shrubs in the
    dry country of Australia" by N. Hall et al., Canberra, Australia :
    Australian Government Publishing Service for the Department of National
    Development   Bd, AA900/78
Les/54   LESHER, C. (1954)   Beekeeping regions of North Dakota.   Glean.
    Bee Cult. 82 : 408-409   Bj, see AA70/56
Lid/81   LINDEN, J.O. VAN DER (1981)   Soybean honey production in Iowa.
    Am. Bee J. 121(10) : 723-725, 731   Bj, AA1361/82
Lie/72   LIEUX, M.H. (1972)   A melissopalynological study of 54 Louisiana
    (U.S.A.) honeys.   Rev. Palaeobot. Palynol. 13 : 95-124   Ba, AA474/74
Lin/77   LIN, Y.-C.; SHEU, S.-Y.; KONG, H.-H. (1977)   Physicochemical
    characteristics of certain Taiwan honeys.   Formosan Sci. 31(1) : 34-39
    Chinese summary   Ba, AA1328/78
Lit/54   LITTLE, L.H. (1954)   Tennessee honey producing areas.   Glean. Bee
    Cult. 82 : 535   Bj, see AA70/56
Liv/81   LIVINGSTON, H. (1981)   W.A.S. 1981 spring quarterly report:
    Alaska.   West. apic. Soc. J. 4(3) : 135-136   Bj
Liv/84   LIVINGSTON, H. (1984)   Personal communication
Liz/76   LITZENBERGER, S.C. (EDITOR) (1976)   Guide for field crops in the
    tropics and the subtropics.   Washington, DC, USA : Office of Agriculture
    Technical Assistance Bureau, Agency for International Development; Peace
    Corps Program & Training J., Reprint Ser. No. 10 : vii + 321 pages   Bd

Lom/77   LOMBARD, A.; BELLIARDO, F.; VIDANO, C.; PATETTA, A. (1977)
   Comparative studies on the trisaccharide fraction of Tilia cordata
   leaves, Eucallypterus tiliae honeydew and honeydew honey.   Pages 163-171
   from "Honey plants - basis of apiculture: International Symposium on
   Melliferous Flora, Budapest, September 1976", Bucharest, Romania :
   Apimondia Publishing House   Bd, AA1247/78
Loo/70   LOOCK, E.E.M. (1970)   Eucalyptus species suitable for the produc-
   tion of honey.   Bull. Dep. For., Pretoria No. 46 : 8 pages Bc, AA920L/71
Loo/83   LOOCK, E.E.M. (1983)   Eucalypts suitable for honey production.
   S. Afr. Bee J. 55(3) : 42-46   Bj
Lor/79   LORD, W.G. (1979)   Beekeeping in the North Carolina coastal
   plain.   Am. Bee J. 119(2) : 112-113   Bj
Lou/66   LOUVEAUX, J. (1966)   Essai de caractérisation des miels de callune
   (Calluna vulgaris Salisb.).   Annls Abeille 9(4) : 351-358   In French;
   English summary   Ba, AA385/67
Lou/77   LOUVEAUX, J. (1977)   Les bruyères et leur miel.   Bull. tech.
   apic. 4(2) : 31-36   Fiche technique, CIDA No. 70 In French Bc, AA972/78
Lou/80   LOUVEAUX, J. (1980)   Les abeilles et leur élevage.   Paris, France
   : Hachette   235 pages   In French   Bd, AA1249/81
Lou/81   LOUVEAUX, J. (1981)   Personal communication
Lov/54   LOVELL, H.B. (1954)   Let's talk about honey plants.   Glean. Bee
   Cult. 82 : 426-427; (a) 479, 505; (b) 606, 638; (c) 677, 701
                                                                Bj, AA137/56
Lov/55   LOVELL, H.B. (1955)   Let's talk about honey plants.   Glean. Bee
   Cult. 83 : 28-29; (a) 101, 125; (b) 228; (c) 296-297; (d) 422-423;
   (e) 750-751   Bj, AA137/56
Lov/56   LOVELL, H.B. (1956)   Honey plants manual.   Medina, OH, USA : A.I.
   Root Co.   64 pages   Bd, AA417/57
Lov/56a-f LOVELL, H.B. (1956a-f)   Let's talk about honey plants.   Glean.
   Bee Cult. 84 : (a) 112-113, 124; (b) 173, 188; (c) 233, 252; (d) 423,
   447; (e) 616-617; (f) 39, 63   Bj, AA416/57
Lov/57   LOVELL, H.B. (1957)   Let's talk about honey plants.   Glean. Bee
   Cult. 85 : 37, 60; (a) 99, 123; (b) 228, 249, 253; (c) 366-367; (d)
   434, 443; (e) 479, 510; (f) 742, 759   Bj, AA121/59
Lov/58   LOVELL, H.B. (1958)   Let's talk about honey plants.   Glean. Bee
   Cult. 86 : 101, 125; (a) 164-165; (b) 678-679; (c) 742-743 Bj, AA232/59
Lov/59   LOVELL, H.B. (1959)   Let's talk about honey plants.   Glean. Bee
   Cult. 87 : 38-39; (a) 166-167; (b) 291-292; (c) 355-356; (d) 419-
   420; (e) 483-484; (f) 611-612   Bj, AA94/61
Lov/60   LOVELL, H.B. (1960)   Let's talk about honey plants.   Glean. Bee
   Cult. 88 : 419-420; (a) 547, 574; (b) 611, 633; (c) 675-676
                                                                Bj, AA370/61
Lov/61   LOVELL, H.B. (1961)   Let's talk about honey plants.   Glean. Bee
   Cult. 89 : 35, 57; (a) 163-164; (b) 289; (c) 355-356; (d) 419-420;
   (e) 547-548; (f) 611, 634   Bj, AA271L/62
Lov/62   LOVELL, H.B. (1962)   Let's talk about honey plants.   Glean. Bee
   Cult. 90 : 35-36; (a) 99-100, 122; (b) 227-228; (c) 291, 317; (d) 483
                                                                Bj, AA148L/68
Lov/63   LOVELL, H.B. (1963)   Let's talk about honey plants.   Glean. Bee
   Cult. 91 : 163, 183; (a) 611-612   Bj, AA149L/68
Lov/64   LOVELL, H.B. (1964)   Let's talk about honey plants.   Glean. Bee
   Cult. 92 : 483-484, 505; (a) 675-697   Bj, AA150L/68
Lov/65   LOVELL, H.B. (1965)   Let's talk about honey plants.   Glean. Bee
   Cult. 93 : 419-420; (a) 483-484   Bj, AA151L/68

Lov/66   LOVELL, H.B. (1966)   Let's talk about honey plants.   Glean. Bee
   Cult. 94 : 35, 54-55; (a) 675-676   Bj
Lov/67   LOVELL, H.B. (1967)   Let's talk about honey plants.   Glean. Bee
   Cult. 95 : 35-36, 58; (a) 291-292   Bj
Lov/77   LOVELL, H.B. (rev. GOLTZ, L.R.) (1977)   Honey plants manual.
   Medina, OH, USA : [A.I. Root Company]   96 pages   Bd
Lun/71   LUNDER, R. (1971)   Håndbok i birøkt.   Oslo, Norway : Fabritius &
   Sønner   374 pages   2nd ed.   In Norwegian   Bd, AA87/73
Lut/63   LUTSCHER, A.M. (1963)   Cincuenta plantas utiles para las abejas y
   otros empleos, en la Provincia de Buenos Aires.   Gac. Colmen. 25 : 194-
   199, 228-231, 233   In Spanish   Bj
Maa/73   MALAN, C.E.; MARLETTO, O.I.O. (1973/74)   Elastomiceti da mieli di
   diverse origini e provenienze.   Annali Accad. Agric. Torino 116 : 1-18
   In Italian; English summary   Ba, AA258/76
Mac/80   MACFARLANE, R.P. (1980)   Bees, bee forage and pollination on the
   Chatham Islands.   N.Z. Beekpr 41(1) : 5-6   Bj, AA1492/81
Mad/81   MA DEH-FENG; HUANG WEN-CHENG (1981)   Apiculture in the new
   China.   Bee Wld 62(4) : 163-166   Bj, AA452L/82
Mae/75   MATHEW, T.J. (1975)   The rubber tree as a source of honey.
   Glean. Bee Cult. 103(6) : 190   Bj, AA1086L/76
Mag/78   MARKGRAF, V.; D'ANTONI, H.L. (1978)   Pollen flora of Argentina.
   Tucson, AZ, USA : University of Arizona Press   x + 208 pages Bd, AA716/80
Mah/78   MAHINDRE, D.B. (1978)   Emherzal forecasts good honey flow.
   Indian Bee J. 40(2) : 39   Bj, AA211L/81
Mai/52   MARIOLOPOULOU, N. (1952)   Some general information about bee-
   keeping in Greece.   Bee Wld 33(7) : 118-119   Bj
Mak/78   MAURIKOS, P.I.; DARATSIANOS, I.N.; KATSOULIS, M.T.; MARKETOS,
   D.G. (1978)   [Studies on Greek honey. 1. Adulteration with invert sugar.
   2. Content of free amino acids.]   Chimika Chronika 7(1) : 33-37   In
   Greek; English summary   Ba, AA608/80
Mal/77   MARLETTO, F.; FERRAZZI, P.; PATETTA, A.; MANINO, A. (1977)
   Caratterizzazione dei mieli.   Ind. Aliment. 16(3) : 111-115   In Italian
                                                           Bc, AA605/80
Mal/79   MARLETTO, F.; MANINO, A.; PATETTA, A.; LOMBARD, A.; BUFFA, M.
   (1979)   Indagini sui carboidrati della "manna del larice" e del relativo
   miele.   Apicolt. mod. 70(2) : 35-41   In Italian; English summary
                                                           Ba, AA913/80
Mao/82   MATHESON, A.G. (1982)   Nectar and pollen sources - summer/autumn/-
   early winter.   Fm Prod. Pract. Minist. Agric. Fish. N.Z. No. 529 : 4 pages
                                                           Bc, AA550L/83
Mao/82a MATHESON, A.G. (1982a)   Nectar and pollen sources - winter/spring/-
   early summer.   Fm Prod. Pract. Minist. Agric. Fish. N.Z. No. 530 : 4 pages
                                                           Bc, AA551L/83
Mar/81   MARCHENAY, P. (1981)   Places of beekeeping interest in France.
   Bee Wld 62(3) : 118-122   Bj, AA449L/82
Mas/81   MARSHALL, A.P. (1981)   Beekeeping in the Marshall Islands.
   Glean. Bee Cult. 109(8) : 430, 433-434   Bj, AA816L/82
Mat/49   MATHIS, M. (1949)   Impressions sur l'apiculture au Maroc.   Gaz.
   apic. (516) : 272-273   In French   Bj, AA33/50
Mau/71   MATSUKA, M.; SHIMIZU, Y. (1971)   [Enzyme activities in domestic
   and imported honey samples.]   Bull. Fac. Agric. Tamagawa Univ. No. 11 :
   73-78   In Japanese; English summary   Ba, AA1054/72

Maz/59   MAURIZIO, A. (1959)   Papierchromatographische Untersuchungen an
    Blütenhonigen und Nektar.   Annls Abeille 2(4) : 291-341   In German;
    French summary   Ba, AA451/61
Maz/64   MAURIZIO, A. (1964)   Das Zuckerbild blütenreiner Sortenhonige.
    Annls Abeille 7(4) : 289-299   In German; English, French summaries
                                                            Bj, AA379/67
Maz/75   MAURIZIO, A.; DUISBERG, H.; EVENIUS, J.; FOCKE, E.; VORWOHL, G.
    (1975)   Der Honig: Herkunft, Gewinnung, Eigenschaften und Untersuchung
    des Honigs.   Stuttgart, German Federal Republic : Verlag Eugen Ulmer
    212 pages   2nd ed.   In German   Bd, AA543/76
Maz/82   MAURIZIO, A.; GRAFL, I. (1982)   Das Trachtpflanzenbuch. Nektar
    und Pollen - die wichtigsten Nahrungsquellen der Honigbiene.   Munich,
    German Federal Republic : Ehrenwirth Verlag   368 pages   3rd ed.   In
    German   Bd, AA1270L/82
Mcc/58   MCCUTCHEON, D.M. (1958)   Beekeeping in Saskatchewan.   Glean. Bee
    Cult. 86(6) : 329-331, 380   Bj, AA357/59
Mcg/59   MCGREGOR, S.E.; ALCORN, S.M.; KURTZ, E.B., JR; BUTLER, G.D., JR
    (1959)   Bee visitors to saguaro flowers.   J. econ. Ent. 52(5) :
    1002-1004   Bb, AA366/60
Mcg/76   MCGREGOR, S.E. (1976)   Insect pollination of cultivated crop
    plants.   Agric. Handb. U.S. Dep. Agric. No. 496 : viii + 411 pages
                                                            Bd, AA1088/77
Mea/76   MEAD, P. (1976)   Beekeeping in the Gambia 9/9/76.   Unpublished
    report : 7 pages   Bs
Med/54   MEAD, R.M. (1954)   Beekeeping areas of Vermont.   Glean. Bee Cult.
    82 : 470-471   Bj, see AA70/56
Mei/71   MELLINGER, M.B. (1971)   Sourwood honey.   Foxfire 5(3) : 142-
    144   Bb
Mel/30   MELLOR, J.E.M. (1930)   A brief note on beekeeping in Cyprus (July
    1929).   Bull. Soc. ent. Egypte 14(2/3) : 65-67 + 5 plates   Bb
Met/66   METCALFE, J.R. (1966)   Report on experiments to assess the extent
    of honey bee poisoning after spraying sugar cane with malathion.   Tech.
    Bull. Sug. Manufact. Ass. No. 4/66 : 28 pages   Bd, AA387/68
Mic/54   MITCHELL, T.J.; DONALD, E.M.; KELSO, J.R.M. (1954)   An examina-
    tion of Scottish heather honey.   Analyst, Lond. 79(940) : 435-442
                                                            Bb, AA132/55
Mic/55   MITCHELL, T.J.; IRVIN, L.; SCOULAR, R.H.M. (1955)   An examina-
    tion of Scottish heather honey. Part II.   Analyst, Lond. 80(953) :
    620-622   Bb, AA281/56
Mis/79   MISHRA, R.C.; DOGRA, G.S.: GUPTA, P.R. (1979)   Apicultural
    activities in Himachal Pradesh.   Indian Bee J. 41(3/4) : 29-31
                                                            Bj, AA152L/81
Mit/49   MITCHENER, A.V. (1949)   Nectar and pollen producing plants in
    Manitoba.   West. Can. Beekpr 12(11) : 3-4   Bj, AA202/51
Miz/78   MIZE, M. (1978)   Biču ganyklu vertinimas Latvijos TSR.   Liet.
    žemdir. moks. tyrimo Inst. Darb. 22 : 125-128   In Lithuanian; English,
    Russian summaries   Bd, AA533/80
Moa/55   MORAL, G. (1955)   Beekeeping in Israel.   Rep. Ia St. Apiar. for
    1954 : 82-88   Bj
Mod/52   MODERN BEEKEEPING (1952)   Beekeeping in Holland.   Mod. Beekeep.
    36(6) : 232-236   Bj
Mof/74   MOFFETT, J.O.; RODNEY, D.R.; SHIPMAN, C.W. (1974)   Consistency
    of honeybee visits to flowering citrus trees.   Am. Bee J. 114(1) : 21-23
                                                            Bj, AA104/75

Mof/81   MOFFETT, J.O.;   STANDIFER, L.N.;   SHIPMAN, C.W. (1981)   Nectar
   flow in a Tucson apiary given minimum management: 1973-1978.   Am. Bee J.
   121(5) : 329-331, 335   Bc, AA176/82
Mog/58   MONTGOMERY, B.E. (1958)   Preliminary studies of the composition of
   some Indiana nectars.   Proc. Indiana Acad. Sci. 68 : 159-163
                                                                 Ba, AA365/63
Moh/82   MOHAMED, M.A.;   AHMED, A.A.;   MAZID, M.M. (1982)   Studies on
   Libyan honies.   J. Fd Qual. 4(3) : 185-201   Ba
Moi/57   MORRISON, W.C. (1957)   Woody honey plants for roadside planting in
   New Jersey.   Circ. New Jers. Dep. Agric. No. 403 : 23 pages   Ba, AA48/59
Moo/65   MOSTOWSKA, I. (1965)   Aminokwasy w nektarach i jednogatunkowych
   miodach.   Zesz. nauk. wyższ. Szk. roln. Olsztyn. 20(3) : 417-432   In
   Polish;   English, Russian summaries   Ba, AA679/66
Mor/56   MORSE, R.A. (1956)   Florida beekeeping.   Bull. Plant Bd Fla No.
   10 : 113 pages   Ba, AA107/59
Mor/58   MORSE, R.A. (1958)   The pollination of birdsfoot trefoil (Lotus
   corniculatus L.) in New York State.   Proc. X int. Congr. Ent. 4 : 951-953
                                                                           Ba
Mor/66   MORSE, R.A. (1966)   The beekeeping potential in the Philippines.
   Glean. Bee Cult. 94(10) : 592-595   Bj, AA54L/67
Mor/68   MORSE, R.A.;   LAIGO, F.M. (1968)   Beekeeping in the Philippines.
   Fm Bull. Univ. Philippines Coll. Agric. No. 27 : 56 pages   Ba, AA462L/69
Mos/81   MOSQUERA QUIJANO, R.;   MARIN QUINTERO, A.;   BAUTISTA ANGEL, J.
   (1981)   Heterosis and increase of beekeeping production in the zone of
   coffee plantations in Valle del Cauca, Colombia.   Proc. XXVIII int.
   Beekeep. Congr. 263-264   Bd
Mot/64   MORTON, J.F. (1964)   Honeybee plants of south Florida.   Proc. Fla
   St. Hort. Soc. 77 : 415-436   Bb, AA921L/71
Mot/78   MORTON, J.F. (1978)   Brazilian pepper - its impact on people,
   animals and the environment.   Econ. Bot. 32(4) : 353-359   Bb, AA1352/81
Mou/72   MOUNTAIN, P. (1972)   Honey production in South Africa.   Am. Bee
   J. 112(11) : 408-410   Bj, AA565L/74
Mou/75   MOUNTAIN, P. (1975)   Yellow box - Eucalyptus melliodora.   S. Afr.
   Bee J. 47(2) : 10-11   Bj
Mue/82   MURRELL, D.C.;   SHUEL, R.W.;   TOMES, D.T. (1982)   Nectar produc-
   tion and floral characteristics in birdsfoot trefoil (Lotus corniculatus
   L.).   Can. J. Pl. Sci. 62 : 361-371   French summary   Bb, AA913/83
Mul/78   MULZAC, H.C. (1978)   Beekeeping for the good of Haiti.   Am. Bee
   J. 118(5) : 360-362   Bj, AA1154L/78
Mul/79   MULZAC, H.C. (1979)   Beekeeping in Belize.   Pages 155-163 from
   "Beekeeping in rural development: unexploited beekeeping potential in the
   tropics with particular reference to the Commonwealth" ed. Commonwealth
   Secretariat and International Bee Research Association, London, UK :
   Commonwealth Secretariat   Bd, AA1212/80
Mun/53   MUNRO, J.A. (1953)   Beekeeping in Bolivia: the land of pleasant
   contrasts.   Glean. Bee Cult. 81(4) : 204-208, 253   Bj
Mun/54   MUNRO, J.A. (1954)   Sidelights on beekeeping in Brazil.   Am. Bee
   J. 94(10) : 380-382   Bj
Mur/76   MURKO, D.;   PAŠIĆ, T.;   RAMIĆ, S. (1976)   Istraživanje sastava
   raznih vrsta pčelinjeg meda.   Hemijska Industrija 30(3) : 113-115   In
   Croat;   English summary   Bc, AA594/78
Naa/70   NARAYANA, N. (1970)   Studies in Indian honeys and bees waxes.
   Poona, India : Central Bee Research Institute   13 pages   Ba, AA420/73

Nai/76   NAIM, M.; PHADKE, K.G. (1976)   Bee flora and seasonal activity of
   Apis cerana indica at Pusa (Bihar).   Indian Bee J. 38(1/4) : 13-19
                                                            Bj, AA126/80
Nak/65   NAIR, P.K.K. (1965)   Pollen grains of Western Himalayan plants.
   London, UK : Asia Publishing House   viii + 102 pages + 1 table + 15
   plates   Bd, AA844/72
Nak/74   NAIR, P.K.K.; SINGH, K.N. (1974)   A study of two honey plants,
   Antigonon leptopus Hook. and Moringa pterigosperma Gaertn.   Indian J.
   Hort. 31(4) : 375-379   Bb, AA211/78
Nap/83°° NAIR, P.K.R.; FERNANCES, E.C.M.; WAMBUGU, P.N. (1983)   Multi-
   purpose leguminous trees and shrubs for agroforestry.   Int. Symp.
   Nitrogen-Fixing Trees for the Tropics, Rio de Janeiro   Bs
Nar/81   NAIR, K.S. (1981)   Beekeepers of Muzaffarpur (Bihar).   Indian Bee
   J. 43(2) : 47-49   Bj
Ndi/74   N'DIAYE, M. (1974)   L'apiculture au Sénégal.   Thèse Docteur
   Vétérinaire d'État, Université de Dakar : 134 pages   In French
                                                            Bd, AA430/78
Nee/78   NEW ZEALAND, MINISTRY OF AGRICULTURE AND FISHERIES (1978)   Papers
   presented at the honeydew seminar, 10 August 1978, Christchurch.
   Christchurch, New Zealand : Ministry of Agriculture and Fisheries   80
   pages   Bd, AA981/79
New/79   NEW SOUTH WALES FORESTRY COMMISSION [1979]   Tree planting for bee
   keeping.   New South Wales, Australia : Forestry Commission   3 pages   Bc
Nez/81   NEW ZEALAND BEEKEEPER (1981)   Potential for New Zealand beekeeper
   in Niue.   N.Z. Beekpr 42(3) : 10   Bj
Nic/55   NICOLAIDIS, N.J. (1955)   Facts about beekeeping in Greece.   Bee
   Wld 36(8) : 141-149   Ba, AA263/56
Nig/68   NIGERIA, MINISTRY OF AGRICULTURE [1968?]   Bee keeping in northern
   Nigeria.   Ext. Bull. Minist. Agric. No. 3 : 10-26   Bb
Nih/83   NIGHTINGALE, J. (1983)   A lifetime's recollections of Kenya tribal
   beekeeping.   London, UK : International Bee Research Association   37
   pages + 1 map   Bd
Nor/52   NORDBERG, M. (1952)   Notes sur l'apiculture en Finlande.   Revue
   fr. Apic. 3(78) : 175-176   In French   Bj, AA110/53
Nsu/77   NSUBUGA-NVULE, E.C.B. (1977)   Bee-keeping annual report ending
   December, 1974.   Annual Report, Apiary Farming Industry Animal Produc-
   tion : 5 pages   Bs
Nye/71   NYE, W.P. (1971)   Nectar and pollen plants of Utah.   Monograph
   Ser. Utah St. Univ. 18(3) : ii + 81 pages   Bd, AA922L/71
Ord/44   ORDETX, G.S. (1944)   Plantas melíferas de Cuba.   Revta Agric.,
   Habana 27(24) : 5-160   In Spanish   IBRA translation E122   Bd, AA27/52
Ord/54   ORDETX, G.S. (1954)   Coral vine a perennial honey source.   Glean.
   Bee Cult. 82 : 656-657   Bj
Ord/56   ORDETX, G.S. (1956)   Beekeeping in Cuba.   Am. Bee J. 96(1) : 27-
   28;   Glean. Bee Cult. 85(3) : 168-171 (1957)   Bj, AA46/58
Ord/63   ORDETX, G.S. (1963)   Flora apícola de Honduras.   Tegucigalpa,
   Honduras : Banco Nacional de Fomento   74 pages   In Spanish Bd, AA334/64
Ord/63a ORDETX, G.S. (1963a)   Informe sobre los recursos apícolas de
   Nicaragua.   Managua, Nicaragua : Instituto de Fomento Nacional   58
   pages   In Spanish   Bd, AA552/65
Ord/64   ORDETX, G.S. (1964)   Estudio apibotanico de la República Domini-
   cana.   Santo Domingo, Dominican Republic : published by the author   102
   pages   In Spanish   Bd, AA553/65

Ord/66   ORDETX, G.S.;   ESPINA PÉREZ, D. (1966)   La apicultura en los
   tropicos.   Mexico D.F. : Bartolome Trucco   412 pages   In Spanish
                                                            Bd, AA53/67
Ord/72   ORDETX, G.S.;   ZOZAYA RUBIO, J.A.;   MILLAN, W.F. (1972)   Estudio
   de la flora apícola nacional.   Chapingo, Mexico : Dirección General de
   Extensión Agrícola   95 pages   In Spanish   Bd, AA804/76
Ord/83, also Ord/82; refer to the book published as Esp/83
Orm/63   ORMSBY, E. (1963)   A calendar of eucalypts.   S. Afr. Bee J. 35(1)
   : 10-11   Bj
Orz/58   ORZHEVSKII, M.D. (1958)   [Honeydew, honeydew honey, bees.]
   Moscow : Sel'khozgiz   84 pages   In Russian   IBRA translation E985
                                                            Bd, AA2/60
Ota/56   OTANES, F.Q. [1956?]   The prospects of modern beekeeping in the
   Philippines.   Publ. Dep. Agric. Philippines Republic : 3 pages   Bc
Oti/79   OTIS, G.W.;   TAYLOR, O.R., JR (1979)   Beekeeping in the Guianas.
   Pages 145-154 from "Beekeeping in rural development: unexploited beekeep-
   ing potential in the tropics with particular reference to the Common-
   wealth", ed. Commonwealth Secretariat and International Bee Research
   Association, London : Commonwealth Secretariat   Bd, AA1217/80
Pae/77   PATETTA, A.;   FERRAZZI, P.;   MANINO, A. (1977)   Caratteristiche
   fisico-chimiche di mieli di Robinia (Robinia pseudo-acacia L.) piemon-
   tesi.   Apicolt. mod. 68(5) : 144-149   In Italian;   English summary
                                                            Bj, AA263/79
Pae/83   PATETTA, A.;   MANINO, A.;   CORRADO, S. (1983)   Osservazioni
   preliminari sull'interesse apistico di afidi produttori di melata.   Atti
   XIII Congr. naz. ital. Ent. : 721-728   In Italian;   English summary
                                                            Bd, AA234/84
Pag/77°° PALGRAVE, K.C. (1977)   Trees of southern Africa.   Cape Town,
   South Africa : C. Struik   959 pages   Bd
Pak/77   PAKISTAN AGRICULTURAL RESEARCH COUNCIL (1977)   Research on honey-
   bee management in Pakistan.   1st Annual Report 1976/77.   Islamabad,
   Pakistan : Pakistan Agricultural Research Council   iii + 58 pages   Bj
Pak/78   PAKISTAN AGRICULTURAL RESEARCH COUNCIL (1978)   Research on honey-
   bee management in Pakistan.   2nd Annual Report 1977/78.   Islamabad,
   Pakistan : Pakistan Agricultural Research Council   iv + 41 pages   Bj
Pal/49   PALMER, S. (1949)   Argentine rape as a honey crop.   West. Can.
   Beekpr 12(7) : 8-10   Bj
Pal/59   PALMER, S. (1959)   A nectar source par excellence.   Glean. Bee
   Cult. 87(8) : 460-461   Bj, AA95/61
Pam/77°° PALMER, V.C. (EDITOR) (1977)   Reforestation in arid lands.   USA :
   Action/Peace Corps/Volunteers in Technical Assistance   x + 248 pages   Bd
Pan/59   PANGAUD, C. (1959)   Dosage de l'acidité d'un miel.   Bull. apic.
   Inf. Docum. scient. tech. 2(2) : 11-14   In French   Bj, AA383/61
Pan/60   PANGAUD, C. (1960)   Revue bibliographique des intoxications de
   l'abeille.   Bull. apic. Inf. Docum. scient. tech. 3(2) : 109-180   In
   French   Bs, AA420/61
Pap/69   PAPADOPOULO, P. (1969)   Bulletin on some Eucalyptus spp. suitable
   for honey and pollen production in Rhodesia.   Salisbury, Rhodesia :
   Department of Conservation & Extension   4 pages   rev.   Ba, AA654L/71
Pap/70   PAPADOPOULO, P. (1970)   Rhodesian indigenous trees and shrubs.
   [Salisbury, Rhodesia] : Department of Conservation & Extension   15 pages
                                                            Bc, AA708L/72

Pap/73   PAPADOPOULO, P. (1973)   Rhodesian cultivated fruit trees, orna-
    mental trees, shrubs, crops and flowers.   [Salisbury, Rhodesia] :
    Department of Conservation and Extension   16 pages   Bc, AA419L/74
Par/77   PARKER, C.H. (1977)   Tasmania's unique honey flora.   Proc. XXVI
    int. Beekeep. Congr. : 529-533   Bd, AA596L/79
Pat/77   PATERSON, P.D. (1977)   Report and evaluation of the progress of
    the beekeeping project of the Christian Rural Service Programme in
    Burundi, with recommendations on the future development.   Unpublished
    report : 20 pages   Bc, AA130L/78
Pat/82   PATERSON, P.D. (1982)   Evaluation report of beekeeping aspects of
    the Salvation Army Agriculture and Rural Technology Outreach Project.
    Unpublished report : 10 pages   Bs
Pay/78   PATTY, G.E. (1978)   The honey industry of Mexico. Situation and
    prospects.   Publ. Foreign agric. Serv. U.S. Dep. Agric. No. FAS M-285 :
    iv + 18 pages   Bc, AA1435/79
Pec/61   PERCIVAL, M.S. (1961)   Types of nectar in Angiosperms.   New
    Phytol. 60 : 235-281   Bb, AA844/64
Pec/65   PERCIVAL, M.S. (1965)   Floral biology.   Oxford, UK : Pergamon
    Press   243 pages   Bd, AA650/65
Peh/76   PECHHACKER, H. (1976)   Zur Vorhersage der Honigtautracht von
    Physokermes hemicryphus Dalm. (Homoptera, Coccidae) auf der Fichte
    (Picea excelsa).   Apidologie 7(3) : 209-236   In German; English,
    French summaries   Ba, AA214/78
Peh/77   PECHHACKER, H. (1977)   Neue Ergebnisse der Honigtauforschung.
    Anz. Schädlingsk. Pflanz. Umweltschutz 50(3) : 45-47   In German;
    English summary   Bc, AA1081/79
Pek/77   PETKOV, V. (1977)   [Investigation of the nectar-producing quali-
    ties of some forage plants.]   Rast. Nauki Sofia 14(3) : 123-133   In
    Bulgarian; English, Russian summaries   Ba, AA214/81
Pek/78   PETKOV, V. (1978)   [Investigation of the nectar-producing quali-
    ties of some oil-containing and medicinal plants.]   Rast. Nauki Sofia
    15(3) : 63-70   In Bulgarian; English, Russian summaries   Ba, AA213/81
Pek/80   PETKOV, V. (1980)   [Study on the melliferous characters of plants
    from various biological groups. Part I.]   Rast. Nauki Sofia 16(1) : 55-
    63   In Bulgarian; English, Russian summaries   Ba
Pek/80a PETKOV, V. (1980a)   [Nectar-yielding properties of plants from
    various botanical groups. Part II.]   Rast. Nauki Sofia 17(8) : 49-57
    In Bulgarian; English, Russian summaries   Ba, AA880/82
Pel/49   PELLETT, F.C. (1949)   Honey plants for roadside and wasteland.
    Rep. Ia St. Apiar. for 1948 : 26-33   Bj
Pel/76   PELLETT, F.C. (1976)   American honey plants.   Hamilton, IL, USA :
    Dadant & Sons   467 pages   Reprint of 1947 ed.   Bd, AA205L/77
Pem/65   PEL'MENEV, V.K. (1965)   Factors influencing nectar secretion of
    the far eastern lime-trees and Phillodendron amurensis.   Proc. XX int.
    Beekeep. Congr. : 345-347   Bd
Pem/77   PEL'MENEV, V.K. (1977)   Anthecology of introduced honey plants.
    Pages 67-69 from "Honey plants - basis of apiculture: International
    Symposium on Melliferous Flora, Budapest, September 1976", Bucharest,
    Romania : Apimondia Publishing House   Bd, AA1228L/78
Pen/61°° PENFOLD, A.R.; WILLIS, J.L. (1961)   The eucalypts.   London, UK :
    Leonard Hill   550 pages   Bd, AA539/62
Pen/70   PETROV, V. (1970)   Mineral constituents of some Australian honeys
    as determined by atomic absorption spectrophotometry.   J. apic. Res.
    9(2) : 95-101   Bj, AA750/71

Peo/71   PETROV, V. (1971)   Qualitative determination of amino acids in some Australian honeys, using paper chromatography.   J. apic. Res. 10(3) : 153-157   Bj, AA799/72

Peo/72   PETROV, V. (1972)   Qualitative and quantitative determinations of sugars by chromatography in Australian honey.   Am. Bee J. 112(1) : 10-12, 15   Bj, AA943/73

Peo/72a   PETROV, V. (1972a)   Quantitative determination of amino acids in some Australian honeys by means of an amino analyser.   Am. Bee J. 112(5) : 171-173, 175   Bb, AA682/73

Peo/74   PETROV, V. (1974)   Quantitative determination of amino acids in some Australian honeys.   J. apic. Res. 13(1) : 61-66   Bb, AA174/75

Peo/74a   PETROV, V. (1974a)   Biological origin of honey.   Apiacta 9(2) : 53-58, 72   Bj, AA369L/75

Per/80   PERSANO, A.L. (1980)   Apicultura practica.   Buenos Aires, Argentina : Editorial Hemisferio Sur   xviii + 300 pages   In Spanish
                                                                       Bd, AA810L/82

Pes/80   PERSANO ODDO, L.; ACCORTI, M.; PIAZZA, M.G. (1980/81)   I mieli monoflora italiani. I. Conducibilita elettrica, ceneri e PK di 8 tipi di miele.   Annali Ist. sper. zool. Agr. 7 : 61-75   In Italian; English summary   Bb

Pet/71   PÉTER, J. (1971)   Florális nektárszekréciós vizsgálatok szántóföldi növényeken.   Mosonmagyaróvári Mezőgazdaságtud. Kar Közl. 14(8) : 5-35   In Hungarian; English, German, Russian summaries   Bb, AA207/77

Pet/72   PÉTER, J. (1972)   Nem gyümölcstermő fák és cserjék florális nektárszekréciójának vizsgálata.   Mosonmagyaróvári Mezőgazdaságtud. Kar Közl. 15(18) : 3-46   In Hungarian; English, German, Russian summaries
                                                                       Ba, AA491/76

Pet/72a   PÉTER, J. (1972a)   A gyümölcsfák mézelési értékelése nektártermelésük alapján.   Mosonmagyaróvári Mezőgazdaságtud. Kar Közl. 15(8) : 5-31   In Hungarian; English, German, Russian summaries   Ba, AA208/77

Pet/77   PÉTER, J. (1977)   Importance of crops, fruit trees and ornamental shrubs for apiculture.   Pages 61-66 from "Honey plants - basis of apiculture: International Symposium on Melliferous Flora, Budapest, September 1976", Bucharest, Romania : Apimondia Publishing House
                                                                       Bd, AA1206L/78

Peu/80   PERUM PERHUTANI (1980)   Proyek perlebahan, Gunung Arca. [Apiary project, Gunung Arca.]   Jakarta, Indonesia : Direski Perum Perhutani 24 pages (folded leaflet)   In Indonesian and English   Ba

Pey/82°°   PERRY, F.; HAY, R. (1982)   Tropical and subtropical plants: how to recognize them.   London, UK : Ward Lock   136 pages

Pha/62   PHADKE, R.P. (1962)   Physico-chemical composition of major unifloral honeys from Mahabaleshwar (Western Ghats).   Indian Bee J. 24(7/9) : 59-65   Bj, AA907/64

Pha/64   PHADKE, R.P. (1964)   Nectar concentration in Carvia callosa Bremk.   Indian Bee J. 26(2) : 22-25   Bb, AA274/66

Pha/65   PHADKE, R.P. (1965)   Nectar concentration in Thelapaepale ixiocephala Bremk.   Indian Bee J. 27(2) : 73-76   Bb, AA163/68

Pha/67   PHADKE, R.P. (1967)   Studies on Indian honeys. 2. Proximate composition and physico-chemical characteristics of unifloral honeys of Mahabaleshwar.   Indian Bee J. 29 : 33-46   Bj, AA479/72

Pha/70   PHADKE, R.P.; NAIR, K.S.; NANDEDKAR, K.U. (1970)   Studies on Indian honeys. IV. Minor constituents.   Indian Bee J. 32(1/2) : 28-35
                                                                       Ba, AA1056/72

Phd/74   PHADKE, K.G.;   NAIM, M. (1974)   Observations on the honeybee
    visitation to the litchi (Nephelium litchi) blossoms at Pusa (Bihar,
    India).   Indian Bee J. 36(1/4) : 9-12   Bj, AA774/7
Phi/51   PHILLIPS, G.W. (1951)   Jamaica beekeeping fifty years after.
    Glean. Bee Cult. 79(11) : 656-659, 700-701   B
Phl/14   PHILLIPS, E.F. (1914)   Porto Rican beekeeping.   Bull. Porto Rico
    agric. Exp. Stn No. 18 : 24 pages   B
Pia/81   PIANA, G.;   RICCIARDELLI D'ALBORE, G.;   ISOLA, A. (1981)   Il
    miele.   Bologna, Italy : Edagricole   x + 59 pages   In Italian
                                                             Bd, AA273/8
Pig/77   PIGGIN, C.M. (1977)   The nutritive value of Echium plantagineum L.
    and Trifolium subterraneum L.   Weed Res. 17 : 361-365   B
Pla/52   PLANCKH, E. (1952)   Beekeeping in Austria.   Mod. Beekeep. 36(2) :
    58-62   B
Pol/65°° POLUNIN, O.;   HUXLEY, A. (1965)   Flowers of the Mediterranean.
    London, UK : Chatto & Windus   257 pages   B
Pol/69°° POLUNIN, O. (1969)   Flowers of Europe. A field guide.   London, UK
    : Oxford University Press   662 pages + 192 coloured plates
                                                             Bd, AA595L/6
Poo/65   POOS, J. (1965)   Apiculture in the Grand Duchy of Luxembourg.
    Proc. XX int. Beekeep. Congr. : 696-697   B
Pop/79   POPESKOVIĆ, D.;   DAKIĆ, M. (1979)   Antibacterial effect of three
    honey types after heat treatment.   Proc. XXVII int. Beekeep. Congr. :
    491-495   Bd, AA278/8
Pop/79a   POPESKOVIĆ, D.;   DIMITRIJEVIĆ, M.;   SAVOVIĆ, M. (1979a)   Lipid
    composition of various honey types.   Proc. XXVII int. Beekeep. Congr. :
    495-498   Bd, AA1434/8
Por/74   PORTUGAL-ARAÚJO, V. DE (1974)   Apiáros e instalações apícolas na
    extensão rural (tecnologia apícola) planalto central de Angola.   Nova
    Lisboa, Angola : Estado de Angola   ix + 110 pages + 17 plates   In
    Portuguese; English, French summaries   Bd, AA404/7
Pos/72   POSZWINSKI, L. (1972)   Spektrograficzne oznaczanie wybranych
    pierwiastków w miodach rzepakowych i wrzosowych pochodzących z okreś-
    lonych rejonów glebowych województwa Warszawskiego.   Pszczel. Zesz.
    nauk. 16 : 173-180   In Polish;   English, Russian summaries Bj, AA840/7
Pou/70   POURTALLIER, J.;   TALIERCIO, Y. (1970)   Les caractéristiques
    physico-chimiques des miels en fonction de leur origine florale. 1.
    Application à un projet de normes pour les grandes variétés de miels.
    Bull. apic. Doc. sci. tech. Inf. 13(1) : 58-68   In French;   English,
    German, Spanish summaries   Bj, AA483/7
Pre/79   PREHN, D. (1979)   Development proposal for improved services to
    beekeepers on Ko Samui.   Unpublished report : 12 pages   B
Pre/80   PREHN, D. (1980)   Report about the beekeeping project on Ko Samui
    island.   Unpublished report : 12 pages   B
Pre/82   PREHN, D. (1982)   Beekeeping project at the University of Khon
    Kaen, North-East Thailand: volunteer's report.   Unpublished report : 6
    pages   B
Pry/50   PRYCE-JONES, J. (1950)   The composition and properties of honey.
    Bee Wld 31(1) : 2-6   Bc, AA219/5
Pry/52   PRYCE-JONES, J. (1952)   "Stringiness" in honey and in sugar syrup
    fed to bees.   Bee Wld 33(9) : 147-150, 154-155   Ba, AA226/5
Pur/68   PURDIE, J.D. (1968)   Honey and pollen flora of South Australia.
    J. Agric. S. Aust. 71 : 207-216;   also Leafl. Dep. Agric. S. Aust. No.
    3870 : 11 pages   Bb, AA942L/7

Pus/68°° PURSEGLOVE, J.W. (1968) Tropical crops. Dicotyledons, Volumes 1 and 2. London, UK : Longmans   xii + 719 pages

R.B/38 R.B. (1938) Características de algunas mieles chilenas. Apicultura chil. (2) : 45-46 In Spanish   Bj

Raa/82 RAMSAMY, M.P. ET AL. (1982) Honey plants in Rodrigues (provisional list, January 1982). Unpublished : 1 page   Bs

Rad/61 RADOEV, L.; BOZHINOV, M. (1961) [A study on the nectar production of cotton and the role of bees in its pollination.] Izv. kompl. sel. Inst., Chirpan 1 : 87-108 In Bulgarian; English, Russian summaries IBRA translation E691   Ba, AA669/64

Rae/80 RAMESH, B. (1980) A visit to the cultivated forest area of Peepalpada, U.P. Indian Bee J. 43(3) : 86   Bj

Rah/40 RAHMAN, K.A.; SINGH, S. (1940) Beekeeping in the Punjab. Publ. Dep. Agric. Punjab : 23 pages   Bb

Rah/41 RAHMAN, K.A.; SINGH, S. (1941) Nectar and pollen plants of the Punjab. Indian Bee J. 4(3/4) : 32-35   Ba

Raj/70 RAJEBHONSALE, M.R.; KAPADNIS, D.G. (1970) Viscosity of honey under different ambient conditions. Indian Bee J. 32(3/4) : 58-61
                                                                  Bj, AA484/72

Ram/37 RAMACHANDRAN, S. (1937) Bee-keeping in South India. Bull. Dep. Agric. Madras No. 37 : 78 pages 2nd ed.   Bb

Rap/69 RAPHAEL, T.D.; CUNNINGHAM, D.G. (1969) Beekeeping in Tasmania. Bull. Tasm. Dep. Agric. No. 42 : 64 pages   Bd, AA612L/70

Ras/78 RASHAD, S.E.; EL-SARRAG, M.S. (1978) Beekeeping in the Sudan. Bee Wld 59(3) : 105-111   Bj, AA899/79

Rav/75 RAVN, V.; HAMMER, B.; BARTELS, H. (1975) En sukkerkemisk og pollenanalytisk undersøgelse af nogle danske honningtyper. Tidsskr. PlAvl 79(1) : 13-36 + 1 table In Danish; English summary
                                                                  Ba, AA1005/77

Raw/80 RAWAT, B.S. (1980) Anand Sinh Mehta: an ideal commercial beekeeper. Indian Bee J. 42(3) : 89-90   Bj

Ray/16 RAYMENT, T. (1916) Money in bees in Australia. Melbourne, Australia : Whitcombe & Tombs Ltd   293 pages   Bd

Raz/80 RAZMADZE, I.P.; REBENOK, V.A. (1980) [Beekeeping in Vietnam.] Pchelovodstvo (9) : 28-29 In Russian   IBRA translation E1563
                                                                  Bj, AA847L/81

Rea/74 REA, J. (1974) Some beekeeping observations in Ethiopia. Bee Wld 55(2) : 61-63   Bj, AA298L/75

Ric/75 RICCIARDELLI D'ALBORE, G.; D'AMBROSIO, M.; PERSANO, L. (1975) Raccolta di polline e di nettare in un aranceto del Lazio da parte delle api. Italia agric. 112(4) : 103-109 In Italian   Bb, AA160/79

Ric/77 RICCIARDELLI D'ALBORE, G. (1977) Secondary Italian honey sources. Pages 180-182 from "Honey plants - basis of apiculture: International Symposium on Melliferous Flora, Budapest, September 1976", Bucharest, Romania : Apimondia Publishing House   Bd, AA1219L/78

Ric/78 RICCIARDELLI D'ALBORE, G.; PERSANO ODDO, L. (1978) Flora apistica Italiana. Florence, Italy : Istituto Sperimentale per la Zoologia Agraria   286 pages In Italian   Bd, AA965/79

Ric/80 RICCIARDELLI D'ALBORE, G.; VORWOHL, G. (1980) Sortenhonige in Mittelmeergebiet: Dokumentation mit Hilfe der mikroskopischen Honiguntersuchung. Riv. Agric. subtrop. trop. 74(1/2) : 89-118   Bb, AA284/82

Ric/83 RICCIARDELLI D'ALBORE, G. (1983) Problemi relativi alla conoscenza della flora apistica nel bacino del Mediterraneo. Riv. Agric. subtrop. trop. 77(1) : 93-121 In Italian   Bb

Rih/64   RIHAR, J. (1964)   Novejša dognanja pri zazimovanju čebel.
   Socialist. Kmet. Gos. (16) : 403-404   In Slovene   Ba, AA571/65
Rih/77   RIHAR, J. (1977)   Examination of a survey made for 49 years in
   Vipava (Slovenia) on the fir tree honeydew.   Pages 188-190 from "Honey
   plants - basis of apiculture: International Symposium on Melliferous
   Flora, Budapest, September 1976", Bucharest, Romania : Apimondia Publi-
   shing House   Bd, AA1250/78
Rin/79   RINDFLEISCH, J.K. (1979)   Beekeeping in Barbados, West Indies.
   Am. Bee J. 119(2) : 131, 135   Bj, AA154L/80
Rob/56   ROBERTS, D. (1956)   Sources and qualities of New Zealand honey.
   N.Z. Jl Agric. 92(3) : 285, 287, 289-290   Bb, AA196/59
Roc/68   ROFF, C.; RHODES, J. (1968)   Colour grades of some Queensland
   honeys.   Qd agric. J. 94(3) : 137-139   Ba, AA599L/68
Rod/59   RODRIGUES YCART, F. (1959)   Beekeeping in Uruguay.   Glean. Bee
   Cult. 87(12) : 711-715, 717-719   Bj, AA117/61
Roe/71   ROBERTS, E. (1971)   A survey of beekeeping in Uganda.   Bee Wld
   52(2) : 57-67   Ba, AA320/73
Rof/75   ROBINSON, F.A.; OERTEL, E. (1975)   Sources of nectar and pollen.
   Pages 283-302 (Chapter IX) from "The hive and the honey bee" ed. Dadant
   & Sons, Hamilton, IL, USA : Dadant & Sons   rev. ed.   Bd, see AA859/77
Roi/81   ROBINSON, W.S. (1981)   Beekeeping in Jordan.   Bee Wld 62(3) :
   91-97   Bj, AA803/82
Roj/39   ROJAS U., R.; RUSSI, O. (1939)   La miel de abejas en Chile.
   Apicultura chil. (12) : 354-358   In Spanish   Bj
Ron/37°° ROBINSON, D.H. (1937)   Leguminous forage plants.   London : Edward
   Arnold & Co.   119 pages   Bd
Roo/74   ROOT, A.I. (rev. ROOT, E.R.; ROOT, H.H.: ROOT, J.A.) (1974)   The
   ABC and XYZ of bee culture ...   Medina, OH, USA : A.I. Root Company
   vi + 712 pages + xiv pages index   35th ed.   Bd
Ros/60   ROSÁRIO NUNES, J.F.; TORDO, G.C. (1960)   Prospecções e ensaios
   experimentais apícolas em Angola.   Lisbon, Portugal : Junta de Investi-
   gações do Ultramar   186 pages   In Portuguese; English, French sum-
   maries   Bd, AA595/62
Row/76   ROWLEY, F.A. (1976)   The sugars of some common Philippine nectars.
   J. apic. Res. 15(1) : 19-22   Bb, AA807/76
Rut/76   RUTTNER, F. (1976)   Apiculture in the past and present in Crete -
   model of a successful development project.   Apiacta 11(4) : 187-191
                                                            Bj, AA1167L/77
Ryc/65   RYCHLIK, M.; FEDOROWSKA, Z. (1965)   Polskie miody gryczane.
   Pszczel. Zesz. nauk. 9(1/2) : 92-100   In Polish; English, Russian
   summaries   Ba, AA574/66
Saa/39   SANTA CRUZ, A. (1939)   Flora melífera de Concepción y Arauco.
   Apicultura chil. (10) : 300-303   In Spanish   Bj
Sab/82   SABOT, J. (1982)   Eucalyptus - un choix restreint.   Revue fr.
   Apic. (410) : 342-343   In French   Bj, AA544L/83
Sac/55   SACCHI, R. (1955)   Mieli umbri e analisi rifrattometrica dell'-
   umidita.   Annali Sper. agr. 10 : 1029-1042   In Italian   Ba, AA259/57
Sad/60   SĂNDULEAC, E.; LĂZĂRESCU, C. (1960)   Lime and black locust as
   important honey sources in the Romanian People's Republic.   Bee Wld
   41(9) : 225-228   Bj, AA369/61
Saf/73   SANFORD, M.T. (1973)   A geography of apiculture in Yucatán,
   Mexico.   University of Georgia, Athens, OA, USA : M.A. Thesis   02
   pages   Bs, AA818/78

Sai/66   SARIAN, Z.B. (1966)   Beekeeping in the Philippines.   Glean. Bee
  Cult. 94(8) : 486-491   Bj, AA55L/67
Sak/82   SAKAI, T.; MATSUKA, M. (1982)   Beekeeping and honey resources in
  Japan.   Bee Wld 63(2) : 63-71   Bj
San/79   SANTAS, L.A.; BIKOS, A.A. (1979)   The apicultural flora of
  Greece.   Apiacta 14(3) : 120-123   Bj, AA1276L/80
San/80   SANTAS, L.A. (1980)   Marchalina hellenica (Genadius): an important
  insect for apiculture of Greece.   Proc. XXVII int. Beekeep. Congr. :
  419-422   Bd, AA1356/81
San/81   SANTAS, L.A. (1981)   Insects useful to apiculture in Greece.
  Proc. XXVIII int. Beekeep. Congr. : 404-407   Bd
Sao/54   SANTOS, C. FERRAZ DE O. (1954)   Contribuição ao conhecimento dos
  nectários de algumas espécies da flora apícola.   Universidade de São
  Paulo, Escola Superior de Agricultura "Luiz de Queiroz", Piracicaba,
  Brazil : Doctorate thesis   66 pages   In Portuguese   Bs, AA418/57
Sao/61   SANTOS, C.F. DE O. (1961)   Morfologia e valor taxonómico do polen
  das principais plantas apícolas.   Universidade de São Paulo, Escola
  Superior de Agricultura "Luiz de Queiroz", Piracicaba : Tese Docente-
  Livre   109 pages   In Portuguese; English summary   Bs, AA807/63
Sar/72   SARAF, S.K. (1972)   Bee flora of Kashmir.   Indian Bee J. 34(1/2)
  : 1-10   Bj, AA155/74
Sar/73   SARAF, S.K. (1973)   Honey flow in Kashmir Valley.   Indian Bee J.
  35(1/4) : 50-51   Bb, AA1090L/76
Sat/75   SATYANARAYANA, I. (1975)   Sugar concentration in nectar from
  Syzygium cumini Skeels.   Indian Bee J. 37(1/4) : 21-24   Bj, AA1223/78
Sau/82   SAURET, J.[S.] (1982)   El eucalipto (Eucalyptus globulus Labill-
  ardière), Eucaliptus camaldulensis Dehn).   Vida apíc. (3) : 13-14   In
  Spanish   Bj
Sau/82a SAURET, J.S. (1982a)   El espliego (Lavandula spica L.), la al-
  hucema (Lavandula latifolia Villars).   Vida apíc. (2) : 13-14   In
  Spanish   Bj
Sau/82b SAURET, J. (1982b)   El biercol (Calluna vulgaris Salisbury). Vida
  apíc. (4) : 13-14   In Spanish   Bj
Sau/82c SAURET, J. (1982c)   El romeo (Rosmarinus officinalis L.)   Vida
  apíc. (1) : 9-10   Bj
Saw/81   SAWYER, R. (1981)   Pollen identification for beekeepers.   Cardiff, UK : University College Cardiff Press   112 pages + 50 punched
  cards   Bd, AA1271/82
Scd/50   SCHNEIDER-ORELLI, O. (1950)   Das Problem des Blatthonigs.   Beih.
  schweiz. Bienenztg 2(19) : 471-484   In German   IBRA translation E101
                                                                Ba, AA107/52
Sce/66   SCHEPARTZ, A.I. (1966)   Honey catalase: occurrence and some
  kinetic properties.   J. apic. Res. 5(3) : 167-176   Bb, AA789/67
Sch/77   SCHEURER, S. (1977)   Honeydew from Pinus silvestris L. and Pinus
  nigra Arn.: production and forecasting.   Pages 172-177 from "Honey
  plants - basis of apiculture: International Symposium on Melliferous
  Flora, Budapest, September 1976", Bucharest, Romania : Apimondia Publishing House   Bd, AA1248L/78
Sch/80   SCHEURER, S. (1980)   Die Verbreitung der auf Picea, Pinus und
  Larix lebenden Cinarinen im Gebiet der Deutschen Demokratischen Republik.   Acta Mus. reginaehradecensis Suppl. : 90-99   In German
                                                                Bb, AA233L/84

Sch/82   SCHEURER, S. (1982)   Die Verbreitung unserer Honigtauerzeuger auf
    Fichte und Lärche.   Garten u. Kleintierz. C (Imker) 21(23) : 8-9   In
    German   Bj, AA231L/84
Sch/82a SCHEURER, S. (1982a)   Die Verbreitung unserer Honigtauerzeuger auf
    Kiefern.   Garten u. Kleintierz. C (Imker) 21(24) : 8-9   In German
                                                           Bj, AA232L/84
Sci/81   SCHILLING, H. (1981)   31 miele des appellations fantaisistes.
    Revue fr. Apic. (395) : 134-138   In French   Bj, AA279/82
Scn/59   SCHNEIDER, H. (1959)   [Honey crop report, Switzerland.]   Bee Wld
    40(8) : 205   Bj
Scu/66   SCHULZ-LANGNER, E. (1966)   Quantitativer Nachweis kleinster
    Saponinmengen durch Beachten der Hämolysedauer. Untersuchungen am Nektar
    der Rosskastanie Aesculus hippocastanum.   Planta med. 14(1) : 49-56
    In German   Ba, AA539/67
Scu/67   SCHULZ-LANGNER, E. (1967)   Über den Trachtwert der Rosskastanie
    (Aesculus hippocastanum) unter besonderer Berücksichtigung des Saponin-
    gehaltes im Nektar.   Z. Bienenforsch. 9(2) : 49-65   In German
                                                           Ba, AA306/69
See/79   SEETHALAKSHMI, T.S.;   PERCY, A.P. (1979)   Borassus flabellifer
    (Palmyrah palm) - a good pollen source.   Indian Bee J. 41(1/2) : 20-21
                                                           Bj, AA944/81
Sel/49   SELWYN, H.H. (1949)   Save the basswoods.   Can. Bee J. 57(11) : 4-5
                                                           Bj, AA58/50
Sha/72   SHAH, F.A. (1972)   False acacia - a promising bee plant of Kashmir.
    Indian Bee J. 34(1/2) : 34-35   Bj
Sha/75   SHAH, F.A. (1975)   Some facts about beekeeping in Kashmir.   Bee
    Wld 56(3) : 103-108   Bj, AA650L/75
Sha/76   SHAH, F.A.;   SHAH, T.A. (1976)   A note on the bee activity and bee
    flora of Kashmir.   Indian Bee J. 38(1/4) : 29-33   Bj, AA128/80
Sha/79   SHAH, F.A. (1979)   Honeys of Kashmir.   Indian Honey 1(2) : 19
                                                           Bj, AA563L/80
She/63   SHERIFF, J.S. (1963)   Beeswax and honey production - the Nyasaland
    potential.   Zomba, Nyasaland : Ministry of Natural Resources & Surveys
    43 pages   Bd, AA69/64
Shh/81   SHAH, T.A. (1981)   "Hazel nut" - an early pollen source in the
    Kashmir Valley, India.   Indian Bee J. 43(3) : 67   Bj, AA547L/83
Shi/77   SHAHID, M.;   QAYYUM, A. (1977)   Bee flora of the N.W.F.P.
    Pakist. J. For. 27(1) : 1-10   Ba, AA900/80
Shl/81   SHLJAKHOV, P. (1981)   Contribution to the investigations of
    chestnut (Castanea sativa Mil.) as bee pasture.   Proc. XXVIII int.
    Beekeep. Congr. : 407   Bd
Shm/77   SHARMA, D.S. (1977)   Personal communication
Shn/69   SHENDE, S.G. (1969)   Bees visiting rubber tree blossoms do not
    affect latex production.   Indian Bee J. 30(1) : 36   Bj, AA129/71
Shp/79   SHAPIRA, D.K.;   SHAMYATKOŬ, M.F.;   ANIKHIMOŬSKAYA, L.V.;   NARYZH-
    NAYA, T.I.;   GARODKA, YA.S. (1979)   [Study of biologically active
    substances in the pollen of some nectar-yielding plants.]   Vestsi Akad.
    Navuk BSSR sel'.-gas. Navuk No. 2 : 59-63   In Belorussian; English
    summary   Bb, AA344/81
Shp/80   SHAPIRA, D.K.;   SHAMYATKOŬ, M.P.;   ANIKHIMOŬSKAYA, L.V.;   NARYZH-
    NAYA, T.I.;   GARADKO, YA. S. (1980)   [Biochemical characterization of
    the pollen (pollute) of honey plants.]   Vestsi Akad. Navuk. BSSR, Khim.
    Navuk No. 4 : 68-73   In Ukrainian; English summary   Bb

Shr/48   SHARMA, P.L. (1948)   Studies on seasonal activities of Apis indica
    F. at Lyallpur.   Indian Bee J. 10(3/4) : 20-23   Bj
Shr/58   SHARMA, P.L. (1958)   Sugar concentration of nectar of some Punjab
    honey plants.   Indian Bee J. 20(7) : 86-91   Bj, AA296/59
Shr/59   SHARMA, P.L. (1959)   Honey crops of the north-western Himalayas.
    Bee Wld 40(7) : 180   Bj
Shw/50   SHAW, F.R. (1950)   Honey and pollen plants of Massachusetts.
    Spec. Circ. Mass. Ext. Serv. No. 27 : 5 pages   rev.   Bc, AA241/53
Shw/50a   SHAW, F.R. (1950a)   Bee keeping.   Ext. Serv. Leafl. Univ. Mass.
    No. 148 : 44 pages   Ba
Shw/53   SHAW, F.R. (1953)   The sugar concentration of the nectar of some
    New England plants.   Glean. Bee Cult. 81(2) : 88-89   Bj, AA157/55
Sig/48   SINGH, S. (1948)   Some important honey plants of the Punjab
    (India).   Rep. Ia St. Apiar. for 1948 : 34-42   Bj
Sig/62   SINGH, S. (1962)   Beekeeping in India.   New Delhi, India : Indian
    Council of Agricultural Research   214 pages   Bd, AA291/64
Sil/69   SILBERRAD, R. (1969)   The remotest colonies.   Br. Bee J. 97(4192)
    : 126-128   Bj, AA863L/70
Sil/70   SILBERRAD, R.E.M. (1970)   Bee-keeping in Seychelles.   Republic of
    Seychelles : Department of Agriculture   ii + 21 pages + 4 plates
                                                          Bd, AA817L/78
Sil/76   SILBERRAD, R. (1976)   African sweetness and fury: beekeeping in
    Zambia.   Ctry Life, Lond. 160(4143) : 1566, 1568   Bc, AA433L/78
Sil/76a   SILBERRAD, R.E.M. (1976a)   Bee-keeping in Zambia.   Bucharest,
    Romania : Apimondia Publishing House   76 pages   Bd, AA218/77
Sim/65   SIMIDCHEV, T. [SIMIDCHIEV, T.] (1965)   The nectar production of
    some agricultural cultures in Bulgaria.   Proc. XX int. Beekeep. Congr.
    : 270-271   Bd
Sim/75   SIMIDCHIEV, T. (1975)   Problems of the honey flow and the study of
    the melliferous properties of some plants.   Proc. XXV int. Beekeep.
    Congr. : 444-449   Bd, AA521/78
Sim/76   SIMIDCHIEV, T. [Studies of nectar and honey production in raspberry
    (Rubus idaeus L.) and blackberry (Rubus fruticosus L.).]   Gradinar.
    lozar. Nauk. 13(2) : 42-49   In Bulgarian; English, Russian summaries
                                                          Ba, AA607/79
Sim/77   SIMIDCHEV, T. [SIMIDCHIEV, T.] (1977)   Pollen and nectar produc-
    tion of sunflower (Helianthus annuus L.).   Pages 125-135 from "Honey
    plants - basis of apiculture: International Symposium on Melliferous
    Flora, Budapest, September 1976", Bucharest, Romania : Apimondia Publish-
    ing House   Bd
Sim/77a   SIMIDCHEV, T. [SIMIDCHIEV, T.] (1977a)   Pollen production of some
    trees.   Pages 85-90 from "Honey plants - basis of apiculture: Interna-
    tional Symposium on Melliferous Flora, Budapest, September 1976",
    Bucharest, Romania : Apimondia Publishing House   Bd, AA1243/78
Sim/80   SIMIDCHIEV, T.K. (1980)   [Nectar and pollen productivity of fruit
    and other plants, and the role of bee pollination.]   Thesis, "Vasil
    Kolarov" Higher Institute of Agriculture, Plovdiv, Bulgaria : 30 pages
    In Bulgarian; English, Russian summaries   Ba, AA152/82
Sin/62 = Sig/62
Sin/71   SINGH, G.; SINGH, G. (1971)   Plectranthus rugosus Wall., the
    major honey plant of Kashmir Valley.   Indian Bee J. 33(3/4) :58-59
                                                          Bj, AA874L/73
Sip/82   SIMPSON, K. (1982)   Bees and plants.   Small Fmr No. 19 : 9, 11, 13
                                                                Bc

Ske/72   SKENDER, K. (1972)   La situation de l'apiculture algérienne et ses possibilités de développement.   Algiers, Algeria : Mémoire d'Ingénio-rat, Institut National Agronomique   102 pages   In French   Bs, AA810/7{

Smi/72   SMIRL, C.B.;   JAY, S.C. (1972)   Manitoba nectar flows: a review of flows from 1924 to 1954 with an analysis of flows from 1955 to 1971.   Manitoba Ent. 6 : 23-26   Bb, AA199/7(

Smt/56   SMITH, F.G. (1956)   Honey.   Pamph. Tanganyika Bee Div. No. 3 : 14 pages   Bb, AA299/5`

Smt/56a SMITH, F.G. (1956a)   Bee botany in Tanganyika.   University of Aberdeen : Thesis   174 pages + photographs   Bd, AA268/5(

Smt/57   SMITH, F.G. (1957)   Bee botany in East Africa.   E. Afr. agric. J. 23(2) : 119-126   Bb, AA95/6(

Smt/59   SMITH, F.G. [1959?]   Beekeeping in Northern Rhodesia: its pros-pects and recommendations for its development.   Report prepared for N. Rhodesian Government : 21 pages   Ba, see AA520/6`

Smt/60   SMITH, F.G. (1960)   Beekeeping in the tropics.   London, UK : Longmans   265 pages   Bd, AA362/6(

Smt/63   SMITH, F.G. (1963)   An introduction to beekeeping in Western Australia.   Bull. Dep. Agric. West. Aust. No. 3108 : 20 pages   Bb, AA547L/6`

Smt/64   SMITH, F.G. (1964)   Some pollen grains in the Caesalpiniaceae of East Africa.   Pollen Spores 6(1) : 85-98   Bb, AA653/6`

Smt/65   SMITH, F.G. (1965)   The sucrose contents of Western Australian honey.   J. apic. Res. 4(3) : 177-184   Bb, AA731/6(

Smt/67   SMITH, F.G. (1967)   Deterioration of the colour of honey.   J. apic. Res. 6(2) : 95-98   Bj, AA399/6{

Smt/69   SMITH, F.G. (1969)   Honey plants in Western Australia.   Bull. Dep. Agric. West. Aust. No. 3618 : 78 pages   Bd, AA143/7`

Smt/83   SMITH, F.G. (1983)   Australian co-operation with the National Agricultural Research Project, Thailand: apicultural research.   Report, Western Australian Department of Agriculture : 64 pages   B{

Sol/63   SOLOV'EVA, T. YA.;   BAZAROVA, V.I. (1963)   [Investigations of sugars of lime-tree and buckwheat honey by paper chromatography.]   Izv. vyssh. ucheb. Zaved. Ser. pishch. Tekhnol. (6) : 139-140   In Russian   IBRA translation E1230   Ba, AA206/7`

Sou/63   SOUTH AFRICAN BEE JOURNAL (1963)   South African eucalypts and their honey flow seasons.   S. Afr. Bee J. 35(2) : 19-20, iii   B`

Sou/65   SOUTH AFRICAN BEE JOURNAL (1965)   The Cape sugar gum among the best of the eucalypts.   S. Afr. Bee J. 37(5) : 15-16   B`

Sou/77   SOUTH AFRICAN BEE JOURNAL (1977)   "Saligna" a misnomer.   S. Afr. Bee J. 49(4) : 19   B`

Spo/50   SPOTTEL, W. (1950)   Honig und Trockenmilch.   Leipzig, German Democratic Republic : J.A. Barth   323 pages   In German   Bd, AA63/5(

Sta/74   STANLEY, R.G.;   LINSKENS, H.F. (1974)   Pollen: biology - bio-chemistry - management.   Berlin : Springer-Verlag   ix + 307 pages   Bd, AA585/7(

Ste/71   STEJSKAL, M. (1971)   Plantas melíferas de Venezuela.   Turrialba 21(1) : 119-120   In Spanish; English summary   Bc, AA1222/7`

Stm/77   STROEMPL, G. (1977)   Distribution and use of basswood and lindens for honey production.   Am. Bee J. 117(5) : 298-301, 322   Bj, AA892/7{

Stn/81   STRANG, L.A.;   DIMICK, P.S. (1981)   Evaluation of the effect of heating on alfalfa honey.   J. apic. Res. 20(2) : 121-124   Bb, AA291/8`

Sto/82°° STORRS, A.E.G. (1982)   More about trees.   Ndola, Zambia : Forest Department   vi + 127 + 28 plates   Bd, AA548L/8`

Stp/54 STEPHEN, W.A. (1954) Beekeeping today in North Carolina. Am.
   Bee J. 94(10) : 382-383   Bj
Str/82 STRETTON, J. (1982) Personal communication
Stv/69 STEVENS, K. (1969) Beekeeping in the Maltese Islands. Glean.
   Bee Cult. 97(2) : 102-108, 119   Bj, AA74L/70
Sub/79 SUBRAMANIAM, K. (1979) The role of Forest Department in develop-
   ing beekeeping in Western Ghat areas of Maharashtra. Indian Bee J.
   41(3/4) : 91-93   Bj, AA928L/81
Suk/81 SUKARTIKO, B. (1981) Country report on beekeeping in Indonesia.
   Proc. XXVIII int. Beekeep. Congr. : 198-202   Bd
Sur/75 SURYANARAYANA, M.C. (1975) Another chapter in the hlalwane
   story. S. Afr. Bee J. 47(1) : 2 pages   Bj
Sur/78 SURYANARAYANA, M.C. (1978) Bee plants of India. 1. Carvia callosa
   (Nees) Brem. Indian Bee J. 40(1) : 7-10   Bj, AA1277L/80
Sve/79 SVENSSON, B. (1979) Beekeeping in some countries of Africa: Tchad.
   Unpublished report : 3 pages   Bs
Sve/80 SVENSSON, B. (1980) Beekeeping in the Republic of Guiné-Bissau
   and the possibilities for its modernization. Unpublished report : 45
   pages including Appendix 1 on Senegal and Appendix 2 on Gambia Bs
Svo/56 SVOBODA, J. (1956) Vlastnosti medu ze snůšky borovicové medovice.
   Včelařství 9(7) : 103   In Czech   Bj, AA332/57
Svo/58 SVOBODA, J. (1958) Facts about beekeeping in Czechoslovakia.
   Bee Wld 39(6) : 137-150   Ba, AA35/62
Swa/76 SWANSON, R.A. (1976) Beekeeping in Upper Volta (North Central and
   West Africa). Am. Bee J. 116 : 56-57, 72, 104-105, 122 Bj, AA1006L/76
Sza/82 SZABO, T.I. (1982) Phacelia tanacetifolia as a honey plant.
   Can. Beekeep. 9(9) : 151   Bj
Szk/73 SZKLANOWSKA, K. (1973) Bory jako baza pożytkowa pszczół.
   Pszczel. Zesz. nauk. 17 : 51-85   In Polish   Bj, AA662/75
Tak/76 TAKEDA, J. (1976) An ecological study of the honey-collecting
   activities of the Tongwe, western Tanzania, East Africa. Kyoto Univ.
   afr. Stud. 10 : 213-247   Bd, AA814/78
Tat/56 TATE, H. (1956) Beekeeping in Mississippi. Glean. Bee Cult. 84
   : 21, 60   Bj, see AA274/57
Tay/56 TAYLOR, P.V. (1956) Melaleuca trees annoy Florida beekeepers.
   Am. Bee J. 96(11) : 449   Bj, AA223/58
Tea/54 TEASLEY, C.M. (1954) Secrets in sourwood getting. Glean. Bee
   Cult. 82(4) : 204-205   Bj
Tem/57 TEMPLER, C.R. (1957) Beekeeping in the Canary Islands. Bee Wld
   38(7) : 184   Bj, AA256/59
Ten/59 TENNENT, J.N. (1959) [Honey crop report, Scotland, 1958.] Bee
   Wld 40(1) : 16   Bj
Tha/62 THAKAR, C.V.; DIWAN, V.V.; SALVI, S.R. (1962) Floral calendar
   of major and minor bee forage plants in Mahabaleshwar Hills (Western
   Ghats). Indian Bee J. 24(4/6) : 35-48   Bj, AA582/63
Tha/76 THAKAR, C.V. (1976) Practical aspects of bee management in India
   with Apis cerana indica. Pages 51-59 from "Apiculture in tropical
   climates" ed. E. Crane, London, UK : International Bee Research Association   Bd, AA1155/77
The/77 THEAN, J.E.; FUNDERBURK, W.C., JR (1977) Sugars and sugar
   products. High pressure liquid chromatographic determination of sucrose
   in honey. J. Ass. off. analyt. Chem. 60(4) : 838-841   Bb, AA267/79

Thi/82   THIMANN R., R.E.; AYMARD, G. (1982)   Flora apícola de la Mesa de
    Cavacas y sus alrededores.   Guanare, Venezuela : Universidad Nacional
    Experimental de los Llanos Occidentales "Ezequiel Zamora"   30 pages
    In Spanish   Bc, AA217/84
Thy/65   THYRI, H. (1965)   Correlation between colony populations and honey
    yields during the Calluna vulgaris honey flow.   Proc. XX int. Beekeep.
    Congr. : 668-669   Bd
Tod/74   TODD, I. (1974)   The Natal wonder plant.   S. Afr. Bee J. 46(5) : 2
                                                              Bj, AA751L/75
Tou/80   TOURN, M.L.; LOMBARD, A.; BELLIARDO, F.; BUFFA, M. (1980)
    Quantitative analysis of carbohydrates and organic acids in honeydew,
    honey and royal jelly by enzymic methods.   J. apic. Res. 19(2) : 144-146
                                                              Bb, AA581/81
Tow/69   TOWNSEND, G.F. (1969)   Beekeeping in East Africa.   Unpublished
    report : 14 pages   Bs
Tow/76   TOWNSEND, G.F.; BURKE, P.W. (1976)   Beekeeping in Ontario.
    Publ. Ont. Minist. Agric. Fd No. 490 : 38 pages   Ba, AA835L/80
Tri/71   TRIPATHI, K.L. (1971)   Bee-keeping in West Bengal.   Indian Bee J.
    33(1/2) : 23-28   Bj, AA318L/73
Tse/54   TSENG, H.N. (1954)   Beekeeping in China.   Glean. Bee Cult. 82(4)
    : 216-217, 249   Bj
Tsu/74   TSUNEYA, T.; SHIBAI, T.; YOSHIOKA, A.; SHIGA, M. (1974)   [Study
    of shina (Tilia japonica) honey flavour.]   Koryo No. 109 : 29-34   In
    Japanese; English summary   Bb, AA1448/79
Tut/64°° TUTIN, T.G. ET AL. (EDITORS) (1964)   Flora Europaea. Volume 1.
    Lycopodiaceae to Platanaceae.   Cambridge, UK : Cambridge University
    Press   xxxii + 464 pages + 5 maps   Bd
Tut/68°° TUTIN, T.G. ET AL. (EDITORS) (1968)   Flora Europaea. Volume 2.
    Rosaceae to Umbelliferae.   London, UK : Cambridge University Press
    xxvii + 455 pages + 5 maps   Bd, AA396/69
Tut/72°° TUTIN, T.G. ET AL. (EDITORS) (1972)   Flora Europaea. Volume 3.
    Diapensiaceae to Myoporaceae.   London, UK : Cambridge University
    Press   xxxi + 370 pages + 5 maps   Bd, AA740/77
Tut/76°° TUTIN, T.G. ET AL. (EDITORS) (1976)   Flora Europaea. Volume 4.
    Plantaginaceae to Compositae (and Rubiaceae).   Cambridge, UK : Cam-
    bridge University Press   xxxi + 505 pages + 5 maps   Bd, AA741/77
Uni/83   UNION NATIONALE DE L'APICULTURE FRANÇAISE (1983)   La fleur et
    l'abeille.   Paris, France : Union Nationale de l'Apiculture Française
    144 pages   In French   Bd
Usa/75°° USA, NATIONAL ACADEMY OF SCIENCES (1975)   Under-exploited tropical
    plants with promising economic value.   Washington, DC, USA : National
    Academy of Sciences   x + 189 pages   Bd
Usa/79°° USA, NATIONAL ACADEMY OF SCIENCES (1979)   Tropical legumes: resour-
    ces for the future.   Washingtion, DC, USA : National Academy of Sci-
    ences   x + 331 pages   Bd
Usa/80°° USA, NATIONAL ACADEMY OF SCIENCES (1980)   Firewood crops: shrub
    and tree species for energy production.   Washington, DC, USA : National
    Academy of Sciences   xii + 237 pages   Bd
Vah/72   VAN HANDEL, H.; HAEGER, J.S.; HANSEN, C.W. (1972)   The sugars of
    some Florida nectars.   Am. J. Bot. 59(10) : 1030-1032   Bb, AA420/74
Van/26   VANSELL, G.H. (1926)   Buckeye poisoning of the honey bee.   Circ.
    Univ. Calif. agric. Exp. Stn No. 301 . 12 pages   Ba

Van/41   VANSELL, G.H. (rev. VANSELL, G.H.; ECKERT, J.E.) (1941)   Nectar
and pollen plants of California.   Bull. agric. Exp. Stn Univ. Calif.
No. 517 : 76 pages   Ba
Van/42   VANSELL, G.H.; WATKINS, W.G.; BISHOP, R.K. (1942)   Orange nectar
and pollen in relation to bee activity.   J. econ. Ent. 35(3) : 321-323
                                                                     Ba
Van/49   VANSELL, G.H. (1949)   Pollen and nectar plants of Utah.   Circ.
Utah agric. Exp. Stn No. 124 : 28 pages   Ba, AA18/50
Vaq/39   VÁSQUEZ M., O. (1939)   Colmenar de Don Orlando Vásquez: fines de
recolección y resultados de la cosecha.   Apicultura chil. (11) : 351-
352   In Spanish   Bj
Var/70   VARJÚ, M. (1970)   Akácmézek ásványi összetétele és ennek össze-
függései a növénnyel és a talajjal.   Élelmiszerv. Közl. 16(4/5) : 253-
258   In Hungarian; English, French, German, Russian summaries
                                                            Ba, AA429/73
Also published as "Mineralstoffzusammensetzung der ungarischen Akazien-
honigarten und deren Zusammenhang mit der Pflanze und dem Boden." Z.
Lebensmittelunters. u. -Forsch. 144(5) : 308-312 (1970) In German;
English summary   Ba, AA789/72
Vas/67   VASU, H.D. (1967)   Studies on beekeeping at Delhi. I. Possi-
bilities of beekeeping and the plants visited by bees for pollen, nectar
or both at Delhi.   Indian Bee J. 29 : 63-64   Bj, AA135L/70
Vas/69   VASU, H.D. (1969)   Studies on beekeeping at Delhi. II. Studies on
Tamarix dioica Roxb. (Tamaricaceae) as a pollen source.   Indian Bee J.
31 : 45-47   Bj
Vat/66°° VARTAK, V.D. (1966)   Enumeration of plants from Gomartak, India
with a note on botanical excursions around Castlerock.   Poona, India :
Maharashtra Association for the Cultivation of Science   viii + 169
pages + 1 plate   Bd
Ver/65   VERMEULEN, L.; PELERENTS, C. (1965)   Suikerfosfor- en ijzerge-
halte van Belgische honing.   Meded. LandbHoogesch. OpzoekStns Gent
30(2) : 527-541   In Dutch; English summary   Ba, AA380/67
Vit/65   VITEZ, P. (1965)   Facts about beekeeping in Argentina.   Bee Wld
46(1) : 19-22   Ba, AA502L/65
Vod/53   VODNIK, F. (1953)   O šetrajih.   Slov. čeb. 55(1/2) : 1-9   In
Slovene   Bj, AA96/54
Vor/64   VORWOHL, G. (1964)   Die Beziehungen zwischen der elektrischen
Leitfähigkeit der Honige und ihrer trachtmässigen Herkunft.   Annls
Abeille 7(4) : 301-309   In German; English, French summaries
                                                        Bj, AA158/67
Vor/68   VORWOHL, G. (1968)   Charakteristik des Tannenhonigs.   Z. Bienen-
forsch. 9(5) : 222-224   In German   Ba, AA179/68
Waa/56   WATANABE, T.; GOTO, M. (1956)   [Studies on useful components in
natural sources. IX. Studies of Japanese honeys.]   Jap. J. Pharmacogn.
10(2) : 145-148   In Japanese   Bb, AA27/59
Waa/60   WATANABE, T.; ASO, K. (1960)   Studies on honey. II. Isolation of
kojibiose, nigerose, maltose and isomaltose from honey.   Tohoku J.
agric. Res. 11(1) : 109-115; also in Nippon Nogeikagaku Kaishi 33 :
1054 (1959)   In Japanese   Bb, AA151/62
Waa/61   WATANABE, T.; MOTOMURA, Y.; ASO, K. (1961)   Studies on honey and
pollen. V. On the sugar composition of honey (2).   Tohoku J. agric.
Res. 12(2) : 187-190   Bb, AA836/65

Waa/61a WATANABE, T.; MOTOMURA, Y.; ANO, K. (1961a) Studies on honey and pollen. IV. On the sugar composition of nectar and nectar flow from the stomach of honeybees. Tohoku J. agric. Res. 12(2) : 179-185
Bb, AA657/62

Wab/80 WATANABE, K.; ECHIGO, T. (1980) [Detection of flavour components in honey.] Honeybee Sci. 1(2) : 69-72 In Japanese; English summary
Bj, AA269/81

Wae/82 WALLER, G.D. (1982) Hybrid cotton pollination. Am. Bee J. 122(8) : 555-560
Bc, AA1044/83

Waf/51 WAFA, A.K. (1951) Egypt and apiculture. S. Afr. Bee J. 26(4) : 7, 9
Bj, AA94/52

Waf/56 WAFA, A.K. (1956) Contribution to the study of factors affecting the amount of pollen grains gathered by honeybees, Apis mellifera L. Bull. Fac. Agric. Cairo Univ. No. 99 : 24 pages
Ba, AA209/59

Wah/74 WATANABE, H.; WATANABE, K. (1974) [Modern beekeeping.] Gifu : Japan Beekeeping Development Association 726 pages In Japanese
Bd

Wai/75 WALLINGFORD, N. (1975) Nectar sources in South Taranaki. Unpublished : 2 pages
Bs

Wak/81 WAKHLE, D.M.; NAIR, K.S.; RAMESH, B. (1981) Sugar composition in nectars of certain plants. Indian Bee J. 43(1) : 6-8 Bj, AA886/82

Wal/78 WALSH, R.S. (1978) Nectar and pollen sources of New Zealand. Wellington, New Zealand : National Beekeepers' Association of New Zealand 59 pages
Bd, AA1214L/78

Wan/64 WANIC, D.; MOSTOWSKA, I. (1964) Cukrowce v nektarze i miodzie. Zesz. nauk. wyźsz. Szk. roln. Olsztyn. 17(4) : 543-551 In Polish; English, Russian summaries
Ba, AA678/66

Wao/79 WALTON, G.[M.] (1979) Beech honeydew honey – a vast potential. N.Z. Beekpr 40(4) : 6-9
Bj, AA673/81

Wao/82 WALTON, G.M. [1982] Personal communication

War/65 WARREN, L.O. (1965) Sources of nectar and pollen for honey bees. Arkans. Fm Res. 14(1) : 2
Ba, AA303L/65

Wat/73 WALTON, J. (1973) Beekeeping in southern Portugal. Am. Bee J. 113(9) : 335
Bj

Wes/49 WESTERN CANADA BEEKEEPER (1949) Alberta flora and honeys. West. Can. Beekpr 12(4) : 14-16
Bj

Wet/81 WESTERN APICULTURAL SOCIETY JOURNAL (1981) Beekeeping cited as honey of a hobby. West. apic. Soc. J. 4(3) : 134, 137-138 Bj, AA815L/82

Whi/62 WHITE, J.W., JR; RIETHOF, M.L.; SUBERS, M.H.; KUSHNIR, I. (1962) Composition of American honeys. Tech. Bull. U.S. Dep. Agric. No. 1261 : 124 pages
Bd, AA655/63

Whi/63 WHITE, J.W., JR; SUBERS, M.H. (1963) Studies on honey inhibine. 2. A chemical assay. J. apic. Res. 2(2) : 93-100
Bb, AA914/64

Whi/64 WHITE, J.W., JR; SUBERS, M.H. (1964) Studies on honey inhibine. 3. Effect of heat. J. apic. Res. 3(1) : 45-50
Bb, AA915/64

Whi/66 WHITE, J.W., JR (1966) Methyl anthranilate content of citrus honey. Fd Res. 31(1) : 102-104
Bb, AA728/66

Whi/67 WHITE, J.W., JR; KUSHNIR, I. (1967) Composition of honey. VII. Proteins. J. apic. Res. 6(3) : 163-178
Bb, AA403/68

Wht/54 WHITEHEAD, S.B. (1954) Bees to the heather. London, UK : Faber & Faber 96 pages
Bd, AA251/54

Why/53°° WHYTE, R.O.; NILSSON-LEISSNER, G.; TRUMBLE, H.C. (1953) Legumes in agriculture. Rome, Italy : Food & Agriculture Organization of the United Nations 367 pages
Bd

Wid/72°° WILD, H. (rev. BIEGEL, H.M.; MAVI, S.) (1972)   A Rhodesian
    botanical dictionary of African and English plant names.   Salisbury,
    Rhodesia : Government Printer   xii + 281 pages   English and African
    languages   2nd ed, enlarged and revised   Bd, AA554/75
Wie/80   WIESE, H. (EDITOR) (1980)   Nova apicultura.   Porto Alegre, Brazil
    : Livraria e Editora Agropecuária Ltda   486 pages   2nd ed.   In Portu-
    gese   Bd, AA503L/81
Wih/79   WILLE, H. (1979)   Promoting bee culture in Democratic Republic of
    the Sudan.   Unpublished report : 7 pages   Bs
Wil/59   WILSON, A. (1959)   Beekeeping in New Brunswick.   Glean. Bee Cult.
    87(1) : 24-26   Bj
Win/70   WINSLOW, W.C. (1970)   There's honey in Turkey ... with a 300%
    profit.   Glean. Bee Cult. 98(10) : 618-620, 630   Bj
Wio/58   WILSON, W.T.; MOFFETT, J.O.; HARRINGTON, H.D. (1958)   Nectar and
    pollen plants of Colorado.   Bull. Colo. St. Univ. Exp. Stn No. 503-S :
    72 pages   Ba, AA295/59
Wio/65   WILSON, W.T. (1965)   Beekeeping in Colorado.   Bull. Colo. St.
    Univ. agric. Exp. Stn No. 418-A : 57 pages   rev.   Bb, AA65L/66
Wir/79   WILSON, R.D. (1979)   The analysis of carbohydrates by high-
    pressure liquid chromatography.   Rep. Chem. Div. Dep. scient. ind. Res.
    N.Z. No. CD 2280 : 15 pages   Bc, AA282/82
Wis/53   WILLSON, R.B. (1953)   Beekeeping in Mexico.   Glean. Bee Cult. 81
    : 79-82, 143-146   Ba
Wis/55   WILLSON, R.B. (1955)   Meet the champions: Miel Carlota!   Glean.
    Bee Cult. 83 : 329-332, 408-410, 447, 473-476   Ba, AA69/56
Woo/76   WOOTTON, M.; EDWARDS, R.A.; FARAJI-HAREMI, R.; JOHNSON, A.T.
    (1976)   Effect of accelerated storage conditions on the chemical
    composition and properties of Australian honeys. I. Colour, acidity and
    total nitrogen content.   J. apic. Res. 15(1) : 23-28   Bb, AA1158/76
Woo/76a WOOTTON, M.; EDWARDS, R.A.: FARAJI-HAREMI, R. (1976a)   Effect of
    accelerated storage conditions on the chemical composition and pro-
    perties of Australian honeys. 2. Changes in sugar and free amino acid
    contents.   J. apic. Res. 15(1) : 29-34   Bb, AA1159/76
Woo/78   WOOTTON, M.; EDWARDS, R.A.; ROWSE, A. (1978)   Antibacterial
    properties of some Australian honeys.   Fd Technol. Aust. 30(5) : 175-176
                                                                 Bc, AA1002/80
Woo/78a WOOTTON, M.; EDWARDS, R.A.; FARAJI-HAREMI, R.; WILLIAMS, P.J.
    (1978a)   Effect of accelerated storage conditions on the chemical
    composition and properties of Australian honeys. 3. Changes in volatile
    components.   J. apic. Res. 17(3) : 167-172   Bb, AA1079/79
Woy/81   WOYKE, J. (1981)   Flora apicola salvadoreña.   San Salvador, El
    Salvador : Ministerio de Agricultura y Ganaderia   14 pages   In Spanish
                                                                 Bc, AA157L/82
Wri/48   WRIGHT, P.H. (1948)   Raspberries as honey plants.   West Can.
    Beekpr 11(5) : 8-9   Bj
Wyk/52   WYKES, G.R. (1952)   An investigation of the sugars present in the
    nectar of flowers of various species.   New Phytol. 51(2) : 210-215
                                                                 Bb, AA149/53
Wyk/53   WYKES, G.R. (1953)   The sugar content of nectars.   Biochem. J.
    53(2) : 294-296   Ba, AA227/54
Yak/73   YAKOVLEVA, L.P.; ZAURALOV, O.A. (1973)   [Variability in sugar
    content of nectar.]   Pchelovodstvo (9) : 20-21   In Russian
                                                                 Bj, AA754/75

Yaz/53   YAZBECK, R. (1953)   Ancienneté de l'apiculture libanaise.   Revue
    fr. Apic. 3(92) : 597-599   In French   Bj, AA116/55
Yes/80   YESUVADIAN, M.S.; ARULDHAS, G.; CHRISTOPHER, M. (1980)   Strategy
    for apicultural extension in tropical climates.   Indian Honey 2(5/6) :
    15-33   Bb, AA487/81
Zam/79   ZAMBIA, FOREST DEPARTMENT (1979)   An introduction to frame hive
    beekeeping in Zambia.   Ndola, Zambia : Forest Department   60 pages
                                                             Bd, AA155/81
Zbo/68   ZBOROWSKI, J. (1968)   Badania nad występowaniem związków flawonoi-
    dowych w odmianowych miodach pszczelich rutyna i kwercetyna. Część II.
    Pszczel. Zesz. nauk. 12(1/2) : 75-83   In Polish; English, Russian
    summaries   Ba, AA1009/70
Zev/75°° ZEVEN, A.C.; ZHUKOVSKY, P.M. (1975)   Dictionary of cultivated
    plants and their centres of diversity. Excluding ornamentals, forest
    trees and lower plants.   Wageningen, Netherlands : Centre for Agricul-
    tural Publishing and Documentation   219 pages   Bd, AA1142/77
Zie/79   ZIEGLER, H.; MAURIZIO, A.; STICHLER, W. (1979)   Die Charak-
    terisierung von Honigen nach ihrem Gehalt an Pollen und an stabilen
    Isotopen.   Apidologie 10(4) : 301-311   In German; English, French
    summaries   Bj, AA666/81
Zma/80   [ZMARLICKI, C.B.] [1980]   Evaluation of honey plants in Burma.
    FAO Project BUR/78/013 Field Document.   Unpublished report : i + 30
    pages   Bs

7. PLANTS WITH SPECIAL CHARACTERISTICS

EXPLANATION OF LISTS BELOW

The lists of plants in this Section were obtained from programmed searches
of some of the 51 coded search fields (see pages 11-14).   Any further
information from the references used may be found in the MAIN ENTRIES (001-
452 and 01D-15D).

(a) IMPORTANT WORLD HONEY SOURCES THAT ARE DROUGHT TOLERANT

Plants marked * are reported to be very drought tolerant.

```
001   Acacia berlandieri Benth.;   Leguminosae
002*  Acacia caffra (Thunb.) Willd.;   Leguminosae
004   Acacia greggii A. Grey; Leguminosae
005   Acacia mellifera (Vahl) Benth.;   Leguminosae
008*  Acacia senegal (L.) Willd.;   Leguminosae
009   Acacia seyal Del.;   Leguminosae
010   Acacia tortilis (Forssk.) Hayne;   Leguminosae
022   Agave americana L.;   Agavaceae
024   Aloe dichotoma Masson;   Liliaceae
026   Aloysia gratissima (Gill. & Hook.) Troncoso;   Verbenaceae
030   Anacardium occidentale L.;   Anacardiaceae
039   Azadirachta indica A. Juss.;   Meliaceae
066   Caesalpinia coriaria (Jacq.) Willd.;   Leguminosae
067   Cajanus cajan (L.) Millsp.;   Leguminosae
070   Calliandra calothyrsus Meissn.;   Leguminosae
076*  Carnegiea gigantea (Engelm.) Britton & Rose;   Cactaceae
078   Cassia siamea Lam.;   Leguminosae
084   Centaurea solstitialis L.;   Compositae
085   Cercidium floridum Benth.;   Leguminosae
086   Cicer arietinum L.;   Leguminosae
093   Citrus limon (L.) Burm. f.;   Rutaceae
106   Combretum celastroides Laws.;   Combretaceae
122   Dalbergia sissoo DC.;   Leguminosae
126   Dialium engleranum Henriques;   Leguminosae
139   Echium lycopsis L.;   Boraginaceae
149   Eriobotrya japonica (Thunb.) Lindl.;   Rosaceae
153   Eucalyptus anceps (Maiden) Blakely;   Myrtaceae
154   Eucalyptus caleyi Maiden;   Myrtaceae
156   Eucalyptus camaldulensis Dehnh.;   Myrtaceae
158   Eucalyptus cladocalyx F. Muell.;   Myrtaceae
160   Eucalyptus crebra F. Muell.;   Myrtaceae
167   Eucalyptus gomphocephala DC.;   Myrtaceae
168   Eucalyptus gracilis F. Muell.;   Myrtaceae
170   Eucalyptus incrassata Labill.;   Myrtaceae
```

172 Eucalyptus leucoxylon F. Muell.; Myrtaceae
176* Eucalyptus melliodora A. Cunn. ex Schauer; Myrtaceae
178* Eucalyptus oleosa F. Muell. ex Miq.; Myrtaceae
180 Eucalyptus paniculata Smith; Myrtaceae
181 Eucalyptus platypus Hook.; Myrtaceae
182 Eucalyptus polyanthemos Schauer; Myrtaceae
185 Eucalyptus rubida Deane & Maiden; Myrtaceae
187 Eucalyptus sideroxylon A. Cunn. ex Woolls; Myrtaceae
191 Eucalyptus wandoo Blakely; Myrtaceae
205 Gleditsia triacanthos L.; Leguminosae
207 Glycine max (L.) Merr.; Leguminosae
208 Gmelina arborea Roxb.; Verbenaceae
217 Gymnopodium antigonoides (Robinson) Blake; Polygonaceae
220 Hedysarum coronarium L.; Leguminosae
221 Helianthus annuus L.; Compositae
237 Ipomoea batatas (L.) Lam.; Convolvulaceae
244 Jacquemontia nodiflora G. Don; Convolvulaceae
246 Julbernardia paniculata (Benth.) Troupin; Leguminosae
272 Lotus corniculatus L.; Leguminosae
280 Mahonia trifoliata (Moric.) Fedde; Berberidaceae
290 Medicago sativa L.; Leguminosae
296 Melilotus alba Desr.; Leguminosae
297 Melilotus officinalis (L.) Pall.; Leguminosae
313 Olea africana Mill.; Oleaceae
314 Onobrychis viciifolia Scop.; Leguminosae
315 Opuntia engelmanii Salm-Dyck; Cactaceae
317 Paliurus spina-christi Mill.; Rhamnaceae
319 Parkinsonia aculeata L.; Leguminosae
330 Pithecellobium dulce (Roxb.) Benth.; Leguminosae
335 Pongamia pinnata (L.) Pierre; Leguminosae
336* Prosopis cineraria (L.) Druce; Leguminosae
338* Prosopis glandulosa Torrey; Leguminosae
339* Prosopis juliflora (Sw.) DC.; Leguminosae
340* Prosopis pallida (Humboldt & Bonpl. ex Willd.) Kunth; Leguminosae
349 Rhigozum trichotomum Burch.; Bignoniaceae
354 Robinia pseudacacia L.: Leguminosae
397 Tamarindus indica L.; Leguminosae
405 Thymus capitatus (L.) Hoffm. & Link; Labiatae
426 Trifolium alexandrinum L.; Leguminosae
441 Viguiera helianthoides Kunth; Compositae
448* Ziziphus mauritania Lam.; Rhamnaceae
450 Ziziphus nummularia (Burm. f.) Wight & Arn.; Rhamnaceae
452* Ziziphus spina-christi (L.) Desf.; Rhamnaceae

(b) IMPORTANT WORLD HONEY SOURCES THAT ARE SALT TOLERANT

This list includes plants reported to show any degree of salt tolerance.

003 Acacia decurrens (Wendl.) Willd.; Leguminosae
019 Aegiceras corniculatum (L.) Blanco; Myrsinaceae
022 Agave americana L.; Agavaceae
037 Avicennia germinans (L.) L.; Avicenniaceae
038 Avicennia marina (Forssk.) Vierh. var. resinifera (Forst.) Bakh.;
     Avicenniaceae
063 Bucida buceras L.; Combretaceae
067 Cajanus cajan (L.) Millsp.; Leguminosae
071 Callistemon citrinus (Curt) Skeels; Myrtaceae
101 Coccoloba uvifera L.; Polygonaceae
104 Cocos nucifera L.; Palmae
122 Dalbergia sissoo DC.; Leguminosae
156 Eucalyptus camaldulensis Dehnh.; Myrtaceae
159 Eucalyptus cornuta Labill.; Myrtaceae
167 Eucalyptus gomphocephala DC.; Myrtaceae
180 Eucalyptus paniculata Smith; Myrtaceae
184 Eucalyptus robusta Smith; Myrtaceae
205 Gleditsia triacanthos L.; Leguminosae
221 Helianthus annuus L.; Compositae
272 Lotus corniculatus L.; Leguminosae
291 Melaleuca leucadendron (L.) L.; Myrtaceae
297 Melilotus officinalis (L.) Pall.; Leguminosae
298 Metrosideros excelsa Sol. ex Gaertn.; Myrtaceae
319 Parkinsonia aculeata L.; Leguminosae
330 Pithecellobium dulce (Roxb.) Benth.; Leguminosae
335 Pongamia pinnata (L.) Pierre; Leguminosae
336 Prosopis cineraria (L.) Druce; Leguminosae
337 Prosopis farcta (Sol. ex Russell) J.F. Macbride; Leguminosae
340 Prosopis pallida (Humboldt & Bonpl. ex Willd.) Kunth; Leguminosae
350 Rhizophora mangle L.; Rhizophoraceae
361 Sabal palmetto (Walt.) Lodd. ex Schultes; Palmae
377 Scaevola frutescens (Mill.) Krause; Goodeniaceae
379 Schinus terebinthifolius Raddi; Anacardiaceae
382 Serenoa repens (Bartr.) Small; Palmae
422 Tournefortia argentea L.f.; Boraginaceae
426 Trifolium alexandrinum L.; Leguminosae
427 Trifolium fragiferum L.; Leguminosae
438 Vicia faba L.; Leguminosae

(c) IMPORTANT WORLD HONEY SOURCES THAT PRESENT A PROBLEM TO BEEKEEPERS

An alert to beekeepers is included in MAIN ENTRIES for the following plants
because they may present a problem in bee management or in honey handling;
see the MAIN ENTRIES for details.

020 Aesculus hippocastanum L.; Hippocastanaceae
022 Agave americana L.; Agavaceae
023 Aloe davyana Schonl.; Liliaceae
025 Aloe mutans Reynolds; Liliaceae
035 Asclepias syriaca L.; Asclepiadaceae
056 Brassica campestris L. var. sarson Prain; Cruciferae

060 Brassica napus L. var. oleifera DC.;   Cruciferae
072 Calluna vulgaris (L.) Hull;   Ericaceae
077 Carvia callosa (Nees) Brem.;   Acanthaceae
097 Citrus sinensis (L.) Osb.;   Rutaceae
098 Citrus unshiu (Mak.) Marc.;   Rutaceae
104 Cocos nucifera L.;   Palmae
130 Diospyros virginiana L.;   Ebenaceae
139 Echium lycopsis L.;   Boraginaceae
152 Eucalyptus albens Benth.;   Myrtaceae
154 Eucalyptus caleyi Maiden;   Myrtaceae
156 Eucalyptus camaldulensis Dehnh.;   Myrtaceae
161 Eucalyptus diversicolor F. Muell.;   Myrtaceae
162 Eucalyptus drepanophylla F. Muell. ex Benth.;   Myrtaceae
164 Eucalyptus fasciculosa F. Muell.;   Myrtaceae
165 Eucalyptus ficifolia F. Muell.;   Myrtaceae
167 Eucalyptus gomphocephala DC.;   Myrtaceae
168 Eucalyptus gracilis F. Muell.;   Myrtaceae
169 Eucalyptus grandis W. Hill ex Maiden;   Myrtaceae
172 Eucalyptus leucoxylon F. Muell.;   Myrtaceae
173 Eucalyptus loxophleba Benth.;   Myrtaceae
175 Eucalyptus maculata Hook.;   Myrtaceae
176 Eucalyptus melliodora A. Cunn. ex Schauer;   Myrtaceae
177 Eucalyptus moluccana Roxb.;   Myrtaceae
180 Eucalyptus paniculata Smith;   Myrtaceae
182 Eucalyptus polyanthemos Schauer;   Myrtaceae
187 Eucalyptus sideroxylon A. Cunn. ex Woolls;   Myrtaceae
197 Eupatorium odoratum L.;   Compositae
206 Gliricidia sepium (Jacq.) Walp.;   Leguminosae
218 Haematoxylum campechianum L.;   Leguminosae
221 Helianthus annuus L.;   Compositae
223 Hevea brasiliensis Muell. Arg.;   Euphorbiaceae
259 Leptospermum scoparium J. & G. Forst.;   Myrtaceae
260 Lespedeza bicolor Turcz.;   Leguminosae
267 Liriodendron tulipifera L.;   Magnoliaceae
282 Malus domestica Borkh.; Rosaceae
290 Medicago sativa L.;   Leguminosae
309 Nicotiana tabacum L.;   Solanaceae
316 Oxydendron arboreum (L.) DC.;   Ericaceae
354 Robinia pseudacacia L.:   Leguminosae
359 Rubus ulmifolius Schott.;   Rosaceae
361 Sabal palmetto (Walt.) Lodd. ex Schultes;   Palmae
363 Salix alba L.;   Salicaceae
364 Salix caprea L.;   Salicaceae
365 Salix nigra Marshall;   Salicaceae
387 Sinapis alba L.;   Cruciferae
398 Taraxacum officinale Weber;   Compositae
410 Tilia cordata Mill.;   Tiliaceae
415 Tilia platyphyllos Scop.;   Tiliaceae
417 Tilia tomentosa Moench;   Tiliaceae
430 Trifolium pratense L.;   Leguminosae
437 Vernonia poskeana Vatke & Hildebrandt;   Compositae

01D Abies alba Miller;   Pinaceae
06D Larix decidua Miller;   Pinaceae
12D Quercus robur L.;   Fagaceae
13D Quercus suber L.;   Fagaceae

(d) IMPORTANT WORLD SOURCES OF HONEY THAT GRANULATES SLOWLY

Honey from these plants is reported not to granulate completely until after one year.

005 Acacia mellifera (Vahl) Benth.; Leguminosae
014 Acer pseudoplatanus L.; Aceraceae
031 Anchusa officinalis L.; Boraginaceae
034 Antigonon leptopus Hook. & Arn.; Polygonaceae
035 Asclepias syriaca L.; Asclepiadaceae
044 Berchemia scandens (Hill) K. Koch; Rhamnaceae
045 Bidens pilosa L.; Compositae
072 Calluna vulgaris (L.) Hull; Ericaceae
077 Carvia callosa (Nees) Brem.; Acanthaceae
080 Castanea sativa Mill.; Fagaceae
097 Citrus sinensis (L.) Osb.; Rutaceae
124 Daniellia oliveri (Rolfe) Hutch & Dalz.; Leguminosae
158 Eucalyptus cladocalyx F. Muell.; Myrtaceae
160 Eucalyptus crebra F. Muell.; Myrtaceae
165 Eucalyptus ficifolia F. Muell.; Myrtaceae
175 Eucalyptus maculata Hook.; Myrtaceae
176 Eucalyptus melliodora A. Cunn. ex Schauer; Myrtaceae
179 Eucalyptus panda S.T. Blake; Myrtaceae
180 Eucalyptus paniculata Smith; Myrtaceae
182 Eucalyptus polyanthemos Schauer; Myrtaceae
197 Eupatorium odoratum L.; Compositae
198 Euphoria longan (Lour.) Steud.; Sapindaceae
200 Faurea saligna Harv.; Proteaceae
216 Guizotia abyssinica Cass.; Compositae
230 Ilex glabra (L.) A. Gray; Aquifoliaceae
242 Isoglossa deliculata C.B. Clarke; Acanthaceae
245 Julbernardia globiflora (Benth.) Troupin; Leguminosae
246 Julbernardia paniculata (Benth.) Troupin; Leguminosae
250 Knightia excelsa R.Br.; Proteaceae
259 Leptospermum scoparium J. & G. Forst.; Myrtaceae
309 Nicotiana tabacum L.; Solanaceae
310 Nyssa aquatica L.; Nyssaceae
311 Nyssa ogeche Bartram; Nyssaceae
316 Oxydendron arboreum (L.) DC.; Ericaceae
321 Persea americana Mill.; Lauraceae
325 Phaseolus multiflorus Lam.; Leguminosae
328 Piscidia piscipula (L.) Sarg.; Leguminosae
340 Prosopis pallida (Humboldt & Bonpl. ex Willd.) Kunth; Leguminosae
344 Psoralea pinnata L.; Leguminosae
351 Rhus glabra L.; Anacardiaceae
354 Robinia pseudacacia L.: Leguminosae
357 Rubus spp [R. fruticosus L.]; Rosaceae
367 Salvia leucophylla Greene; Labiatae
368 Salvia mellifera Greene; Labiatae
370 Salvia officinalis L.; Labiatae
384 Sesamum indicum L.; Pedaliaceae
391 Symphoricarpos albus (L.) S.F. Blake; Caprifoliaceae
395 Syzygium cuminii (L.) Skeels; Myrtaceae
407 Thymus vulgaris L.; Labiatae
426 Trifolium alexandrinum L.; Leguminosae

447 Ziziphus jujuba Mill.;   Rhamnaceae
448 Ziziphus mauritania Lam.;   Rhamnaceae

04D Calocedrus decurrens (Torr.) Florin;   Cupressaceae
08D Picea abies (L.) Karsten;   Pinaceae
09D Pinus halepensis Miller;   Pinaceae

(e) IMPORTANT WORLD SOURCES OF HONEY THAT GRANULATES RAPIDLY

Honey from these plants is reported to granulate completely within two weeks.

008 Acacia senegal (L.) Willd.;   Leguminosae
019 Aegiceras corniculatum (L.) Blanco;   Myrsinaceae
020 Aesculus hippocastanum L.;   Hippocastanaceae
023 Aloe davyana Schonl.;   Liliaceae
025 Aloe mutans Reynolds;   Liliaceae
026 Aloysia gratissima (Gill. & Hook.) Troncoso;   Verbenaceae
028 Ampelopsis arborea (L.) Koehne;   Vitaceae
037 Avicennia germinans (L.) L.;   Avicenniaceae
042 Banksia serrata L.f.;   Proteaceae
054 Brassica campestris L.;   Cruciferae
060 Brassica napus L. var. oleifera DC.;   Cruciferae
084 Centaurea solstitialis L.;   Compositae
118 Cucurbita pepo L.;   Cucurbitaceae
139 Echium lycopsis L.;   Boraginaceae
144 Epilobium angustifolium L.;   Onagraceae
146 Erica cinerea L.;   Ericaceae
152 Eucalyptus albens Benth.;   Myrtaceae
154 Eucalyptus caleyi Maiden;   Myrtaceae
156 Eucalyptus camaldulensis Dehnh.;   Myrtaceae
159 Eucalyptus cornuta Labill.;   Myrtaceae
167 Eucalyptus gomphocephala DC.;   Myrtaceae
169 Eucalyptus grandis W. Hill ex Maiden;   Myrtaceae
172 Eucalyptus leucoxylon F. Muell.;   Myrtaceae
177 Eucalyptus moluccana Roxb.;   Myrtaceae
187 Eucalyptus sideroxylon A. Cunn. ex Woolls;   Myrtaceae
189 Eucalyptus tereticornis Smith;   Myrtaceae
201 Fuchsia excorticata (J. & G. Forst.) L.f.;   Onagraceae
207 Glycine max (L.) Merr.;   Leguminosae
221 Helianthus annuus L.;   Compositae
223 Hevea brasiliensis Muell. Arg.;   Euphorbiaceae
265 Lippia nodiflora (L.) Michx.;   Verbenaceae
269 Litsea stocksii Hook. f.;   Lauraceae
272 Lotus corniculatus L.;   Leguminosae
277 Mackenziea integrifolia (Dalz.) Brem.;   Acanthaceae
290 Medicago sativa L.;   Leguminosae
291 Melaleuca leucadendron (L.) L.;   Myrtaceae
293 Melaleuca quinquenervia (Cav.) S.T. Blake;   Myrtaceae
296 Melilotus alba Desr.;   Leguminosae
297 Melilotus officinalis (L.) Pall.;   Leguminosae
298 Metrosideros excelsa Sol. ex Gaertn.;   Myrtaceae
299 Metrosideros robusta A. Cunn.;   Myrtaceae
300 Metrosideros umbellata Cav.;   Myrtaceae

314 Onobrychis viciifolia Scop.; Leguminosae
324 Phacelia tanacetifolia Benth.; Hydrophyllaceae
338 Prosopis glandulosa Torrey; Leguminosae
342 Prunus x yedoensis Matsum.; Rosaceae
355 Rosmarinus officinalis L.; Labiatae
359 Rubus ulmifolius Schott.; Rosaceae
382 Serenoa repens (Bartr.) Small; Palmae
387 Sinapis alba L.; Cruciferae
388 Sinapis arvensis L.; Cruciferae

398 Taraxacum officinale Weber; Compositae
403 Thelepaepale ixiocephala (Benth.) Bremk.; Acanthaceae
420 Tithonia tubaeformis Cass.; Compositae
424 Trichostema lanceolatum Benth.; Labiatae
428 Trifolium hybridum L.; Leguminosae
430 Trifolium pratense L.; Leguminosae
438 Vicia faba L.; Leguminosae
440 Vicia villosa Roth; Leguminosae
441 Viguiera helianthoides Kunth; Compositae

01D Abies alba Miller; Pinaceae
06D Larix decidua Miller; Pinaceae

(f) IMPORTANT WORLD SOURCES OF HONEY WITH A HIGH SUCROSE CONTENT

Out of the 467 honeys represented in the MAIN ENTRIES, chemical analyses
were available for only 108, and the sucrose content for only 74 of them.
The 10 below contained more than 5% sucrose by weight, the maximum allowed
in the FAO/WHO Draft Codex (Cod/83).

Of the 108 honeys analysed, 77 (71%) were from plants that grow in temperate
regions.

039 Azadirachta indica A. Juss.; Meliaceae
077 Carvia callosa (Nees) Brem.; Acanthaceae
097 Citrus sinensis (L.) Osb.; Rutaceae
139 Echium lycopsis L.; Boraginaceae
206 Gliricidia sepium (Jacq.) Walp.; Leguminosae
223 Hevea brasiliensis Muell. Arg.; Euphorbiaceae
252 Lavandula angustifolia Miller; Labiatae
362 Saccharum officinarum L.; Gramineae
373 Sapindus mukorossi Gaertn.; Sapindaceae
422 Tournefortia argentea L.f.; Boraginaceae

(g) IMPORTANT WORLD SOURCES OF HONEY WITH A HIGH ASH CONTENT

Of the 68 honeys whose ash content was reported, only the 2 below contained
more than 1% by weight, the maximum allowed in the FAO/WHO Draft Codex
(Cod/83).

079 *Castanea pubinervis* (Hassk.) C.K. Schn.; Fagaceae
04D Calocedrus decurrens (Torr.) Florin; Cupressaceae

8. INDEXES TO MAIN ENTRIES

(a) HONEYDEW-PRODUCING INSECTS

In this index and in the entry, bold type indicates current names; others are synonyms. The authority and family are given in the MAIN ENTRY (01D–15D following 001–452). Names have been verified by the Commonwealth Institute of Entomology, London.

(b) SYNONYMS OF PLANT NAMES

The MAIN ENTRY (001-452, preceding 01D-15D) gives the authority for each synonym.

Abies pectinata	01D	Lippia urticoides	027
Acacia verek	008	Madhuca latifolia	278
Aloysia ligustrina	026	Malus communis	282
Archangelica decurrens	033	Malus sylvestris	
Archangelica officinalis	033	var. domestica	282
Avicennia nitida	037	Medicago sativa	
Bidens chaetodonta	113	subsp. falcata	287
Bombax malabaricum	046	Melaleuca parviflora	292
Brassica alba	387	Metrosideros lucida	300
Butea frondosa	064	Metrosideros tomentosa	298
Canthium parviflorum	074	Mikania parkeriana	301
Castanea crenata	079	Moringa pterigosperma	305
Cedrela toona	421	Nephelium litchi	268
Cercidium torreyanum	085	Nephelium longana	128
Chamaenerion angustifolium	144	Nephelium long-yan	128
Combretum trothae	106	Nyssa uniflora	310
Cordia dentata	110	Olea chrysophylla	313
Coreopsis abyssinica	113	Olea europaea subsp.	
Coridothymus capitatus	405	africana	313
Dialium simii	126	Onobrychis sativa	314
Dipsacus sylvestris	132	Picea excelsa	08D
Echium plantagineum	139	Picea vulgaris	08D
Erica carnea	147	Plectranthus coetsa	346
Erica verticillata	148	Plectranthus rugosus	347
Eucalyptus falcata var.		Polygonum fagopyrum	199
ecostata	163	Pongamia glabra	335
Eucalyptus incrassata var.		Pyrus malus	282
costata	170	Quercus pedunculata	12D
Eucalyptus redunca		Randia dumetorum	081
var. elata	191	Rivea corymbosa	434
Eucalyptus rostrata	156	Salmalia malabarica	046
Eugenia caryophyllus	393	Sapindus detergens	373
Eugenia jambolana	395	Sesamum orientale	384
Eugenia jambos	396	Strobilanthes perfoliatus	277
Heliconia brevispatha	222	Strobilanthus ixiocephalus	403
Ilex integrifolia	231	Symphoricarpos racemosus	391
Ipomoea leari	236	Symphoricarpos rivularis	391
Jussiaea nervosa	273	Syzygium jambolanum	395
Lagerstroemia lanceolata	251	Taraxacum dens-leonis	398
Larix europaea	06D	Tilia sylvestris	410
Lavandula officinalis	252	Tipuana speciosa	419
Lavandula spica	252	Trifolium sativum	430
Lavandula vera	252	Vitis parthenocissus	
Libocedrus decurrens	04D	quinquefolia	320
Lippia citriodora	266	Xeromphis spinosa	081
Lippia repens	265	Ziziphus jujuba	448

(c) <u>COMMON NAMES OF PLANTS</u>

MAIN ENTRIES 001-452 precede 01D-15D.   Generally, names consisting of two or more words are indexed only under the noun, e.g. red clover is indexed only under clover, red.   A few names are indexed under both, e.g. honey locust, also locust, honey.

www.ingramcontent.com/pod-product-compliance
Lightning Source LLC
Chambersburg PA
CBHW081044220326
41598CB00038B/6974